T0321516

Synchrotron Radiation

Learn about the properties of synchrotron radiation and its wide range of applications in physics, materials science and chemistry with this invaluable reference. This thorough text describes the physical principles of the subject, its source and methods of delivery to the sample, as well as the different techniques that use synchrotron radiation to analyse the electronic properties and structure of crystalline and non-crystalline materials and surfaces. It explains applications to study the structure and electronic properties of materials on a microscopic, nanoscopic and atomic scale. This book is an excellent resource for current and future users of these facilities, showing how the available techniques can complement information obtained in users' home laboratories, and is perfect for graduate and senior undergraduate students taking specialist courses in synchrotron radiation, in addition to new and established researchers in the field.

D. Phil Woodruff is Emeritus Professor of Physics at Warwick University. He has been awarded a number of prizes for his work in the UK, USA and Germany and is a fellow of the Royal Society. His previous works include *Modern Techniques of Surface Science*, 3rd ed. (Cambridge University Press, 2016).

Synchrotron Radiation

Sources and Applications to the Structural and Electronic Properties of Materials

D. PHIL WOODRUFF
University of Warwick

CAMBRIDGE
UNIVERSITY PRESS

University Printing House, Cambridge CB2 8BS, United Kingdom

One Liberty Plaza, 20th Floor, New York, NY 10006, USA

477 Williamstown Road, Port Melbourne, VIC 3207, Australia

314–321, 3rd Floor, Plot 3, Splendor Forum, Jasola District Centre, New Delhi – 110025, India

103 Penang Road, #05–06/07, Visioncrest Commercial, Singapore 238467

Cambridge University Press is part of the University of Cambridge.

It furthers the University's mission by disseminating knowledge in the pursuit of
education, learning, and research at the highest international levels of excellence.

www.cambridge.org
Information on this title: www.cambridge.org/9781107189805
DOI: 10.1017/9781316995747

First published 2021

A catalogue record for this publication is available from the British Library.

Library of Congress Cataloging-in-Publication Data
Names: Woodruff, D. P., author.
Title: Synchrotron radiation : sources and applications to the structural and electronic properties of materials /
    D. Phil Woodruff, University of Warwick.
Description: Cambridge, UK ; New York, NY : Cambridge University Press, 2021. I Includes
    bibliographical references and index.
Identifiers: LCCN 2021025004 (print) I LCCN 2021025005 (ebook) I ISBN 9781107189805 (hardback) I
    ISBN 9781316995747 (epub)
Subjects: LCSH: Synchrotron radiation. I BISAC: TECHNOLOGY & ENGINEERING / Materials Science /
    General I TECHNOLOGY & ENGINEERING / Materials Science / General
Classification: LCC QC793.5.E627 W66 2021 (print) I LCC QC793.5.E627 (ebook) I DDC 539.7/35–dc23
LC record available at https://lccn.loc.gov/2021025004
LC ebook record available at https://lccn.loc.gov/2021025005

ISBN 978-1-107-18980-5 Hardback

Contents

Preface

It is now more than 50 years since the first experiments exploiting synchrotron radiation were undertaken on electron synchrotrons designed and operated for experiments in particle physics. These early experiments were mostly conducted by physicists studying photoabsorption, but since that time the range and number of synchrotron radiation facilities, of experimental techniques exploiting this radiation, and of different types of users in chemistry, materials science, engineering, biology and medicine have expanded enormously. Increasingly advanced accelerator physics designs of facilities specifically designed to provide synchrotron radiation, each typically able to accommodate at least 20 different experiments simultaneously, operate as national and international facilities, satisfying the needs of users drawn from this wide range of disciplines. Some of these facility users have, as a particular focus of their research, the development and application of synchrotron radiation techniques, but for the great majority the facility is a (very large) 'black box', delivering to them a capability to measure certain properties of their samples that complement other techniques based in their home laboratories.

My own involvement with synchrotron radiation started in the late 1970s through parasitic use of the NINA electron synchrotron at Daresbury in the UK, followed by use of the dedicated synchrotron radiation facilities in Madison, Wisconsin (Tantalus) and near Paris in France (ACO) before returning to Daresbury when the SRS was commissioned. Subsequently, I became a major user of the facilities in Berlin (BESSY and BESSY II) and the ESRF in Grenoble, also running experiments at Brookhaven (NSLS) and Berkeley (ALS) in the USA. During this period I also followed the developments in Trieste (Electra) in Italy and in Lund (from MAX 1 to MAXIV) in Sweden through committee work. My current activities are anchored at the UK's Diamond facility. This involvement with different facilities, including some aspects of beamline design, and even whole source design (most notably for a facility that was never funded!) convinced me that, for my purposes at least, some understanding of the constraints and capabilities of the storage ring, the insertion devices, and the beamline optics and detectors has helped me to optimise my use of these facilities. I am inclined to believe that many users could benefit from seeing a synchrotron radiation source as rather more than a black box. This was one issue that motivated me to prepare this monograph.

Throughout the book I have tried to focus on the physical principles that underlie all aspects of synchrotron radiation, the actual radiation sources (Chapters 1 and 2);

the beamline optics and other components (Chapter 3); structure determination methods including those that explicitly exploit crystalline periodicity (Chapter 4) and those that do not (Chapter 5); probes of electronic and vibrational structure (Chapters 6 and 7); and finally the broad topic of imaging and micro- and nano-scale analysis that touches on all of these techniques (Chapter 8). I have consciously not covered applications in biological and medical sciences in any depth. These fall too far outside my field of competence for me to understand the significance of the clearly very important applications in these field, although I have included a very small number of examples of applications of general techniques that fall into these areas.

Inevitably, in trying to cover such a wide range of techniques I have benefitted from discussions and correspondence from a number of individuals with expertise complementary to my own. I would particularly like to thank Mike Poole, latterly under difficult circumstances, for much information and advice on accelerator physics and its practical implications, both before and during the writing of this book. I would also like to thank (roughly in the order of the topics as presented in the book) Richard Walker, Elaine Seddon, Ian Robinson, Andy Dent, Christian Morawe, Kevin Cowtan, Laura Gunn, Chiu Tang, Katrin Amann-Winkel, Anders Nilsson, Nick Terrill, Andy Smith, Nathan Cowieson, Phil King, Gerrit van der Laan, Malcolm Cooper, Jon Duffy, George King, Peter Gardner, Brian Tanner and Ian Robinson again, who have kindly provided me with figures from their work and/or improved my understanding of various techniques through further discussions. I hope I have represented their advice fairly. If I failed, the fault is mine, not theirs.

Abbreviations and Acronyms

AFM	atomic force microscopy
APPLE	Advanced Planar Polarised Light Emitter
ARPES	angle-resolved photoelectron spectroscopy
ATR	attenutated total reflection
BESSY	Berliner Elektronenspeicherring-Gesellschaft für Synchrotronstrahlung (Berlin electron storage ring for synchrotron radiation)
BXCDI	Bragg coherent X-ray diffraction imaging
CCD	charge-coupled device
CD	circular dichroism
CHESS	Cornell High Energy Synchrotron Source
Cryo-SXT	cryo-soft X-ray tomography
CT	computed tomography
CXDI	coherent X-ray diffraction imaging
DBA	double bend achromat
DESY	Deutsches Elektronen-Synchrotron (German electron synchrotron facility in Hamburg)
DFT	density functional theory
DOS	density of states
EDAX	energy-dispersive analysis of X-rays
EM	electromagnetic
EPU	elliptically polarising undulator
ERL	energy recovery linac
ESCA	electron spectroscopy for chemical analysis
ESRF	European Synchrotron Radiation Facility (Grenoble, France)
EXAFS	extended X-ray absorption fine structure
FEL	free electron laser
FELIX	Free Electron Laser for Infrared eXperiments
FPA	focal plane array
FTIR	Fourier-transform infrared
GIM	grazing incidence monochromator
HAXPES	hard X-ray photoelectron spectroscopy
HREELS	high resolution electron energy loss spectroscopy
INS	inelastic neutron scattering
LCLS	Linear Coherent Light Source (Stanford, USA)

LEED	low energy electron diffraction
Linac	linear accelerator
MAC	multiple analysing crystal (detector)
MAD	multiple-energy anomalous diffraction
MBA	multi-bend achromat
MIR	multiple isomorphous replacements
MX	macromolecular X-ray crystallography/diffraction
NEXAFS	near-edge X-ray absorption fine structure
NIM	normal incidence monochromator
NIXSW	normal incidence X-ray standing waves
OPD	optical path difference
PCXDI	plane-wave coherent X-ray diffraction imaging
PDF	pair distribution function
PDOS	partial density of states
PEEM	photoelectron emission microscopy
PEPICO	photoelectron–photoion coincidence
PES	photoelectron spectroscopy
PGM	plane grating monochromator
PhD	photoelectron diffraction (energy scan mode)
PSD	position-sensitive detector
PSL	photon-stimulated luminescence
PX	protein X-ray crystallography/diffraction
QEXAFS	quick EXAFS
RAIRS	reflection-absorption infrared spectroscopy
RF	radiofrequency
RIXS	resonant inelastic X-ray scattering
RXES	resonant X-ray emission spectroscopy
SASE	self-amplified spontaneous emission
SAXS	small angle X-ray scattering
SEXAFS	surface EXAFS
SGM	spherical grating monochromator
SLAC	originally Stanford Linear Accelerator Center but now known as SLAC National Accelerator Laboratory
SLS	Swiss Light Source
SNOM	scanning near-field optical microscopy
SPEM	scanning photoelectron microscopy
SRS	Synchrotron Radiation Source (Daresbury, UK)
SSRL	Stanford Synchrotron Radiation Laboratory
STM	scanning tunneling microscopy
SXR	soft X-ray
SXRD	surface X-ray diffraction
TEM	transmission electron microscopy
TGM	toroidal grating monochromator

TOF	time-of-flight
TPES	threshold photoelectron spectroscopy
TPESCO	threshold photoelectron spectroscopy coincidence
TTF	TESLA Test Facility (at DESY)
UPS	ultraviolet photoelectron spectroscopy
UV	ultraviolet
VUV	vacuum ultraviolet
WAXS	wide angle X-ray scattering
XANES	X-ray absorption near-edge structure
XFEL	X-ray free electron laser
XFH	X-ray fluorescence holography
XLD	X-ray linear dichroism
XMCD	X-ray magnetic circular dichroism
XPD	X-ray photoelectron diffraction (angle-scan mode)
XPEEM	X-ray photoelectron emission microscopy
XPS	X-ray photoelectron spectroscopy
XRF	X-ray fluorescence
XRT	X-ray topography
XSW	X-ray standing waves

1 Introduction

1.1 New Lamps for Old ...

Electromagnetic radiation is well-established as an important probe of matter, be it in the form of free atoms or molecules, crystalline solids or soft matter including biological materials. In the long wavelength range, notably from the far infra-red to the extreme ultra-violet and beyond, spectroscopic studies reveal many aspects of the energetics that govern physical and chemical properties. In the short wavelength X-ray range, scattering experiments provide detailed structural information on an atomic scale. All of these studies have been transformed by one or both of two key developments initiated in the second half of the twentieth century, namely the discovery and development of facilities to exploit synchrotron radiation, and the invention and development of the laser. These two new light sources are highly complementary in character, leading in large part to different areas of exploitation.

Synchrotron radiation provides a continuum of wavelengths from the far infrared to hard X-rays/γ-rays from $\sim10^6$ Å (100 µm) to $\sim10^{-1}$ Å (10 pm), with associated photon energies from ~10 meV to ~100 keV (see Fig. 1.1, which also shows the approximate ranges of some alternative radiation sources). It has a very high spectral brightness (number of photons per unit area per unit solid angle) due to the intrinsically narrow beam of the emitted radiation. When combined with suitable monochromators this produces a tuneable radiation source over a vast wavelength or energy range. Although it is strictly pulsed in time, the repetition rate is so high and the pulse length is sufficiently long that for very many studies it can be considered as quasi-continuous in time.

Laser radiation is also in the form of a narrow bright source, but is typically provided at a fixed wavelength, although limited tunability (even of up to a factor of ~10 in some cases, but usually much less) can be achieved in some systems. The wavelengths offered by a selection of different lasers are largely restricted to the range from the infrared to the relatively near ultra-violet, although in special cases even 'table-top' lasers (albeit highly specialised ones) have achieved soft-X-ray energies of ~100 eV. Although some lasers have a continuous time structure, many are pulsed, and indeed very short pulses in the fs range have opened up a whole new area of exploration of fast time-resolved studies and, by virtue of the associated huge peak power output, render multi-photon phenomena accessible. One further important difference is coherence; lasers are sources of fully coherent radiation,

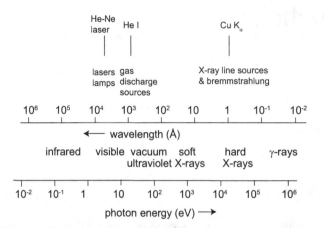

Figure 1.1 The electromagnetic spectrum covered by synchrotron radiation together with some laboratory sources. The range indicated for lasers corresponds to that of conventional 'off-the-shelf' table-top devices.

synchrotron radiation is not, although as discussed in Chapter 2 (and further in Chapter 8) certain types of experiments requiring coherence can be performed using synchrotron radiation. Of course, another key difference between lasers and synchrotron radiation is their physical scale and cost. Most conventional 'table-top' lasers are single-user devices located in the home laboratory of the experimentalist at a cost largely consistent with standard laboratory instruments. By contrast, synchrotron radiation is delivered to many (~20–30) simultaneous users in a central – typically national – facility with a capital cost as much as 10^5 or more times larger. This difference is almost certainly a major reason why the early development and exploitation of lasers was much more rapid than that of synchrotron radiation. The subsequent very significant developments in synchrotron radiation sources, firstly in creating dedicated user facilities, but subsequently in the creation of novel magnet structures that tailor the radiation output to the users' requirements, are described in Chapter 2.

The complementary nature of the two different types of sources in terms of the range of tunability and time structure, but also the readily accessible wavelengths, means that they have been mostly exploited in quite different experimental techniques. Recent developments in free-electron lasers (FELs) with abilities to produce tuneable coherent radiation with extremely high instantaneous power at energies up to the true X-ray range, do occupy a middle ground that somewhat blurs the distinction between these two types of sources in terms of their underlying mechanisms, but it seems likely that many applications of FELs and synchrotron radiation sources will continue to be rather distinct. In this book conventional lasers, based on a lasing medium, will not be discussed further, and the focus will be on synchrotron radiation. However, a brief discussion of FELs and their applications is included in view of some significant similarities in the structure of the sources and some overlap in their fields of exploitation.

1.2   Particle Accelerators: A Brief Historical Introduction

The motivation for the initial development of particle accelerators (and particularly proton, rather than electron, accelerators) has its origins in particle physics, investigating the consequences of high-energy collisions between fundamental particles. The objective was to accelerate the charged particles to high energies and then extract them from the accelerator and direct them to a target. Of course the simplest form of charged particle accelerator simply involves allowing the particles to gain energy in a 'single push' by passage across a large voltage difference. In this sense a simple electron gun, installed in cathode ray tubes (for many years the basis of televisions and oscilloscopes), is a particle accelerator, leading to electron energies of up to a few tens of keV. Much higher voltage single-push accelerators using Van de Graaf generators and Cockroft/Walton cascade generators can achieve energies of hundreds of keV or even a few MeV.

To achieve the highest energies, however, one must increase the energy in multiple sequential steps. Two essentially different designs to achieve this emerged, namely linear accelerators ('linacs') in which the charged particles gain energy while travelling along a linear trajectory, and somewhat more compact devices in which the charged particles perform circular (or approximately circular) orbits, constrained by magnetic fields. Conceptually the simplest versions of these devices, that rely on the charged particles gaining energy by passing through a succession of appropriately applied voltage differences, are primarily of interest for the acceleration of ions (particularly hydrogen ions - protons) rather than the much lower-mass electrons, for reasons that will become clear below. In the earliest type of linac shown in Fig. 1.2(a) the particles pass through a sequence of conducting drift tubes. Within the drift tubes the particles experience no electric field but if the applied voltages to the drift tubes are periodic in time, and are phased such that each time the particle emerges from one tube it 'sees' an attractive voltage on the next tube, then it is accelerated in each gap. Notice that as the speed of the particles increases the length of the drift tubes must increase to ensure the passage between the tubes is periodic in time. However, as the particle energy becomes relativistic this effect becomes less significant. In fact for electrons their speed becomes extremely close to that of light, c, at quite modest energies (e.g. an electron with an energy of only 10 MeV has a speed of ~$0.999c$, whereas a proton of the same energy has a speed of only ~$0.14c$).

In the earliest circular accelerator, the cyclotron (Fig. 1.2(b)), the particles travel in a split conducting cavity comprising two 'Dees' located between the pole pieces of a large fixed-field electromagnet; the magnetic field forces the ions to travel in a circular orbit while at a fixed energy. In this case, too, a periodically-varying voltage difference is applied to the two Dees, phased such that each time the particle passes from one Dee to the other it is accelerated. As the energy increases, the bending radius of the trajectory in the fixed magnetic field increases proportionately, so the overall trajectory follows a spiral until the particles are deflected out of the accelerator. For protons at non-relativistic energies the transit duration of each full orbit is constant (i.e. phase-locked), so the accelerating field frequency is also fixed, but relativistic

(a)

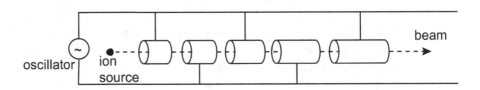

(b)

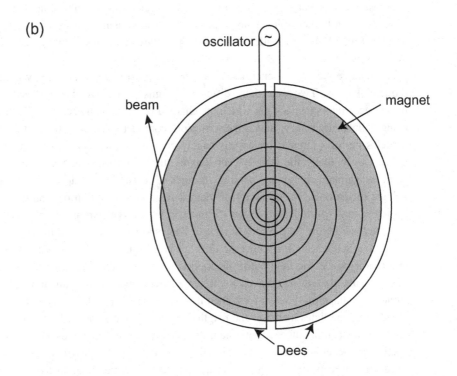

Figure 1.2 Basic features of two ion accelerators based on (a) linear and (b) circular geometries.

effects do limit performance eventually. While the cyclotron is far more compact than the linac, the energy to which the particles can be accelerated is also limited by the physical size of the device: specifically, the size of the magnetic pole pieces and the associated magnetic field strength.

Accelerators based on the circular motion of electrons are the devices of primary interest in the context of synchrotron radiation, but before focusing on this aspect it is appropriate to explain in slightly more detail how a linac is able to accelerate electrons at relativistic energies. Linacs are widely used to inject high (relativistic) energy electrons into the circular path of electron storage rings that are now the basis of synchrotron radiation sources, and are the essential ingredient of free electron lasers, both described in Chapter 2. Notice that the frequency of the oscillators in both linacs and circular accelerators, when operated with electrons, is in the (micro-wave) GHz range, with associated electromagnetic wavelengths of a few cms, and the way these radiofrequency (RF) waves propagate along a linac structure provides a rather different mode of acceleration structure to that described for ions in the context of Fig. 1.2(a). Specifically, in an electron linac the electrons 'surf' the RF travelling wave in a fashion somewhat similar to that of a surfer riding water waves. Fig. 1.3 shows in a simplified fashion how this can occur. A positive value of the longitudinal electric field E is defined here as one that accelerates the electrons in the direction of travel. The acceleration will be greatest if a travelling electron is located, relative to the travelling wave, near a peak in E but will experience less acceleration if it is located closer to the zero value of E, and indeed will be decelerated if it experiences a negative value of E. This leads to two consequences, namely a net acceleration of the 'surfing' electrons, but also a bunching of the electrons into groups located at similar relative positions on the travelling wave. Of course, in order for this to be effective, the travelling wave must match the average speed of the electrons.

As remarked above, for energies above a relatively low threshold value (at which the electrons are injected into this travelling wave), the electrons have a speed very close to c, and will gain energy mostly by gaining relativistic mass. Nevertheless, it is clear that the travelling wave must have a *phase velocity* that is close to, but slightly less than, c. The simplest situation would seem to be when the waves travel down a conducting tube, but in such a hollow tube waveguide, their phase velocity, $v_{ph} = \omega/k$ is actually greater than c, although their group velocity $(d\omega/dk)$, the speed at which information can be transferred, is less than c and thus consistent with the requirements of relativity. In order to create a linear electron accelerator based on this travelling RF wave approach it is therefore necessary to modify the waveguide to slow down the phase velocity. This is achieved by 'loading' the waveguide cavity with a series of periodic obstacles in the form of discs with central apertures or 'irises' (Fig. 1.4).

Electron linacs based on this principle use a succession of these cavities, coupled to individual RF power sources, to achieve the multi-GeV energies required for injection into synchrotrons or electron storage rings (see below), and also for FELs. Notice that

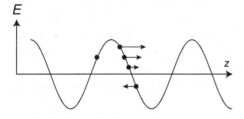

Figure 1.3 Schematic diagram showing the instantaneous force felt by electrons (represented by black discs) located at different positions relative to a travelling RF wave.

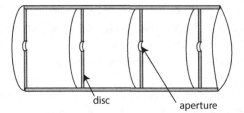

Figure 1.4 Schematic cutaway diagram of a disc-loaded cavity.

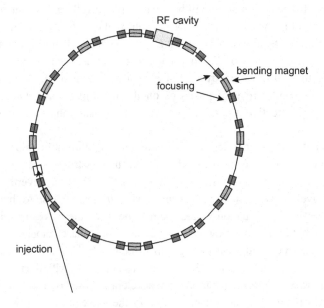

Figure 1.5 Schematic view of a synchrotron showing the main components in a simplified fashion.

essentially the same device is also used prior to high energy acceleration as a 'buncher' to produce electron bunches with the desired time structure and energy spread.

The fact that accelerated electrons quickly achieve highly relativistic speeds means that the simplest circular accelerator, the cyclotron, cannot maintain phase stability to high energies if operated at a fixed frequency. The synchrocyclotron, a development of essentially the same device but operated on single bunches of electrons at variable frequency, addressed this problem, but was still restricted in ultimate energy by the need for huge magnetic pole pieces to accommodate the spiralling trajectory of the accelerating electrons in the fixed magnetic field. The solution to these problems was the synchrotron, in which the electron trajectories are no longer strictly circular; the huge single magnet is replaced by a series of bending magnets located at the 'corners' of a polygonal racetrack (Fig. 1.5). The electron trajectory in such a device (through a very large but thin doughnut-shaped vacuum vessel) is nominally independent of energy as it increases; this fixed trajectory is maintained constant as the energy is increased by increasing the field strength of the bending magnets. Acceleration is achieved by installing one or more RF cavities in straight sections between the bending (dipole) magnets. Unlike the mechanism of the acceleration by travelling waves in a linac described above, in this case the RF cavity produces a standing wave, but the phasing of the time variation of the RF field with the arrival of an electron bunch is such as to achieve acceleration and maintain a near constant energy of electrons within the bunch. Additional magnetic structures (particularly quadrupoles – see Chapter 2) ensure the electrons all stay within a narrow path around the ring. In particular, the synchrotron exploits the idea of 'strong' or 'alternating gradient' focusing, in which the circulating beam is locally refocused at many points around the ring. It transpires that this technique allows the construction of much smaller aperture accelerators than earlier ones exploiting weak focusing. Initial injection of the electrons is from a lower-energy accelerator, most commonly nowadays a linac.

1.3   Synchrotron Radiation Basics

The key characteristics of synchrotron radiation are its extreme brightness or brilliance, its broad spectral continuum and its polarisation. In this section the basic physics leading to these properties is explained and quantified briefly.

Any acceleration or deceleration of a charged particle leads to the emission of electromagnetic radiation. It is this bremsstrahlung ('braking radiation') that provides the continuum background of X-radiation in a conventional laboratory X-ray source; energetic electrons (typically ~40 keV) strike a metallic target and emit bremsstrahlung as they lose energy in the solid. A rather special case of electromagnetic radiation from an accelerated charged particle occurs if the particle is forced into a circular orbit by a central force that constantly accelerates the particle along the direction of this central force. In this case the radiation emitted has the frequency of the circular motion and within the frame of reference of the charged particle one has dipolar emission; the angular dependence of the amplitude of the emission is given by the cosine of the angle between the instantaneous direction of travel and the direction of observation (see Fig. 1.6). It is this emission of radiation by an electron orbiting in a central Coulomb potential, and the resulting continuous energy loss, that leads to the downfall of a purely classical 'planetary' model of the hydrogen atom; the additional postulates of the Bohr theory were the first steps towards the proper quantum mechanical description. However, a quite different mechanism for producing circular motion of a charged particle is the force experienced by the moving particle due to the presence of a constant magnetic field perpendicular to the direction of motion. This is exactly the situation in a cyclotron or a synchrocyclotron as described in the previous section. It also occurs in astronomical plasmas of ionised gases and electrons in the proximity of sources of magnetic fields, the emission being known as cyclotron radiation. This magnetism-induced central-force acceleration is also a feature of the bending magnets used in a synchrotron to guide the electrons around the accelerator, although the nature

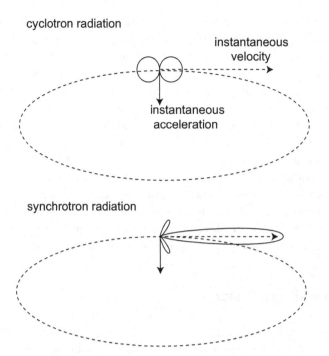

cyclotron radiation

instantaneous velocity

instantaneous acceleration

synchrotron radiation

Figure 1.6 Schematic diagram showing the angular dependence of the radiation emitted from an electron travelling in a circular trajectory in a magnetic field at sub-relativistic energies (cyclotron radiation) and at relativistic energies (synchrotron radiation).

of the radiation emitted is determined by the bending radius in these magnets, rather than the much larger radius of the whole synchrotron. Nevertheless, this bending radius is ~10 m, so if the speed of the electrons is effectively the speed of light, c, $(3 \times 10^8$ ms$^{-1})$ then the frequency of the cyclotron radiation would be ~10^7 Hz (in the radio frequency range) and of little interest as a photon source.

However, because the energy of the electrons is such that their speed is very close to c, there are very strong relativistic effects that modify the properties of the emission, which is then known as synchrotron radiation rather than cyclotron radiation. In describing these relativistic effects it is sometimes convenient to express the electron velocity v relative to the speed of light in a vacuum, c by the parameter $\beta = v/c$ although even more useful is the parameter, γ, defined as

$$\gamma = 1 \Big/ \sqrt{(1 - \beta^2)} = E/m_0 c^2, \qquad (1.1)$$

where m_0 is the rest mass of the electron and E is the total energy, mc^2, where m is the relativistic mass of the electron. Notice that as $m_0 c^2$ is 0.511 MeV, an approximate numerical expression for γ for an electron is simply $\gamma \approx 2,000\ E$ with E expressed in GeV.

One relativistic effect is the consequence of applying the Lorentz transformation to transfer the dipolar angular distribution of the radiation in the rest frame of the electron, into the laboratory frame (Fig. 1.6). The angular spread of the radiation is reduced to a total angle, $\Delta\theta$, with a value of approximately $2/\gamma$.[1] For example, with an electron energy of 5 GeV, this angular spread is only $\sim 10^{-4}$ rad or 0.1 mrad (0.0006°). This is the origin of the highly directional character of synchrotron radiation and a major factor leading to its extreme brightness.

High electron energies also have a major impact on the amount of radiated power. In the non-relativistic case the total power radiated when an electron is accelerated was derived by Larmor at the end of the nineteenth century (Larmor, 1897) as

$$S = \frac{e^2}{6\pi\varepsilon_0 m_0 c^3}\left(\frac{dp}{dt}\right)^2, \tag{1.2}$$

where e is the electronic charge, ε_0 is the dielectric constant of the vacuum and $p = m_0 v$ is the electron momentum. At highly relativistic energies, however, one needs a Lorentz invariant form of Equation (1.2). The time transforms according to $dt \rightarrow d\tau = \frac{1}{\gamma}dt$ while the momentum p must be replaced by the momentum 4-vector P_μ of relativistic electrodynamics

$$\left(\frac{dP_\mu}{d\tau}\right)^2 \rightarrow \left(\frac{dp}{d\tau}\right)^2 - c^2\left(\frac{dE}{d\tau}\right)^2. \tag{1.3}$$

In the case of an electron in a circular trajectory, with the acceleration directed to the centre of the circle, the energy of the electron is constant, so the second term in Equation (1.3) is zero. The total power radiated then becomes

$$S = \frac{e^2 c}{6\pi\varepsilon_0 (m_0 c^2)^2}\left(\frac{dp}{d\tau}\right)^2 = \frac{e^2 c\gamma^2}{6\pi\varepsilon_0 (m_0 c^2)^2}\left(\frac{dp}{dt}\right)^2, \tag{1.4}$$

while for circular motion one can also write

$$\frac{dp}{dt} = p\omega = p\frac{v}{R}, \tag{1.5}$$

where R is the bending radius. Moreover, for large values of γ, $E = pc$, so

$$S = \frac{e^2 c}{6\pi\varepsilon_0 (m_0 c^2)^4}\frac{E^4}{R^2}, \tag{1.6}$$

a result first obtained by Liénard (1898). Of course, Liénard's work predated the theory of relativity, but the equations of electromagnetic theory are invariant under transformations between moving systems so his original equation remains valid at relativistic energies This equation highlights two key results. Firstly, the radiated energy scales as E^4, so while the radiated energy is very small in a low

---

[1] In some presentations the whole angular spread is defined as $1/\gamma$. In fact the exact value of this angle depends on the wavelength of the radiation emitted but also on how the angle is defined, for example as the Gaussian width σ or, indeed, 2σ (see Equation (1.18) and the associated discussion). Here $1/\gamma$ is taken to be the half-angle. The arguments based on either definition are necessarily approximate.

energy cyclotron, it can be very significant at the higher energies found in a synchrotron. Indeed, in a synchrotron one can show from Equation (1.6) that the energy loss of an electron passing just once around the synchrotron is, in keV, equal to 88.5 E^4/R with E expressed in GeV and R in m. For example, if E is 3 GeV and R is 10 m the energy loss by a single electron in each circuit around the ring is 717 keV. The second significant result contained in Equation (1.6) is that the radiated power scales as $1/m_0^4$. This clearly favours a low mass charged particle as the optimum source of synchrotron radiation. Specifically, as the mass of a proton is 1,836 times that of an electron, the radiated power from a proton at the same total energy as an electron is 1.13×10^{13} times smaller. Machines designed to deliver synchrotron radiation to users have thus always used accelerated electrons or positrons.

The second key feature of synchrotron radiation is its frequency spectrum. To understand this property one must note that because the emission from electrons orbiting with a frequency ω_0 has a very narrow angular spread of only $\Delta\theta \approx 2/\gamma$, the radiation illuminates a detector ('observer') for a duration of only $\Delta t \approx \Delta\theta/\omega_0 = 2/\gamma\omega_0$, so the Fourier spectrum of this pulse of radiation must contain frequencies of at least $\sim 1/\Delta t$ or $\gamma w_0/2$. We must, however, also take proper account of relativistic time dilation and the Doppler effect. A particularly simple way of taking account of both of these effects is to consider the trajectory of the electron and the emitted radiation within the time period during which an observer is illuminated (Fig. 1.7). The time interval between the two ends of the detected electromagnetic pulse is just the difference in the time of flight of the electron, speed v, and the photon, speed c, as the electron travels from A to B

$$\Delta t = t_{el} - t_{ph} = \frac{R\Delta\theta}{v} - \frac{R\sin\theta}{c} = \frac{R}{c}\left(\frac{\Delta\theta}{\beta} - \Delta\theta + \frac{\Delta\theta^3}{3!}\ldots\ldots\right) \qquad (1.7)$$

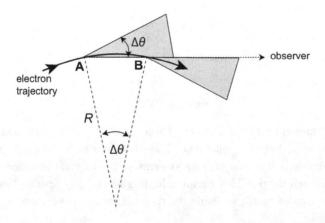

Figure 1.7 Schematic figure showing the range of the electron trajectory in a circular orbit from **A** to **B** during which a fixed observer can observe the (shaded) fan of synchrotron radiation with an angular width of $\Delta\theta$.

and with $\Delta\theta \approx 2/\gamma$ and $\gamma\beta \approx \gamma - 1/2\gamma$ one obtains

$$\Delta t \approx \frac{R}{c}\left(\frac{2}{\beta\gamma} - \frac{2}{\gamma} + \frac{8}{6\gamma^3}\cdots\right) \approx \frac{4R}{3c\gamma^3}. \tag{1.8}$$

This short pulse generates a broad spectrum with some characteristic frequency

$$\omega_{char} \approx \frac{1}{\Delta t} \approx \frac{3c\gamma^3}{4R}. \tag{1.9}$$

A more formal treatment of this problem was presented by Schwinger (1949) who showed that the power radiated by an electron in a unit angular frequency range centred on ω is

$$P(\omega) = \frac{3\sqrt{3}}{16\pi^2\varepsilon_0}\frac{e^2}{R}\gamma^4\frac{\omega_0}{\omega_c}\frac{\omega}{\omega_c}\int_{\omega/\omega_c}^{\infty} K_{5/3}(y)dy \tag{1.10}$$

Where $K_{5/3}(y)$ is a modified Bessel function and ω_c is the critical frequency, leading to a critical energy

$$hv_c = \hbar\omega_c = \hbar\frac{3}{2}\gamma^3\omega_0 = \hbar\frac{3c}{2R}\gamma^3. \tag{1.11}$$

Notice that the associated critical frequency corresponds (within a factor of 2) to the rather roughly estimated value of ω_{char} in Equation (1.9). Equation (1.10) may also be written as

$$P(\omega) = \frac{1}{6\pi\varepsilon_0}\frac{e^2\gamma^4 c}{R^2\omega_c}S_s(\omega/\omega_c) \tag{1.12}$$

in terms of the spectral function $S_s(y)$

$$S_s(y) = \frac{9\sqrt{3}}{8\pi}y\int_y^{\infty} K_{5/3}(\xi)d\xi \tag{1.13}$$

which is normalised such that

$$\int_0^{\infty} S_s(y)dy = 1 \tag{1.14}$$

An important property of this normalised spectral function is that

$$\int_1^{\infty} S_s(y)dy = 0.5 \tag{1.15}$$

The limits of this integral correspond to the range of harmonic values, n (or frequencies ω) from the critical harmonic (frequency) to infinity. Equation (1.15) shows that half the total radiated power corresponds to photon energies greater than the critical energy and half corresponds to lower photon energies. Essentially the same function without this normalisation

$$G_1(\omega/\omega_c) = \left(\frac{\omega}{\omega_c}\right) \int\limits_{\omega/\omega}^{\infty} K_{5/3}(\xi)\,d\xi \tag{1.16}$$

is also referred to as the synchrotron radiation universal function. From this one can produce the so-called universal curve of synchrotron radiation from bending magnets, shown in Fig. 1.8. The abscissa is expressed as the photon energy (or photon wavelength) in units of the critical energy (wavelength) while the ordinate shows the number of photons emitted per second, per mrad of horizontal divergence, per mA of circulating electron current, per GeV of electron energy and per 0.1% bandwidth. As Schwinger pointed out, because the emitted spectrum is from an electron in periodic motion it must consist of harmonics of the circulation frequency ω_0, but the harmonic numbers are very large; at the critical energy the harmonic number (ω/ω_0) is $(3/2)\gamma^3$, so for a machine energy of 3 GeV the harmonic number is $\sim 10^{10}$. Moreover, there must be a small spread in the exact energies of the individual electrons in the synchrotron (in particular, due to

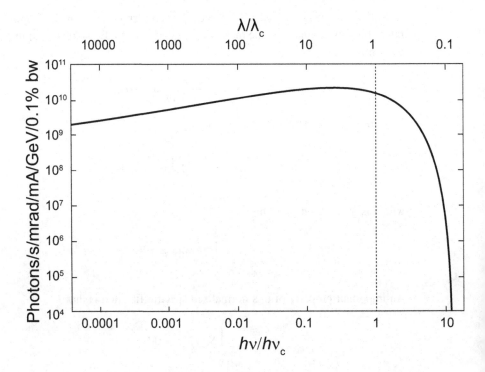

Figure 1.8 Universal curve of synchrotron radiation from bending magnets.

the energy lost to synchrotron radiation during a single circulation of the ring), so the harmonics are smeared and the emitted spectrum really is a continuum as implied by Fig. 1.8. Notice, incidentally, that because Equation (1.12) has a factor ω_c (proportional to γ^3) in the denominator, power $P(\omega)$ is linear in the machine energy, γ, as implied by the ordinate labelling of Fig. 1.8.

The other key property of synchrotron radiation concerns its state of polarisation. Within the circulating plane of the electron the radiation is plane polarised (within this circulating plane) but out of this plane there is a component of radiation polarised perpendicular to the plane of circulation. Equation (1.10) gives the radiation power integrated over all angles perpendicular to the circulating plane, but as a function of the angle, ψ, out of this plane the radiated power is

$$P(\omega, \psi) = \frac{e^2}{16\pi^3 \varepsilon_0 c} \gamma^2 \left(\frac{\omega}{\omega_c}\right)^2 \left(1 + \gamma^2 \psi^2\right)^2 \left[K_{2/3}^2(\xi) + \frac{\gamma^2 \psi^2}{1 + \gamma^2 \psi^2} K_{1/3}^2(\xi)\right] \quad (1.17)$$

where $\xi = \dfrac{\omega}{\omega_c} \dfrac{\left(1 + \gamma^2 \psi^2\right)^{3/2}}{2}$.

Equation (1.17) comprises two terms associated with the two modified Bessel functions in the square brackets. The first of these corresponds to the intensity of radiation polarised in the circulating plane and peaks in this plane at $\psi = 0$. The second term corresponds to the intensity of radiation polarised in the plane perpendicular to the circulating plane; this is zero at $\psi = 0$ and peaks at larger values of ψ. Examples of these functions at three different photon energies (relative to the critical energy) are shown in Fig. 1.9.

Notice that earlier in this chapter the opening angle of the fan of radiation was assumed to be $2/\gamma$. Fig. 1.9 shows that the vertical opening angle actually depends quite strongly on the photon energy relative to the critical energy. If the intensity spread within the total radiation fan is approximated by a Gaussian profile with a

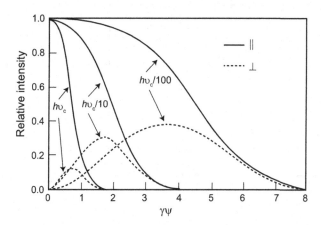

Figure 1.9 Angular distribution out of the electron circulating plane, of synchrotron radiation polarised in this plane (‖) and perpendicular to this plane (⊥), as determined by Equation (1.17).

standard deviation σ_θ, (68% of the intensity lying within $\pm\sigma_\theta$ of the central value), the value of this parameter is

$$\sigma_\theta = \frac{0.565}{\gamma}(h\nu/h\nu_c)^{-0.425} \qquad (1.18)$$

Thus, at the critical energy the full width at half maximum, $2.35\sigma_\theta$, has a value of $1.33/\gamma$. However, if one defines the total width in terms of $\pm2\sigma_\theta$ (with ~95% of the intensity within this range) then the half-width is $1.13/\gamma$, close to the earlier assumed value of $1/\gamma$. At lower photon energies the radiation fan can be much wider, while at higher energies it is narrower. Of course, all of these derivations are effectively based on a single orbiting electron. In a useful synchrotron radiation source there are many electrons circulating and not all have exactly the same orbit. The impact on the emitted radiation of the properties of the electron beams in practical sources will be discussed in Chapter 2.

Above and below the circulation plane there is, respectively, a constant phase difference of $+\pi/2$ and $-\pi/2$ between the parallel and perpendicular components of the electric field, resulting in right and left elliptically polarised radiation. Nevertheless, even integrating the full out-of-plane fan of radiation the degree of linear polarisation, P_{lin},

$$P_{lin} = \frac{I_\parallel - I_\perp}{I_\parallel + I_\perp} \qquad (1.19)$$

can be quite high at energies significantly above the critical energy, although this degree of linear polarisation falls to a value of 0.5 at low energies, as shown in Fig. 1.10.

It is these properties of synchrotron radiation emitted from the electron beam passing through bending magnets that formed the basis of all early experiments

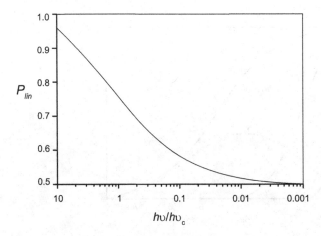

Figure 1.10 Degree of linear polarisation at different photon energies when accepting the full vertical spread of bending magnet radiation. Based on data presented by Green (1976).

seeking to use this radiation, and the design of the first machines designed specifically to offer such radiation to users. Subsequently, however, there have been many significant advances in ways to optimise the radiation provided by these sources, and indeed it is the output from so-called insertion devices, comprising arrays of additional magnets inserted in straight sections of the storage ring, rather than from traditional bending magnets, that increasingly dominates the use of these facilities. These developments and their properties are described in Chapter 2.

2 Synchrotron Radiation Sources

2.1 A Brief History

As described in Chapter 1, the basic theory of synchrotron radiation in terms of the energy loss of an electron travelling in a circular path was already published in 1898 by Liénard. Although this work predated the theory of relativity, the equations of electromagnetic theory are invariant under transformations between moving systems, so his original equations are still valid today. Of course, there are relativistic effects in the spectral and angular distribution of the output, and a very detailed treatment including these effects was first published by Schott in 1912. The fact that his equations involved Bessel functions of very high order, for which there were no available tables, probably accounted for this work being largely neglected in the following years. An historical review by Blewett (1998) provides a more detailed description of this and other early developments in the theory of synchrotron radiation. The principal interest in what we now call synchrotron radiation came from accelerator developers, because of its possible influence as a source of energy loss in the accelerated electrons. However, it was initially thought to be a weak effect and indeed, at low energies, it is (as discussed in Chapter 1). The first evidence that the effect may be significant, which subsequently led to the first direct observation of synchrotron radiation, came about through work at the General Electric (GE) Company in Schenectady in the USA. A group there were developing electron accelerators known as betatrons; these devices function on a rather different principle from the machines described in Chapter 1, in that the acceleration is achieved by magnetic induction (Kerst, 1940). A betatron is effectively a transformer in which the secondary winding is the electrons in a toroidal vacuum chamber. Of course, the importance of synchrotron radiation does not depend on the mode of acceleration of the electrons. GE was interested in developing a betatron as a means of creating an X-ray source from the impact of the accelerated electrons onto a suitable target material. Indeed, electrons from the 100 MeV betatron, which was developed by Westendorp & Charlton (1945) at GE (see Fig. 2.1), did lead to an intense and energetic X-ray source that allowed them to produce a radiograph through 10 cm thickness of steel.

The potential influence of synchrotron radiation on the operation of a betatron was explicitly presented in a short letter in *Physical Review* by Ivanenko & Pemeranchuk (1944), drawing attention to their earlier work published only in Russian. This was pursued by Blewett (1946) at GE, who searched for evidence of this effect in the

Figure 2.1 The 100 MeV GE betatron. Reprinted from Westendorp & Charlton (1945) with the permission of AIP

100 MeV betatron. He initially searched for, and failed to find, direct evidence for the emission of synchrotron radiation in the radio frequency range, but subsequently acknowledged that with hindsight the theoretical work of Schwinger (1949), of which he had knowledge before it was published, indicated that most of the synchrotron radiation energy would be in the infrared and visible range, which he had not been able to investigate because the electron containment vessel was not transparent. However, he did show that a shift in the orbit of the circulating electrons could be attributed to the effect of synchrotron radiation losses. Quite soon afterwards, however, the researchers at GE built one of the first synchrotrons (the very first demonstration of such a device was by Goward & Barnes (1946) in the UK), operating at an energy of 70 MeV, and in this case they were able to observe directly, through a glass-walled vessel, the synchrotron radiation in the visible spectral range (Elder et al., 1947, 1948). The events surrounding this discovery, and to whom it should be attributed, have been described by Pollack (1983) and Blewett (1998). Fig. 2.2 shows a photograph of this small (~30 cm radius) GE synchrotron with the emitted visible synchrotron radiation arrowed.

Despite this discovery of synchrotron radiation, a significant period of time passed before the effect started to be exploited. Of course, in the case of low-energy synchrotrons, the spectral range of the resulting synchrotron radiation (from the infrared to the near ultraviolet) did not offer very significant advantages over other, more conventional, sources. However, higher-energy synchrotrons were certainly being built, mostly to conduct particle physics experiments involving particle–particle collisions, and with suitable agreement it was possible to 'make a hole in' these accelerators to extract the synchrotron radiation, with no obvious detriment to the particle physicists. This was first demonstrated by Tomboulian & Hartmann (1956),

Figure 2.2 Photograph of the 70 MeV GE synchrotron with the emitted visible synchrotron radiation arrowed clearly seen. After Blewett (1998); reproduced with permission of the International Union of Crystallography

who reported detailed measurements of the spectral output from the 300 MeV Cornell University synchrotron in the spectral range 60–300 Å (~40–200 eV). Soon afterwards, similar facilities were constructed at many synchrotrons around the world. Although this 'parasitic' mode of operation opened up the possibility of a range of experiments exploiting synchrotron radiation up to quite hard X-ray energies, there were significant underlying limitations. A synchrotron is designed to accelerate electrons to some specified energy, optimised for the particle physicists, but then to extract the accelerated electrons to undergo a collision event to investigate the consequences. This process of injection, acceleration and extraction was typically performed 50–60 times each second. The synchrotron radiation output was therefore not constant but pulsed at 50–60 Hz with a time-varying output within each pulse as the electron energy was ramped up. Moreover, both the machine energy and the circulating electron current were defined by the particle physics experiments; for some of these experiments low currents were required, leading to very low intensities of synchrotron radiation.

Clearly what was needed by synchrotron radiation users were machines in which the electron beam could be accelerated to the desired energy but then *not* extracted; rather, the beam should be 'stored' in the synchrotron, formally turning the electron synchrotron into an electron storage ring. Such a source has a constant machine energy and thus a constant spectral output, with no low-frequency modulation (although high-frequency modulation arises because the electrons in such a device are localised in 'bunches', as described in more detail below). Some storage rings were built for colliding-beam particle physics experiments, and the parasitic use of these devices clearly offered some advantages over the parasitic use of pulsed synchrotrons.

Ultimately, though, the requirement was for electron storage rings designed for use exclusively as dedicated synchrotron radiation sources. There are distinct characteristics in the optimal design properties of a storage ring, in terms of stored beam lateral dimensions and angular divergence, which differ for particle physics and synchrotron radiation sources. In fact, the first storage ring facility dedicated for use as a synchrotron radiation source was not originally designed for this purpose. The 240 MeV electron storage ring at Stoughton, Wisconsin, was first constructed as an accelerator test device, but after its funding for this purpose was terminated in 1967, new funding was achieved to complete its construction and operate it as a dedicated synchrotron radiation facility, first delivering synchrotron radiation in 1968. In view of its troubled funding history, it was named Tantalus after the mythological figure doomed to eternal punishment by having to stand under a fruit tree with fruit that ever eluded his grasp (he was also stood in a pool of water that always receded before he could take a drink!). The first storage ring specifically designed for dedicated synchrotron radiation exploitations was the 300 MeV SOR-RING at the University of Tokyo, commissioned in 1974 (Miyahara et al., 1976), while the first such ring operating at the higher energies required for use in the hard X-ray spectral range was the SRS (Synchrotron Radiation Source) at Daresbury in the UK, commissioned in 1981. This was soon followed by many other 'second generation' synchrotron radiation sources in which, at least initially, the radiation was obtained only from bending magnets as described in Chapter 1. More recently, however, a third generation of sources has been commissioned and built that is designed particularly to exploit 'insertion devices' – novel magnet structures that are located in the straight sections of the storage ring between the bending magnets. These are described more fully later in this chapter.

2.2   Electron Storage Rings

In terms of the basic layout of the machine, electron storage rings are essentially the same as synchrotrons. The key ingredients are bending (dipole) and focusing (quadrupole and sextupole) magnets (sometimes incorporated into the same magnetic structure); an RF cavity for power input; and some form of kicker magnet to align the injection path with that of the electrons travelling around the ring. In pulsed synchrotrons, of course, the RF power is used to accelerate the electrons from the injection energy to the energy that the electrons are extracted. In a storage ring, the RF power is needed to re-accelerate electrons that have lost energy due to the synchrotron radiation losses; it also plays an important role in optimising the beam lifetime, for example by influencing the tolerance to energy spread of the electrons in the ring. The phasing of the time-dependence of the RF standing wave in the cavity, and the arrival of the electron bunches, is such as to accelerate the electrons by differing amounts depending on how much energy they have lost due to synchrotron radiation in each circuit of the storage ring. Fig. 2.3 shows schematically how this works. An electron with the nominal ideal energy to be expected after a single circuit of the ring arrives in the RF cavity at a time that corresponds to it receiving exactly the correct amount of energy gain to compensate its energy loss due to synchrotron radiation in its single passage around the ring. This is referred to as the synchronous particle. An electron with a slightly lower

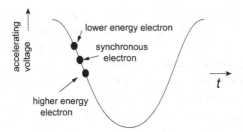

Figure 2.3 Schematic diagram showing how the accelerating voltage received by electrons of different energy entering the RF standing wave cavity depends on the relative time of their arrival.

energy actually arrives in the RF cavity earlier because it has essentially the same speed as the synchronous particle (c), but its lower energy means it is more strongly bent by the bending magnets and so follows a trajectory with a slightly smaller radius around the ring. Arriving earlier means that it experiences a higher accelerating voltage and thus a larger energy gain. Similarly, an electron arriving with a higher energy than the synchronous particle arrives later and gets a smaller accelerating voltage. This 'synchronism', as the name implies, is a key property of a synchrotron and the associated storage ring. Of course, one negative feature of the problem that this aims to address is that there must be oscillations in the energy and in the lateral displacement of the circulating electrons from the nominal ideal trajectory. In fact, the emission of synchrotron radiation can generally lead to the damping of these oscillations. As described in Chapter 1, the energy loss per circuit around the ring scales as the fourth power of the electron energy, so the most energetic electrons lose more energy and the less energetic electrons lose less energy.

In all early synchrotron radiation sources based on storage rings, the energy of the electrons that were initially injected was lower than the final operating energy of the source, so the storage ring was actually used briefly as a synchrotron accelerator immediately after initial injection. Newer sources use full energy injection, overcoming the need for the machine to fulfil this dual function as well as making it possible to 'top up' the current in the ring during operation, as described more fully below. High-energy injection is generally effected from a separate ('booster') synchrotron, a linac or a combination of the two.

Fig. 2.4 is a schematic diagram showing the key elements of a third-generation synchrotron radiation source. The basic magnet structure of the storage ring is similar to that of the synchrotron shown in Fig. 1.5, and in both cases, synchrotron radiation is emitted tangentially from each bending magnet. These radiated beams are located in a sequence of vacuum vessels that also contain components to focus, monochromate and/or filter the radiation delivered to experiments at the end of these beamlines. Within the constraints of this simplified diagram, what distinguishes this as a third-generation (rather than a second-generation) synchrotron radiation source is the elongation of the straight sections between the magnet groups, and the presence of insertion devices in many of these straight sections. From a machine designer's

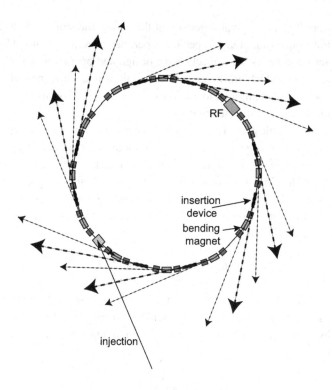

Figure 2.4 Schematic diagram of the electron storage ring layout used in a third-generation synchrotron radiation source. The short-dashed lines represent radiation from bending magnets, the longer-dashed bold lines represent radiation from insertion devices. These radiation beams are contained in the tangential vacuum vessels of the different beamlines.

perspective, a key difference is the magnetic lattice that makes these long straights possible but also ensures the electron beam has appropriate lateral and angular dimensions in these straight sections.

2.3   Insertion Devices: Wigglers and Undulators

2.3.1   Wavelength Shifters and Multipole Wigglers

As described in Chapter 1, synchrotron radiation, the radiation emitted when relativistic-energy electrons are bent through magnetic fields, consists of a very broad spectral continuum that can cover the range from long wavelength far infrared (also known as terahertz radiation) to hard X-rays. The critical wavelength or energy of the radiation delivered by the machine is determined by the electron beam energy and the magnet bending radius, so differences in these parameters at different machines shift the generic spectrum (Fig. 1.8) up or down in photon energy, but this spectrum is the same from all the bending magnets on a given source. The idea of insertion devices, which comprise arrays of additional magnets that perturb the beam within a straight section but restore it to its original trajectory at the end of the straight, is to produce different spectral outputs tailored to specific experimental requirements. Of course, this simple description belies the impact on the design and operation of such a third-generation light source, which must cope with the additional resulting perturbations on the circulating beam, especially due to the introduction of these localised 'errors' that disturb the symmetry of the magnetic lattice.

Conceptually the simplest insertion device is a *wavelength shifter*, which is a type of *wiggler*. Its function is to provide a source with a higher critical energy than that of the bending magnets in the storage ring. This can be achieved (at the fixed electron energy of the ring) by using a higher magnetic field strength than that of the bending magnets, thus reducing the local bending radius. As described in Chapter 1, the critical photon energy in bending magnet radiation is related to the magnet bending radius, R, by

$$h v_c = \hbar \omega_c = \hbar \frac{3}{2} \gamma^3 \omega_0 = \hbar \frac{3c}{2R} \gamma^3. \tag{1.11}$$

It is convenient for the discussion of insertion devices to re-express this in terms of the magnetic field strength. The starting point to achieve this is to equate the Lorenz force experienced by a moving electron, velocity $\mathbf{v}$, in a magnetic field $\mathbf{B}$, with the centripetal force

$$e\mathbf{v} \times \mathbf{B} = \frac{m_0 v^2}{R}, \tag{2.1}$$

but in the highly relativistic case, the rest mass m_0 must be replaced by the relativistic mass γm_0 while v is essentially c, whence

$$ecB = \frac{\gamma m_0 c^2}{R} \tag{2.2}$$

and so

$$R = \frac{\gamma m_0 c}{eB} = \frac{E}{ecB}, \tag{2.3}$$

where E is the electron energy ($\gamma m_0 c^2$). In practical units, this reduces to $R(\mathrm{m}) = E(\mathrm{GeV})/B(\mathrm{T})$; so if B is ~1 T, as in typical bending magnets, and E is a few GeV, then R is a few metres. Using this relationship allows the critical energy (Equation (1.11)) to be re-expressed as

$$h v_c = \hbar \omega_c = \hbar \frac{3eB}{2m_0} \gamma^2, \tag{2.4}$$

so the critical energy of the emitted radiation (at a fixed electron energy) depends linearly on the magnetic field strength, B. Thus, if one can insert a magnet with a field strength of 6 T into the straight section of a storage ring with 1 T bending magnets, the critical energy of radiation from this magnet is increased by a factor of 6 relative to that from the bending magnets; for a 3 GeV storage ring, this shifts the critical energy from ~6 keV to ~36 keV. Fig. 2.5 shows the spectral output for these two cases, assuming a circulating electron current of 200 mA. The shape of the spectrum is unchanged, but the higher field wavelength shifter causes the spectrum to shift to higher energy, greatly increasing the output at higher energies.

Of course, this high-field magnet strongly deflects the electron beam, which must be restored to its original trajectory if it is not to be lost from the storage ring. Fig. 2.6 illustrates how this can be achieved. Fig. 2.6(a) is a schematic diagram of the electron trajectory along the straight section of the storage ring with three superimposed dipole magnets: the central high-field magnet and two smaller magnets that direct the electrons into the central magnet and back on to the original trajectory within the straight. Fig. 2.6(b) is a graphical representation of the variation in magnetic field

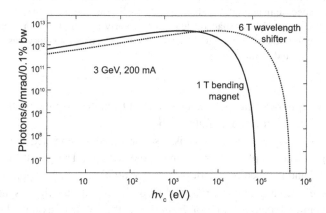

Figure 2.5 Spectral output from a 1 T bending magnet and a 6 T wavelength shifter for a beam energy of 3 GeV and a circulating current of 200 mA.

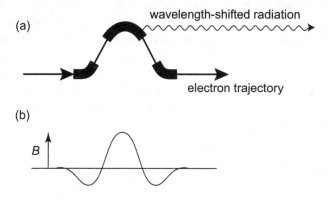

Figure 2.6 (a) Schematic diagram showing the trajectory of electrons in a straight section passing through the three dipole magnets of a wavelength shifter. (b) the magnetic field variation along the length of this device.

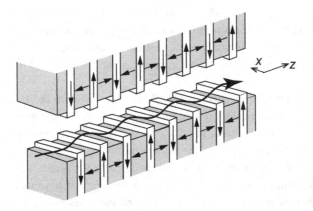

Figure 2.7 Schematic diagram of the general magnet structure of either a multipole wiggler or an undulator. The specific magnet design here is a 'hybrid' type based on permanent magnet blocks (shaded) and iron poles (unshaded) that is particularly favoured in short-period undulators. The amplitude of the 'wiggles' suffered by the electron trajectory passing through it are exaggerated in the diagram.

through this device. Notice that the radiation emitted from the 'wiggle' of the beam in this device peaks in the direction parallel to the straight section.

A rather more advanced device that is also designed to enhance the radiation intensity of higher-energy photons is the multipole wiggler. The general concept is shown in Fig. 2.7 and is based on a periodic array of dipole magnets of opposite sense that causes the electrons passing through it to oscillate left and right, synchrotron radiation being emitted along the axis of the device at the points of extreme lateral left and right displacement. If there are N periods in the magnetic lattice, the radiation emitted along the axis is thus $2N$ times larger than that induced by each 'wiggle'. The strength of the periodic magnetic field in such a device with a reasonably short (~5 cm) period cannot reach the highest values that can be achieved in a single-wiggle wavelength shifter, so the shift of the spectral output to higher energies is smaller, but the multipole device also shifts the whole spectrum up in intensity by a factor of $2N$ (or by only a factor of N if the acceptance angle of the beamline is such that only one side of the 'wiggle' can be 'seen'). The combined effect is thus a large enhancement of the intensity of radiation, this being particularly valuable at the higher energy range of the bending magnet output. Multipole wigglers, some using superconducting magnet technology producing periodic fields of 3 T or more, have thus become the insertion devices of choice for enhancing the intensity of higher-energy radiation on storage ring sources.

2.3.2   Undulators

If the magnetic field strength in such a periodic magnet structure is more modest, the device becomes referred to as an undulator (strictly, a planar undulator – other magnetic structures introducing periodic transverse fields also exist as discussed later), and the mode of operation and the spectral output is fundamentally different. This is because radiation from successive wiggles (undulations) can become coherent. The key factor determining whether such a device operates as a wiggler or an undulator is the amplitude of the angular variation suffered by the electrons in their passage through the device. Defining the lateral shift coordinate of the electron passing through the device as x, and the coordinate along the axis as z, and assuming that the angular deflections in the device are small, the curvature of the path in the local field B can be written as

$$\frac{d^2x}{dz^2} = \frac{1}{R} = \frac{eB}{\gamma m_0 c}. \tag{2.5}$$

So if the magnetic field is assumed to be sinusoidal in distance with a period λ_0,

$$B = -B_0 \sin\left(\frac{2\pi z}{\lambda_0}\right), \tag{2.6}$$

then integrating Equation (2.5) gives the horizontal angular deflection from the z axis

$$\frac{dx}{dz} = \frac{B_0 e}{\gamma m_0 c} \frac{\lambda_0}{2\pi} \cos\left(\frac{2\pi z}{\lambda_0}\right) = \frac{K}{\gamma} \cos\left(\frac{2\pi z}{\lambda_0}\right), \tag{2.7}$$

where K is known as the 'deflection parameter' or 'undulator parameter':

$$K = \frac{B_0 e}{m_0 c} \frac{\lambda_0}{2\pi} = 93.36 B_0 \lambda_0, \tag{2.8}$$

with B_0 in T and λ_0 in m. Equation (2.7) shows that the peak angular deflection is K/γ. However, as described in Chapter 1, the half-angle defining the spread of synchrotron radiation is $\sim 1/\gamma$; so if $K > \sim 1$, the angular width of this spread is less than the angular range of the variation in the central direction of emission along the electron trajectory. In this case, an observer along the axis sees the radiation flashing on and off (Fig. 2.8(a)). By contrast, if $K < \sim 1$, the observer on axis sees the radiation being emitted at all points on the electron trajectory continuously (Fig. 2.8(b)). This simple argument thus leads to the conclusion that a K value of (approximately) unity is the nominal point at which a high-K wiggler becomes a low-K undulator, although in reality the transition between the two regimes is much less sharp than implied by this. As will become clear later in this chapter, the spectral output of such a device shows undulator characteristics at significantly higher values of K. Notice that integrating Equation (2.7) gives the amplitude of the lateral displacements of the electron beam as it passes through the device

$$x(z) = \frac{K}{\gamma} \frac{\lambda_0}{2\pi} \sin\left(\frac{2\pi z}{\lambda_0}\right). \tag{2.9}$$

For example, if $B_0 = 1$ T, $\lambda_0 = 5$ cm and the electron energy is 3 GeV, then $K \approx 5$ and these lateral oscillations have an amplitude of ~ 100 μm.

(a) wiggler

(b) undulator

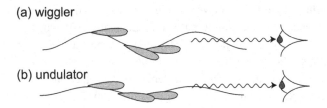

Figure 2.8 Schematic diagram showing the difference in the visibility of emitted synchrotron radiation along the axis of a wiggler and an undulator. The amplitude of the lateral movement of the electron is greatly exaggerated.

As remarked above, the key property of an undulator that distinguishes it from a multipole wiggler in terms of the nature of its radiated spectrum is that the emissions of radiation from equivalent points in the periodic magnetic field are coherent. In order to understand the conditions to achieve this coherent interference, it is necessary to first establish the speed of the electron along the axis of the device as it undergoes its transverse modulations. Expressing the speed of the electron along its (undulating) trajectory in terms of the parameter $\beta = v/c$ and assuming the angular deviation of the electron trajectory, given by Equation (2.7), is small, the transverse velocity can be written as

$$\frac{dx}{dt} = \beta_x c = \beta c \frac{K}{\gamma} \cos\left(\frac{2\pi z}{\lambda_0}\right), \tag{2.10}$$

while

$$\beta_z^2 = \beta^2 - \beta_x^2, \tag{2.11}$$

with $\beta_z = (dz/dt)/c$ so

$$\beta_z^2 = \beta^2 - \frac{K^2}{\gamma^2} \cos^2\left(\frac{2\pi z}{\lambda_0}\right). \tag{2.12}$$

Using the identity $\cos 2A = 2\cos^2 A - 1$ gives

$$\beta_z^2 = \beta^2 - \frac{K^2}{\gamma^2}\left(\frac{1}{2} + \frac{1}{2}\cos\left(\frac{4\pi z}{\lambda_0}\right)\right), \tag{2.13}$$

and using the approximation $(1 - x)^{1/2} \approx 1 - x/2$ for small x leads to

$$\beta_z \approx \beta\left(1 - \frac{K^2}{4\beta^2\gamma^2} - \frac{K^2}{4\beta^2\gamma^2}\cos\left(\frac{4\pi z}{\lambda_0}\right)\right). \tag{2.14}$$

The first two terms on the right-hand side of this equation are constant, while the third term is oscillatory. The average electron velocity along the axis, $\hat{\beta}_z$, is thus given by the first two terms alone, and using $\beta = \sqrt{(1 - (1/\gamma^2))}$ (whence $\beta^2\gamma^2 = \gamma^2 - 1$) leads to

$$\hat{\beta}_z \approx 1 - \frac{1}{2\gamma^2} - \frac{K^2}{4\gamma^2}. \tag{2.15}$$

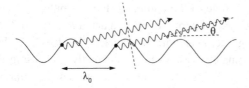

Figure 2.9 Schematic diagram showing the emission of radiation from two equivalent points on the electron trajectory through an undulator at a phase matching condition.

Fig. 2.9 shows schematically the condition required for coherent interference between radiation emitted at two equivalent points in the periodic magnetic field of an undulator. Although the electron is travelling at almost the speed of light, the fact that it has a transverse modulation in its trajectory means that it takes longer than the emitted radiation to cover the distance λ_0 than does the light, so the coherence condition is that this time difference is an exact multiple of the period (the inverse frequency) of the emitted radiation. Specifically, the time the electron takes to cover one period, λ_0, is $\left(\lambda_0 / c\hat{\beta}_z\right)$, and in this time the first wavefront of the radiation emitted in the previous period has travelled a distance $\left(\lambda_0 / \hat{\beta}_z\right)$. The separation between these two wavefronts travelling at an angle θ to the axis is therefore

$$d = \frac{\lambda_0}{\hat{\beta}_z} - \lambda_0 \cos\theta \tag{2.16}$$

and for constructive interference this must equal an integral number of wavelengths of the emitted radiation, namely,

$$n\lambda = \frac{\lambda_0}{\hat{\beta}_z} - \lambda_0 \cos\theta. \tag{2.17}$$

Substituting from Equation (2.15) and using the approximation $1/(1-x) \approx 1+x$ gives

$$n\lambda \simeq \lambda_0 \left(1 + \frac{1}{2\gamma^2} + \frac{K^2}{4\gamma^2}\right) - \lambda_0 \cos\theta \tag{2.18}$$

and using the identity $(1 - \cos A) = 2\sin^2(A/2)$ and the small angle approximation $\sin\theta \approx \theta$ leads to

$$n\lambda \simeq \lambda_0 \left(\frac{\theta^2}{2} + \frac{1}{2\gamma^2} + \frac{K^2}{4\gamma^2}\right) = \frac{\lambda_0}{2\gamma^2}\left(1 + \frac{K^2}{2} + \theta^2\gamma^2\right), \tag{2.19}$$

giving the standard undulator equation for the wavelength of the emitted radiation

$$\lambda = \frac{\lambda_0}{2n\gamma^2}\left(1 + \frac{K^2}{2} + \theta^2\gamma^2\right). \tag{2.20}$$

Several features of this result are immediately striking. Firstly, the dependence of the emitted wavelength on γ^{-2} means that this radiation wavelength is vastly smaller than the period of the magnetic field of the undulator. For example, at an energy of 3 GeV ($\gamma \sim 6{,}000$) an undulator with a period of 50 mm and a K value of 3 leads to a first harmonic ($n = 1$) along the axis with a wavelength of ~ 40 Å. In effect, millimetres transform approximately to ängströms. Secondly, increasing the magnetic field strength increases the value of K and thus increases the emitted wavelength, thereby *reducing* the photon energy. This is the opposite effect to that in bending magnet radiation in which the critical photon energy increases with increasing magnetic field strength. Of course, increasing the magnetic field in an undulator does not reduce the total power output (indeed, this increases, consistent with the foregoing discussion of bending magnet radiation), but it does lower the photon energy of each harmonic. Perhaps even more notable is the fact that the output from an undulator consists of well-separated harmonics and not the broad continuum of bending magnet radiation. This is one of the particular advantages of undulators: the spectral output can be optimised for particular experiments in a specific photon energy range. The photon energy of each harmonic can be tuned by changing the magnetic field strength and thus the value of K, generally achieved by varying the gap between the magnet pole pieces.

Before exploring further the properties of undulator radiation it is instructive to describe an alternative way of deriving the undulator equation by considering the movement of an electron in the frame of reference moving along the undulator with the average speed of $\hat{\beta}_z c$ (Equation (2.15)), alternatively expressed as a relativistic factor $\hat{\gamma}_z$ where

$$\hat{\gamma}_z = \frac{1}{\left(1 - \hat{\beta}_z^2\right)^{1/2}} = \frac{\gamma}{\left(1 - K^2/2\right)^{1/2}}. \tag{2.21}$$

The electron in this reference frame sees the undulator approaching at this speed, so it sees the undulator period Lorentz contracted to a value $\lambda_0/\hat{\gamma}_z$. Because, for a low value of K, the transverse velocity β_x is low (non-relativistic), the resulting oscillations will cause it to emit dipole radiation only at this wavelength (i.e. with no significant transfer of energy into higher harmonics). Transforming into the laboratory frame this wavelength undergoes a relativistic Doppler shift to

$$\lambda = \lambda_0 \hat{\gamma}_z \left(1 - \hat{\beta}_z \cos\theta\right) \approx \frac{\lambda_0}{2\,\hat{\gamma}_z}\left(1 + \hat{\gamma}_z^2\theta^2\right) \tag{2.22}$$

and substituting into this equation the value of $\hat{\gamma}_z$ from Equation (2.21) leads to the undulator Equation (2.20) for the first harmonic. Radiation from a low-K undulator can thus be viewed as relativistically transformed dipole radiation resulting from the transverse oscillations that the electron undergoes as it passes through the device. Notice that this discussion considers only the impact of the transverse oscillations (in the x direction) of the electron in its moving frame. There are also smaller oscillations in the z direction that give rise to weaker radiation with a completely different angular distribution and polarisation, as described later in this chapter.

In view of the harmonic character of undulator radiation an obvious question is: what is the bandwidth of this radiation? Indeed, one might ask if one could use a single harmonic for an experiment without any need for high-resolution monochromators that are an essential feature of most beamlines, as described in Chapter 3 (although one might expect to need to filter out other harmonics in some way, perhaps with a low-resolution monochromator). How useful this might be depends on the harmonic bandwidth. If an undulator contains N_u periods then the general condition for constructive interference is (cf. Equation (2.17))

$$N_u n\lambda = \frac{N_u \lambda_0}{\hat{\beta}_z} - N_u \lambda_0 \cos\theta. \tag{2.23}$$

As the wavelength is changed, the ideal phase matching is lost, until at a wavelength λ' one obtains destructive interference and thus zero emitted intensity. This occurs when

$$N_u n\lambda' + \lambda' = \frac{N_u \lambda_0}{\hat{\beta}_z} - N_u \lambda_0 \cos\theta, \tag{2.24}$$

because the emitted wave from the first source point in the undulator is exactly out of phase with the wave emitted from the centre of the undulator; similar matching pairs of waves are emitted at all points further along the undulator. This means that $Nn\lambda' + \lambda' = Nn\lambda$ so the range over which there is emission is determined by $\Delta\lambda = \lambda - \lambda'$ which leads to a bandwidth of the harmonic line of

$$\frac{\Delta\lambda}{\lambda} \approx \frac{1}{N_u n}. \tag{2.25}$$

Notice, of course, that the undulator equation depends on θ, with the emitted wavelength increasing off-axis, so to benefit from the narrow bandwidth it is necessary to restrict the angular range of radiation that is accepted by an experiment. In practice most experiments require significantly narrower bandwidths than this, so a monochromator between the undulator and the experiment is still required (Chapter 3).

A similar argument to that used to determine the bandwidth allows one to determine the angular width of a single harmonic emission because destructive interference occurs at an angle θ' defined by

$$N_u n\lambda + \lambda = \frac{N_u \lambda_0}{\hat{\beta}_z} - N_u \lambda_0 \cos\theta', \tag{2.26}$$

whence $N_u \lambda_0 \cos\theta' + \lambda = N_u \lambda_0 \cos\theta$. For the case of radiation emitted around the axis of the device ($\theta = 0$), applying the approximation $\cos\theta \approx 1 - \theta^2/2$ leads to an expression for the off-axis angle at which the intensity falls to zero of

$$\Delta\theta = \sqrt{\frac{2\lambda}{N_u \lambda_0}}. \tag{2.27}$$

Taking the example used above of an electron energy of 3 GeV, an undulator with a period of 50 mm, a K value of 3 and 100 periods leads to a first harmonic ($n = 1$) with an angular width of ~40 μrad, much narrower than the value for conventional bending magnet radiation at this energy of ~170 μrad. In the limit of very small K (taking only the first term in Equation (2.20)) the angular width expression of Equation (2.27) reduces to $\Delta\theta = \sqrt{\frac{1}{N_u n \gamma^2}}$ but this approximation is clearly far from valid with $K = 3$. Indeed, notice that, as is clear from the undulator Equation (2.20), the wavelength of the emitted radiation depends on the off-axis angle, so the actual spectral output delivered to a beamline and on to an experiment can depend quite strongly on the acceptance angle of the beamline and its optics. In this regard one further feature of the spectral output of a planar undulator is that along the axis only odd harmonics are detected; radiation in even harmonics is emitted, but this has zero intensity along the axis and peaks in intensity to either side of the axis.

One way to rationalise this absence of axial emission at even harmonics (and some other aspects of the harmonic output) is to consider the Fourier transform of the time

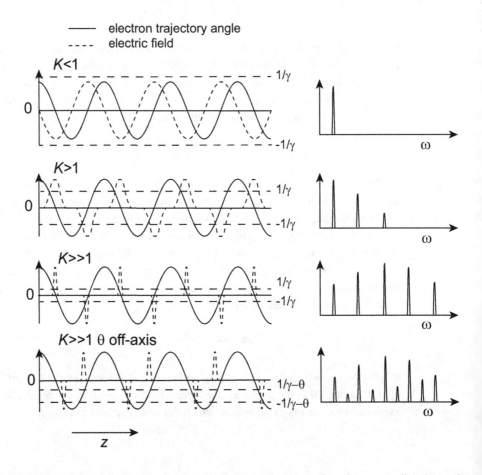

Figure 2.10 Schematic diagram showing the electron trajectory angle and corresponding detected electric field variations together with the resulting radiation spectrum (Fourier transform) for different values of K and detection angle.

dependence of the electric field variation seen by an observer. Fig. 2.10 shows, on the left-hand side, the angular variation of the electron trajectory as is passes along the undulator, together with the electric field 'seen' by an observer, for different values of K, and for observation along and off the undulator axis. On the right hand side is shown schematically the Fourier transform, which corresponds to the harmonic output. If K is very small ($\ll 1$) the angular excursions experienced by the electron ($\sim K/\gamma$) are much smaller than the synchrotron radiation emission cone ($\sim 1/\gamma$) and this observer 'sees' all of the emission cone at all times, experiencing a continuous sinusoidally varying electric field. The Fourier transform is thus just a single frequency – the first harmonic radiation. For larger values of K, however, when the angular excursions of the electron beam are larger than the synchrotron radiation cone, the observer sees only flashes of radiation as the emission cone passes over the axial direction. As these flashes become sharper (shorter) at higher values of K the Fourier transform contains higher and higher frequency components. However, on axis, these flashes are evenly-spaced in time and of alternating amplitude, so the Fourier transform contains only odd harmonics. Off axis, however, the alternating amplitude components are no longer periodic in time, and even harmonics appear in the Fourier transform.

Notice that the very much larger number of harmonics that are excited at high values of K (for a $K \sim 20$ thousands of harmonics contribute to the emitted spectrum) means that at high energies one approaches a continuum of radiation. In this way the description of a high-K undulator becomes compatible with the simpler description ascribed to such a device earlier in the section – namely a multipole wiggler emitting bending-magnet type synchrotron radiation with an intensity enhanced by a factor of $2N_u$ due to the many deflecting magnets. Even at high values of K, however, the spectral output at the lowest energies, corresponding to very low harmonics, is undulator-like, with intense harmonic output and a low continuum background. This actually highlights one limitation of the use of undulators for a wide range of experiments on any particular storage ring. Bending magnet radiation is a continuum covering many decades of useful photon energies. For a given machine energy, however, the useful range of photon energies that can be achieved by undulators (i.e. devices with relatively low values of K) is much narrower. In practice there is a rather small range of periodicities of magnet structures that can be used.

The shortest period achievable is determined by the design of the magnet structures together with the constraint of the aperture required for the circulating electron beam to pass through, which limits the smallest useable gap of the magnet pole pieces. For undulators based on electromagnets the extra space occupied by the electrical windings is a significant constraint on the shortest period device that can be built. For this reason, so-called hybrid magnet structures, with alternating permanent magnet blocks and iron poles (see Fig. 2.7) are used for short-period undulators to overcome this physical constraint. Notice that these magnet structures, with alternating rectangular magnetic pole pieces, do produce a harmonic variation of the field along the axis of the device and thus a harmonic modulation of the electron beam trajectory as has been assumed in the foregoing discussion. A more fundamental lower limit to the period is that it must be greater than about twice the size of the gap between the poles, otherwise there is severe magnetic flux leakage. The smallest achievable value of this gap is determined by the characteristics of the electron beam in the storage ring. Traditionally insertion device magnets were placed outside the vacuum vessel of the ring, but smaller gaps (and higher values of K) can be achieved using in-vacuum magnets. The ultimate limit is determined by the aperture required for the circulating electron beam which in the most modern sources can be as small as a few mm.

The longest useful period is determined by the length of the straight sections. Straight sections of storage rings are typically ~5 m or less in length, so if undulators with at least ~20 periods are to be used the practical values of λ_0 are only in the range of ~25–250 mm (and generally less) – only one decade of values of first harmonic radiation. The resulting useful spectral range is extended by the use of higher harmonics (generally not much beyond $n \sim 9$), so this means the effective range of undulator-derived photon energies at a given storage ring energy is no more than about 2 decades. Moreover, to achieve the lowest photon energies requires operating at high K values at which, of course, most of the output radiation is actually at (unwanted) much higher energies.

The other key question regarding an undulator is how much radiation is emitted into the central axial cone at the various harmonics. This is usually written in terms of the number of photons emitted per unit time, $dn_{ph}/dt = \dot{n}_{ph}$. The number of photons/s/mrad^2/0.1% bandwidth is then given by

$$\frac{d^2\dot{n}_{ph}}{d\omega/\omega d\Omega}\bigg|_{\theta=0} = 4.54 \times 10^7 N_u^2 \gamma^2 F_n(K) I_b, \tag{2.28}$$

where I_b is the average electron beam current in amps. F_n is a function defining the relative contributions to the different harmonics

$$F_n(K) = \frac{n^2 K^2}{\left(1 + K^2/2\right)^2} \left(J_{(n+1)/2}(Y) - J_{(n-1)/2}(Y)\right)^2, \tag{2.29}$$

with $J(Y)$ being a Bessel function of $Y = \dfrac{nK^2}{4\left(1 + K^2/2\right)}$.

F_n is shown in Fig. 2.11. As predicted in the simple Fourier transform arguments described above, for small values of K (significantly less than unity) the spectral output contains only the first harmonic, but as K increases more and more higher harmonics are emitted and the total power output increases very significantly. Indeed, the total power output is proportional to B_0^2, and thus to K^2, but as K increases the (horizontal) angular spread of the radiation increases, so the gain in the on-axis power density is more modest.

Of course a key feature of the intensity emitted from an undulator, as defined by Equation (2.28), is that it is proportional to N_u^2, rather than the linear dependence on N_u that characterises a multipole wiggler. This is a consequence of the coherence in the emission from different parts of the electron trajectory, leading to the emitted

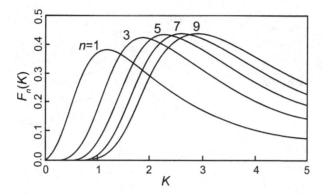

Figure 2.11 On-axis undulator flux density function $F_n(K)$ for several harmonics.

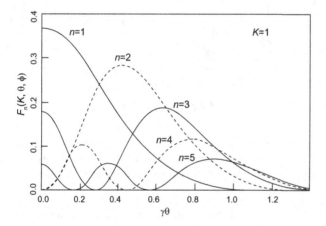

Figure 2.12 Dependence of the flux density on the off-axis angle in the horizontal plane for $K = 1$ (Equation (2.29)). After Walker (1993). Copyright (1993) with permission from Elsevier

amplitudes being added in an undulator, rather than the intensities being added in a (incoherent) multipole wiggler.

The angular distribution of the emitted radiation is given by a similar expression to Equation (2.28) but with $F_n(K)$ replaced by $F_n(K, \theta, \phi)$. The dependence of this function on θ, the off-axis angle in the horizontal plane, is shown for a value of $K = 1$ in Fig. 2.12.

As a more specific example, Fig. 2.13 shows the spectral output, expressed as the central flux density, from an undulator with a length of 2 m and a period of 5 cm on a storage ring operating at 3 GeV with a circulating current of 300 mA, calculated using the SPECTRA code of Tanaka & Kitamura (2001). The main figure in Fig. 2.13 shows the tuning curves of the different harmonics as the K parameter of the undulator is varied (increased magnetic field strength and thus K value being achieved by closing the gap between the opposing magnet structures). The inset shows the spectrum

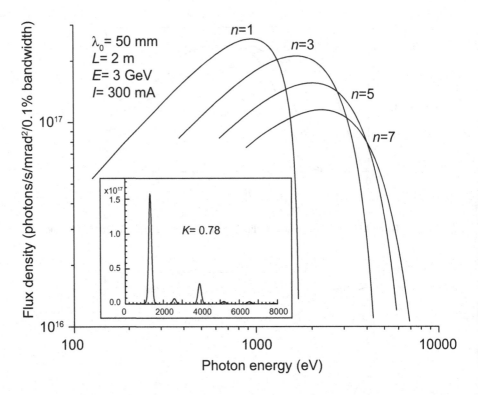

Figure 2.13 Calculated output from an undulator with a 5 cm period operating at an electron energy of 3 GeV. The main figure shows the tuning curves of the lower harmonics achieved by varying K, while the inset shows the spectral output at a K value 0.78.

achieved at a value of $K = 0.78$. Notice that weak even harmonics in this nominally on-axis geometry are visible because the calculation includes a finite size of detector and also a finite range of angles of the electron trajectory.

By contrast, Fig. 2.14 shows the calculated output spectrum from the same undulator, but at a higher value of K of 4.0. At low photon energies, the output shows the dominance of the harmonic output characteristic of undulator behaviour, as also seen at the much lower value of K shown in the inset of Fig. 2.13, although the energy of the first (and higher) harmonic(s) is much reduced when using the higher value of K. However, at the highest energies the spectrum becomes dominated by the continuum expected of a multipole wiggler. The dashed line shows the output calculated if the output were to be simply an incoherent sum of the bending magnet radiation from the successive dipole magnets in the undulator (i.e. if the device were regarded solely as a multipole wiggler).

This example also highlights a problem remarked upon earlier, that to achieve lower energy harmonic radiation from an undulator one may have to increase the value of K well above the nominal value of unity that characterises the idealised undulator situation, leading to a large increase in the generation of (unwanted) higher energy radiation. Recall (Section 1.2) that the critical photon energy in bending

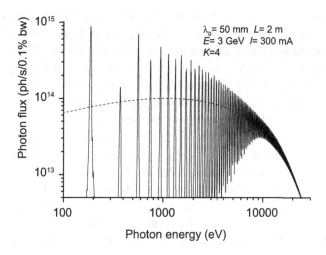

Figure 2.14 Calculated axial flux from the same undulator as in Fig. 2.13, but with a K value of 4.

magnet radiation, which occurs close to the peak in the spectral output (as shown by the dashed line in Fig. 2.14), corresponds to the value at which half the emitted power is at lower photon energies while half is at higher photon energies. Evidently the low harmonics in Fig. 2.14 correspond to photon energies well below this half-power value. In this regard it is interesting to determine the harmonic number that corresponds to the critical energy of the underlying synchrotron radiation continuum. This critical energy was shown in

$$\hbar\omega_c = \hbar\frac{3c}{2R}\gamma^3 \tag{1.11}$$

and using Equations (2.2) and (2.8) one can re-express this in terms of K rather than R as

$$\hbar\omega_c = \hbar\frac{3\pi\gamma^2 cK}{\lambda_0}. \tag{2.30}$$

From the undulator Equation (2.20) the photon energy of the nth harmonic on the axis is

$$\hbar\omega_n = \frac{2\pi c}{\lambda_n} = \frac{4\pi n c\gamma^2}{\lambda_0\left(1 + K^2/2\right)} \tag{2.31}$$

so defining the critical harmonic n_c by $\omega_c = n_c\omega_1$ gives

$$n_c = \frac{3K}{4}\left(1 + \frac{K^2}{2}\right). \tag{2.32}$$

Notice that for $K = 1$ the value of $n_c \approx 1$, consistent with the dominance of a single (first) harmonic at this low value of K, whereas for the case of Fig. 2.14, in which

$K = 4$, the value of n_c is 27 so all the low harmonics are significantly below the critical energy and the majority of the power output is at higher energies. Notice, incidentally, that the total radiated power from an undulator over the full spectra range is exactly the same as if there was no coherent interference and it behaved like a wiggler; the interference simply concentrates most of the power into certain harmonics, in the same way that an optical diffraction grating or an X-ray scattering crystal concentrates the diffracted intensity into specific directions.

In considering the way the output of a high-K undulator or multipole wiggler approaches a true continuum spectrum the acceptance angle of the experiment that is taking the radiation is a relevant factor. As derived above, very close to the axis of an undulator the spectral width of the individual harmonics is (Equation (2.25)) $\Delta\lambda/\lambda \approx 1/N_u n$, but the inverse dependence on the harmonic number means that the lines become narrower and narrower as the harmonic number increases, so this would imply that they never reach a continuum. However, the emitted wavelength differs off the axis, so a finite angular acceptance angle ensures that additional (asymmetric) broadening of the harmonic components occurs. As remarked in discussing Fig. 2.13, the finite angular distributions of the electrons passing through the undulator (and also the distribution of their energies) also contributes to this broadening so that even on axis one sees off-axis effects in the radiation.

Of course, so far all the preceding discussion and formulae (but not the results of the calculations shown in Figs. 2.13 and 2.14) have considered only a single electron trajectory through the bending magnets and undulators. In reality, the 'bunches' of circulating electrons have finite lateral dimensions, angular spread and energy spread. This important aspect of practical synchrotron radiation sources is discussed below in Section 2.5.

2.3.3   Sources of Variable Polarisation

As discussed in Chapter 1 (see also Fig. 1.9) bending magnet radiation is linearly polarised within the (horizontal) plane of the electron circulation, and the same is true for radiation from a planar undulator as described above. However, some experiments using synchrotron radiation *require* circularly polarised radiation, while in others it is useful to have linearly polarised radiation but to be able to vary the angle of the plane of polarisation. As shown in Fig. 1.9, out of the plane of the electron circulation in a storage ring the bending magnet radiation contains a component of vertically-polarised radiation and the mixture of the two linear polarisations gives rise to a component of circular polarisation (it is elliptically polarised), of opposite sense above and below the horizontal plane. In general, if one defines the electric field variation in the horizontal plane as $E_x = E_{x0} \cos \omega t$, and that in the vertical direction as $E_y = E_{y0} \cos (\omega t + \phi)$, then if $\phi = 0$ one has pure linear polarisation at an angle defined by the relative amplitudes of the two components, whereas if $\phi = \pi/2$, and the amplitude of the two components is the same, one has pure circularly polarised radiation. More generally it is usual (and measurable) to define the state of polarisation in terms of the four Stokes parameters:

$$S_0 = I_x + I_y = I_{45°} + I_{-45°} = I_R + I_L$$

$$S_1 = I_x - I_y$$

$$S_2 = I_{45°} - I_{-45°}$$ (2.33)

$$S_3 = I_R - I_L,$$

where the intensity suffices x, y, $45°$, $-45°$, R and L relate to the linearly polarised components in the x and y directions, at the two orthogonal $45°$ orientations to them, and the right and left circularly polarised components, respectively. For a single monochromatic wave these four parameters are related (i.e. reduce to three independent parameters) by the expression for the total intensity, S_0:

$$S_0^2 = S_1^2 + S_2^2 + S_3^2.$$ (2.34)

In practice, however, the finite size of the source due to the characteristics of the electron beam (discussed more fully in Section 2.5 below), as well as the fact that emission occurs from different parts of the magnet structure, mean that there is an unpolarised component, S_4 whence

$$S_0 = \sqrt{S_1^2 + S_2^2 + S_3^2} + S_4$$

$$1 = \sqrt{P_1^2 + P_2^2 + P_3^2} + P_4$$ (2.35)

The second form of this equation defines $P_1 = S_1/S_0$, etc., the different P values describing the degrees of linear and circular polarisation.

Fig. 2.15 shows the different states of polarisation as a function of the normalised out-of-plane angle, $\gamma\theta$, for bending magnet radiation. As indicated by the data of Fig. 1.9, in plane $(\theta = 0)S_3 = 0$ and $S_1 = S_0 = 1$, so one has 100% linearly polarised radiation in the horizontal plane $(P_1 = 1)$. By contrast, at large values of $\theta(> {\sim}1/\gamma)$, $S_3 \approx S_0$, so $P_3 \approx 1$ and one has almost 100% circularly polarised radiation. Of course, at large out-of-plane angles the intensity falls sharply, as

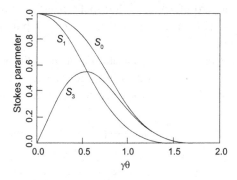

Figure 2.15 Stokes parameter values as a function of out-of-plane angle from bending magnet radiation. After Walker (1998) with permission under CC-BY-3.0 licence

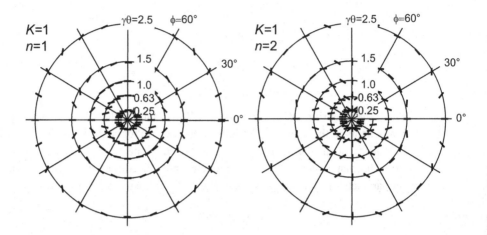

Figure 2.16 Dependence of the direction of linear polarisation on the emission direction for the first and second harmonic radiation from a low-K undulator. After Kitamura (1980). Copyright (1980) The Japan Society of Applied Physics

indicated by the low value of S_0. Notice that $S_2 = 0$ at all angles, a result that can be derived from a simple symmetry argument.

While a planar undulator shares with the bending magnet the fact that the in-plane (on-axis) radiation is horizontally linearly polarised, the situation out of this plane is quite different. Specifically, radiation from a planar undulator is *always* linearly polarised, although the direction of this linear polarisation changes in a complex fashion for emission directions away from the undulator axis. An example is shown in Fig. 2.16, in the form of maps of this polarisation direction for different off-axis angles (scaled in units of $1/\gamma$) and azimuths for the first two harmonics from an undulator with a low value of K of unity. Notice that for the case of $K = 1$, the central cone has a nominal angular width corresponding to $\gamma\theta \approx 1$. In fact the overall pattern of this polarisation behaviour in most directions is quite similar for these odd and even harmonics, with pure horizontal polarisation in the horizontal plane. However, out of the plane of the undulator and at zero offset in the horizontal plane the polarisation direction is horizontal for the odd harmonic but *vertical* (but of zero amplitude exactly on the axis) for the even harmonic.

A return to the arguments used to derive the undulator Equation (2.20) provides an explanation of this feature. Recall that the derivation involved determining the lateral velocity of the electron due to the periodic magnetic field variations, which was found to be

$$\frac{dx}{dt} = \beta_x c = \beta c \frac{K}{\gamma} \cos\left(\frac{2\pi z}{\lambda_0}\right), \tag{2.10}$$

while the velocity of the electron along the axis of the undulator was shown to be

$$\beta_z \approx \beta\left(1 - \frac{K^2}{4\beta^2\gamma^2} - \frac{K^2}{4\beta^2\gamma^2} \cos\left(2\frac{2\pi z}{\lambda_0}\right)\right). \tag{2.14}$$

Thus, not only does the electron have a lateral modulation, but there is also a longitudinal modulation relative to its average speed, which has half the spatial period, and thus twice the frequency, of the lateral modulations. These longitudinal modulations must also give rise to the emission of radiation, but at frequencies that are even harmonics relative to the transverse modulations. However, in the same way that the standard undulator equation can be thought of as a relativistically transformed dipole radiation pattern with a node in the x direction, the modulations in z give rise to a relativistically transformed dipole radiation pattern with a node in the z axis. Within the vertical plane the radiation from these z modulations is linearly polarised, but because of the node along z of its dipole radiation pattern, its intensity is zero exactly along the axis. Notice that a comparison of Equations (2.10) and (2.14) shows that while the magnitude of the transverse velocity term depends linearly on K, the amplitude of the longitudinal velocity (second harmonic) term scales as K^2.

A rather simple pictorial argument provides a rationale for the difference in appearance, or non-appearance, of circularly polarised radiation above and below the horizontal plane from bending magnets and undulators. Fig. 2.17 shows the electron trajectories through these two magnet structures with superimposed arrows showing the direction of instantaneous acceleration and the components of these quantities perpendicular and parallel to the direction of radiation collection. In the case of a bending magnet (Fig. 2.17(a)) an observer exactly along the axis sees only the radiation due to the acceleration along the x direction which is therefore linearly polarised in the horizontal plane. Observing above or below the plane, however, the acceleration along the z direction now has a component in the vertical plane perpendicular to the direction of observation which varies in magnitude and direction as the electron travels through the bending magnet. The sum of these two radiation components seen by the observer leads to an electric field that rotates, resulting in elliptically polarised radiation (i.e. a sum of circular and linear polarisation). Evidently the direction of rotation of this vector is opposite above and below the horizontal plane, giving rise to circularly polarised radiation components of opposite sign. By contrast, in the undulator (Fig. 2.17(b)) different components of

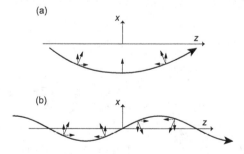

Figure 2.17 Schematic diagram showing the electron trajectories through (a) a bending magnet and in (b) a (planar) undulator. The arrowed continuous lines show the direction of the electron acceleration at several points on the trajectories, while the arrowed dashed lines show the components of these accelerations resolved along the x and z axes.

the trajectory have opposite directions of the acceleration vectors, so in any specific out-of-plane direction one would generate an equal amount of elliptically polarised radiation of opposite sign, leading to no net circular polarisation. Strictly, in the case of the undulator, one should take account of the coherence of the radiation at different points on the trajectory (adding amplitudes rather than intensities) but the final conclusion proves to be the same.

Of course, this raises a further question. If a planar undulator provides only linearly polarised radiation, can one design a different undulator that provides only circularly polarised radiation? The answer is yes, and a number of designs have emerged. A conceptually attractive solution is to fabricate a device that produces a helically-varying magnetic field along its axis. Such a device has the interesting property that along its axis only radiation at the fundamental frequency is observed. The design, however, is not very flexible. Conceptually a simpler approach exploits the fact, introduced right at the beginning of this section, that if one adds horizontal and vertical linearly polarised components with the same amplitude and a phase difference of $\pi/2$, one obtains circularly polarised radiation. Thus, if one uses two planar undulators, one with a periodic vertical magnetic field, the other rotated by 90° having a periodic horizontal magnetic field, then by appropriate control of the relative phase (achieved by insertion of an intermediate magnetic structure to change the phase of the output of the first undulator (Fig. 2.18), one can obtain the required circularly polarised radiation. Moreover, by varying the phase between the two devices, one can switch between left- and right-polarised circularly polarised radiation.

A major disadvantage with this approach is that it means the vacuum vessel (tube) containing the electron beam must be small in both horizontal and vertical dimensions in order that the magnet gaps can be closed to increase K sufficiently, but as all storage rings have very asymmetric beam sizes and corresponding aperture needs, this leads to serious technical difficulties. A range of alternative magnet structural designs have emerged to address this problem, vertical fields being achieved by varying the relative positions of the individual magnet arrays that are only above and below the horizontal plane. These structures mostly allow a wide range of polarisations to be provided under different operating conditions. Specifically, these devices provide the possibility

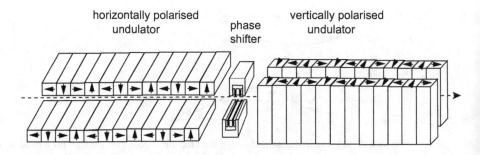

Figure 2.18 Schematic diagram of a device for producing circularly polarised radiation using two orthogonal planar undulators and an intermediate phase shifter.

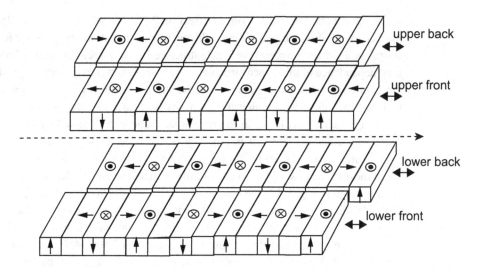

Figure 2.19 Schematic diagram of a section of an APPLE-II (Advanced Planar Polarised Light Emitter) EPU.

of delivering right- and left- circularly polarised radiation but also linearly polarised radiation that can be not only horizontally or vertically polarised, but also at any chosen intermediate orientation. One such arrangement that has proved to be popular is the APPLE-II (Advanced Planar Polarised Light Emitter) elliptically polarising undulator (EPU) (Sasaki, 1994), shown in Fig. 2.19, which comprises four separate periodic magnet structures that can be adjusted in their relative positions.

2.4   Electron Bunches, Time Structure, Lifetimes

So far the discussion of the radiation output from both bending magnets and insertion devices has effectively assumed there is a single electron in the storage ring or that all electrons have exactly the same trajectories. Clearly this is not the case; the electrons must have a spread of trajectories, and this has an important impact on the exact properties of the radiation that emerges from these different sources, as well as the way that the radiation can be delivered to a sample in an experiment. This aspect of the lateral spread in spatial coordinates and angle will be discussed in the following section. Another aspect of the way the electrons are distributed in a storage ring is their spatial distribution *along* their trajectories. As is clear from the description of the properties of linacs in Chapter 1, the way the electrons are accelerated through their interaction with the RF power input actually leads to them being distributed along the accelerator path in 'bunches' that match a particular part of the RF wave. The length and spatial separation of these bunches are related to the wavelength associated with the RF waves, while the temporal distribution is related to the RF frequency but is also influenced by the characteristics of the magnetic lattice and the effects of the synchrotron radiation. As described in Section 2.1 above, electrons are accelerated or reaccelerated (after energy loss through synchrotron radiation) by a similar RF cavity source in a storage ring, so this bunch structure is also present in such a device. The time dependence of this bunch structure, of course, determines the time dependence of the emitted synchrotron radiation. Because the RF frequency is substantially higher than the frequency with which the electrons orbit the ring, the possible number of electron bunches circulating in the ring at any time is determined by this frequency ratio, which has a value that is typically of the order of several hundred or more. Fig. 2.20(a) shows schematically the resulting time structure in a multi-bunch mode when most of the possible 'buckets' are filled, together with the RF voltage variation. As illustrated in Fig. 1.3 in the context of a linac, the electron bunch is located in time at the leading edge of the positive RF voltage excursion. Notice that the widths of individual bunches are relatively much smaller than shown schematically in Fig. 2.20. For example, a typical value of the RF frequency of 500 MHz leads to a bucket separation of 2 ns, but the bunch length can be as short as ~5 ps, although it is generally significantly longer. Different modes of operation can be used with different numbers of buckets filled. For example, some users of the emitted radiation who are interested in time-resolved studies may want the shortest possible bunch length with the longest possible time between bunches. For such experiments the ideal operating condition is with only a single bunch of electrons circulating the storage ring (Fig. 2.20(b)); the interval between bunches is then determined by the time taken for the bunch to circulate the storage ring which is of the order of μs. Of course, single bunch operation leads to much lower total circulating currents, and thus to a much lower time-averaged intensity of the synchrotron radiation; this is clearly unattractive to most users who are not interested in exploiting the time structure. A more common mode of operation that goes some way to accommodating the needs of both types of

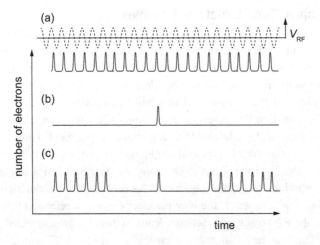

Figure 2.20 Schematic diagram showing the time structure of bunches in a storage ring in (a) multibunch mode, (b) single bunch mode, (c) hybrid mode. Also shown is the time dependence of the RF voltage.

user is a hybrid mode (Fig. 2.20(c)), in which a single bunch is separated from a group of filled bunches by a number of unfilled buckets.

Although the RF cavity tops up the electron energy lost through synchrotron radiation for each circulation of the storage ring, maintaining a constant average energy over many turns, electrons can be lost from the beam, causing the circulating current to decay. One reason for this relates to a fundamental property of the lattice of bending and focusing magnets and RF cavity of the ring, known as the dynamic aperture. This defines the region of phase space (a combination of lateral and angular displacements from the nominal ideal electron trajectory, defined in Section 2.5) in which stable motion occurs. If the phase space region occupied by the beam exceeds the dynamic aperture, then some particles will be pushed to higher and higher displacements and lost from the beam after many circuits around the ring. In addition, however, electrons can be scattered out of the beam. One such mechanism is collisions with residual gas molecules in the vacuum vessel that contains the electrons, an effect that can be minimised by operating the ring under ultra-high vacuum conditions. A second scattering mechanism is large-angle electron-electron scattering within the individual bunches, known as the Touschek effect after Touschek and co-workers first recognised this problem (Bernardini et al., 1963). As described more fully in Section 2.5, an increasing trend in the design of synchrotron radiation sources is the push for smaller and smaller lateral dimensions (and divergence) in the electron beam, but of course this increases the local electron density and thus also increases the probability of this electron-electron scattering. Similarly, making the electron bunches shorter also increases the electron density. Ensuring that the Touschek lifetime is not too short was an important design criterion for many sources. By doing so storage ring beam lifetimes (the inverse of the characteristic exponential decay rate) can still be arranged to be more than ~10 hours. Of course, even with these relatively long lifetimes, users of the synchrotron radiation will notice a significant fall in intensity over a typical operating shift (the time between new injections of electrons into the ring) of ~8–24 hours. This led to the development of a new mode of operation – so-called top-up – that is now a feature of essentially all of the newer sources. In this mode of operation, extra electrons are injected at frequent intervals (of order minutes), replacing the lost electrons to the stored beam and thus maintaining a near-constant (to within ~1% or less) beam current. Of course, top-up necessarily implies that the electrons must be injected into the storage ring at its final operating energy; top-up is incompatible with the earlier mode of operation of many sources in which the storage ring was used as a synchrotron immediately after injection to ramp up the energy from the lower injection energy to the final operating energy.

2.5　Emittance and Spectral Brightness

As remarked above, almost all of the discussion of the properties of synchrotron radiation so far, both from bending magnets and from insertion devices, has assumed

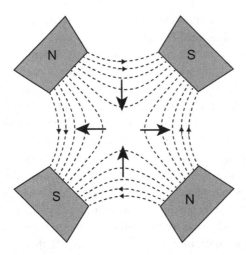

Figure 2.21 Schematic diagram of a quadrupole focusing magnet. The small arrows superimposed on the magnetic field lines indicate the direction of the **B** field. For an electron passing through this device (into the page) the large arrows show the direction of the Lorentz force on an electron travelling into the page. As shown this device focuses in the vertical plane but defocuses in the horizontal plane.

that the electron(s) follow a single trajectory, whereas in reality there are many electrons in each bunch that must, necessarily, have a finite lateral dimension, a finite angular spread, and a finite energy spread (and thus a series of slightly different trajectories). As described in Chapter 1, a synchrotron (which has the same basic design as an electron storage ring) uses 'strong focusing' (also known as 'alternating gradient focusing') in which the electron bunches are refocused at many points around the ring, particularly by the quadrupole magnets. Fig. 2.21 is a schematic diagram of a quadrupole focusing magnet with the direction of the local **B** field lines shown together with the directions of the Lorentz force experienced by electrons passing through it in the direction into the page. Notice that while this device focuses an electron beam in the vertical direction (out of the circulating plane on the storage ring, it actually defocuses in the horizontal (in-plane) direction. Of course, a similar device rotated by 90° about the axis has the opposite effect.

The simplest 'lattice' of focusing devices around a storage ring (and a synchrotron) therefore comprises alternate focusing and defocusing quadrupole magnets, here defining 'focusing' as focusing in the horizontal (circulating) plane (which is defocusing in the vertical direction) and 'defocusing' as defocusing in the horizontal plane (which is focusing in the vertical direction). Fig. 2.22 illustrates this basic idea in the simplest type of lattice, the FODO lattice, where O corresponds to 'nothing' (no focusing component) between the focusing (F) and defocusing (D) quadrupole magnets. The earliest storage rings built as synchrotron radiation sources used this type of lattice. Notice that a combination of focusing and defocusing elements of the same strength does lead to net focusing. This is shown by the usual simple lens

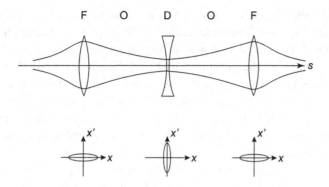

Figure 2.22 Schematic diagram showing a section of a FODO lattice of quadrupole magnets including the envelope of possible electron trajectories. s is the parameter defining the distance along the central (ideal) trajectory around the ring. Below shows the horizontal phase space at different positions.

formulae also used in light optics. The focal length, f, of a pair of lenses of focal lengths f_1 and f_2, separated by a distance d is given by

$$\frac{1}{f} = \frac{1}{f_1} + \frac{1}{f_2} - \frac{d}{f_1 f_2}.$$ (2.15)

So if $f_1 = f_2$ then

$$\frac{1}{f} = \frac{d}{|f_1 f_2|}.$$ (2.16)

Focusing, be it of charged particles or light (electromagnetic radiation), involves changes in the size and divergence of the focused beam, but an important general theorem, the Helmholtz-Lagrange law (which stems from Liouville's theorem), states that the product of the lateral dimension, say σ_x, and the angular divergence, $\sigma_{x'}$, is constant. This product, $\sigma_x \sigma_{x'}$, known as the emittance, describes an ensemble of particles distributed in phase space. Also shown in Fig. 2.21 are sketches of this emittance in phase space (space defined by the two parameters x and x') at different positions along the axis; the area of the ellipse in phase space (defining the emittance) is constant, but its orientation changes around the ring. It is the reduction of this emittance in the circulating electrons, or more exactly these emittances, the horizontal and vertical emittances, which is the goal of modern synchrotron radiation source design. An ultimate objective is to achieve the ability to focus the radiation output to smaller and smaller spots on a sample, leading to large increases in the spectral brightness, the number of photons per unit source size and unit source divergence, typically expressed as the number of photons/s/mrad^2/mm^2/0.1% bandwidth (i.e. it is inversely proportional to the product of the horizontal and vertical emittances). Notice that there is some confusion in the literature over the terminology for this quantity. Particularly in Europe the word brightness is commonly replaced by brilliance (with or without the preceding adjective 'spectral'); this seems to have arisen because the German word for brightness is brillianz, leading to the English word being changed to brilliance (Poole, 2017). An international working group has concluded that 'spectral brightness' best describes the quantity and should therefore be used (Mills et al., 2005). Of course, what is important to the user of synchrotron radiation is strictly the photon emittance, not the electron emittance *per se*, but if the electron emittance is too large it may limit the ability to reduce the photon emittance. It is therefore appropriate to first consider the electron emittance, the quantity that is under the control of the designers of the storage ring.

The key effect that increases emittance in a storage ring is dispersion in the bending magnets, namely the fact that electrons with different energies have different trajectories through the field of a magnet. Electrons passing through bending magnets and insertion devices lose energy through the emission of synchrotron radiation so there is inevitably a spread of energies on their passage around the ring. The energy is eventually returned to its full value on passage through the RF cavity, although this may take several circuits around the ring during which the exact trajectory and energy oscillates. To ameliorate this problem more modern sources insert 'achromats' into the storage ring, the simplest version, which was widely used, being the so-called double bend achromat (DBA). Fig. 2.23 shows the basic idea of this lattice component. The displacement of an off-energy trajectory (in the horizontal plane within which the dipole magnets cause bending) can be written as

$$x(s) = D(s)\frac{\Delta E}{E},$$
(2.17)

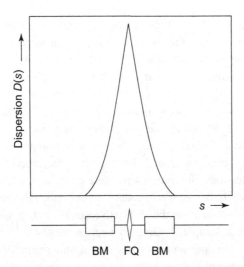

Figure 2.23 Schematic graph showing the way a double bend achromat, comprising two equal bending magnets (BM) and an intermediate focusing quadrupole magnet (FQ) can correct the dispersion induced by the bending magnets.

where s is the distance along the ideal (full energy) reference trajectory, $D(s)$ is the dispersion function and E is the electron energy. As shown in Fig. 2.23, the symmetry of the DBA arrangement ensures that if the electron beam enters the first bending magnet with zero dispersion, the dispersion at the end of the cell beyond the second bending magnet will also be zero, providing that the central focusing quadrupole magnet simply reverses the gradient of the dispersion. Of course, (horizontally) defocusing quadrupoles must also be included in the lattice to maintain focus in the vertical direction, but if the dipolar bending magnets are properly aligned they do not produce any dispersion in this direction. For this reason the vertical emittance of storage rings is generally very much smaller than the horizontal emittance; reducing the horizontal emittance is the major challenge in machine design.

While the horizontal and vertical electron emittances are conserved around the storage ring, the frequent focusing and defocusing means that the individual components, namely the lateral dimension and the angular divergence of the beam, do change at different positions in the ring, such that these two quantities follow an ellipse in the two-dimensional phase space defined by the parameters x and x' in the horizontal direction (and a second ellipse in the phase space defined by y and y' in the vertical direction). The variation in the root-mean-square (rms) amplitude of the lateral displacement parameter in the horizontal (radial) direction around the storage ring, in terms of s, the coordinate in the reference orbit direction, can be written as

$$x(s) = \sqrt{\varepsilon}\sqrt{\beta_x(s)} \sin\left(Q\psi(s) + \phi\right), \tag{2.18}$$

where ε is the emittance in terms of the rms values of x and $x'(\varepsilon = \sigma_x\sigma_x')$, β_x is a modulating function dependent on the specific focusing components around the ring, $\psi(s)$ is related to the azimuthal angle around the ring and Q is the betatron wavenumber defining the number of oscillations the particle executes in one trip around the ring. The oscillations of the electrons around the reference orbit are known as betatron oscillations, a name derived from the earliest (betatron) accelerators of electrons (also known as negatively-charged beta particles). An equivalent expression for $y(s)$ gives the vertical modulation function $\beta_y(s)$. These two β functions define the shape and orientation of the emittance ellipse in phase space. Notice that

$$\sigma_x = \sqrt{\varepsilon_x\beta_x} \quad \text{and} \quad \sigma_y = \sqrt{\varepsilon_y\beta_y}, \tag{2.19}$$

while

$$\sigma_x' = \sqrt{\varepsilon_x/\beta_x} \quad \text{and} \quad \sigma_y' = \sqrt{\varepsilon_y/\beta_y}. \tag{2.20}$$

Fig. 2.24 shows a specific example of the variation of the two β functions and the dispersion in a section of the Diamond storage ring, comprising four double-bend achromats surrounding three straight sections used for insertion devices. At the ends of the range shown are halves of two long straights. Notice that in addition to the quadrupole focusing magnets are additional sextupole magnet structures (with 6 pole pieces arranged at 60° intervals around the beam) to apply corrections for the chromatic behaviour (what is known as chromatic aberration in light optics) of the quadrupole magnet lattice.

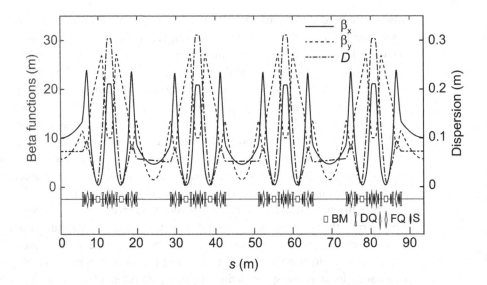

Figure 2.24 Dispersion and beta functions in a section of the Diamond storage ring comprising four double bend achromats with three intermediate straight sections and, at each end of the section, half of a long straight. Below these graphs are representations of the associated magnet structures, namely bending magnets (BM) horizontally defocusing and focusing quadrupoles (DQ and FQ) and sextupoles (S). Courtesy of Richard Walker, Diamond Light Source

  The values of the beta functions are of particular importance in the straight sections to match the characteristics of the insertion devices. Many insertion devices rely on being able to achieve a small gap of only a few mm between the magnetic pole pieces either to achieve very high fields or to achieve reasonably modulated high fields in a device with a short period. Evidently this means that σ_y must be sufficiently small, which implies a sufficiently small value of β_y. More specific issues concerned with matching the beta functions of the straight sections to the intended insertion devices can also be crucial, as described briefly below.

  The push for smaller and smaller emittances has led to an increasing trend to upgrade existing sources, or build new ones, with more complex magnet structures, and particularly multi-bend achromats (MBAs). The motivation for this is that the theoretical minimum horizontal emittance scales as the square of the energy, but is inversely dependent on the cube of the number of bending magnets, N_b (i.e. $\varepsilon_x \propto E^2/N_b^3$), so a large number of weak bending magnets offers the potential advantage of a much smaller emittance. However, a large number of weaker dipole magnets would, in a DBA lattice, imply a much larger (and therefore much more expensive) storage ring. Moving these dipole magnets close together in groups to form MBAs provides a solution to this particular problem. Of course, the lower dispersion associated with the smaller emittance also means that stronger magnets are required to provide the necessary focusing fields very close to the axis of the lattice. The solution adopted at the MAX-IV facility in Sweden, first commissioned in 2016, is a 7-bend achromat with a very small diameter vessel for the beam, allowing magnets to be very close to the axis. This does, however, raise further problems with achieving reliable ultra-high vacuum conditions (to avoid beam loss through electron-molecule scattering) in the long narrow vessels; poor pumping speeds in such vessels precludes reliance of conventional pumps and has led to the use of 'NEG' (Non-Evaporable Getter) thin film coatings on the inside of the vessel, which act as local pumps. One consequence of this strategy of multiple weak bending magnets in compact structures containing other focusing elements means that it is no longer possible to access useful bending magnet radiation and the facility becomes entirely reliant on insertion devices in the straight sections.

Of course, as remarked above, ultimately what is important to the user of synchrotron radiation is the photon emittance, so minimising the electron emittance is only paramount if the electron emittance is so large that it becomes comparable to (or even larger than) the photon emittance, the emittance experienced by the user of the radiation being a convolution of the electron emittance and the photon emittance. Ultimately, the photon emittance is limited by diffraction. For a radiation source of length L, at a wavelength λ, the angular spread due to diffraction can be evaluated in a similar way to the usual treatment of single-slit diffraction in optics, but considering the phase relationship of wavefronts from the two ends of the source rather than the two sides of a slit. Approximating the source distribution by a Gaussian function leads to the diffraction limited divergence being given by $\sigma'_{ph} \approx \sqrt{(\lambda/2L)}$. The diffraction-limited emittance for Gaussian distributions is $\sigma'_{ph}\sigma_{ph} = \lambda/4\pi$[1], so the apparent source size is $\sigma_{ph} = (1/\pi)\sqrt{(\lambda L/8)}$. Evidently if the electron emittance, $\varepsilon = \sigma_e\sigma'_e$ is significantly less than $\lambda/4\pi$, it does not limit the photon emittance; a source with this characteristic is known as diffraction-limited, and diffraction-limited storage rings are increasingly seen as the ultimate goal. It is important to note, however, that this limit depends on the wavelength, λ. At sufficiently low photon energies (large λ), *all* storage ring sources are diffraction limited. At the highest photon energies, however, achieving this presents a major challenge. The widely perceived current goal is to produce a diffraction-limited source at a wavelength of 1 Å, a rather characteristic wavelength used in a wide variety of X-ray diffraction experiments. As yet no source design is capable of achieving this goal in the horizontal emittance, although it is approached in the vertical emittance.

Fig. 2.25 illustrates the extent to which the diffraction limit is achieved in a standard third generation source at different wavelengths, specifically the case of undulator radiation from the ELETTRA storage ring facility in Trieste, Italy. The effective total photon emittance due to both electron emittance and the diffraction limit can be estimated by adding the two components in quadrature, so for example the effective rms source size in the horizontal direction can be written as

---

[1] An estimate of the diffraction-limited emittance can be derived from the single-slit diffraction case in optics. Passage of light through a slit of full width w leads to a central beam characterised by an angle to the first zero intensity of $\Delta\theta$ such that $w\Delta\theta = \lambda$; for Gaussian distributions the product of the Gaussian σ values is clearly smaller. An alternative derivation can be obtained from the Heisenberg Uncertainty Principle, $\Delta x\Delta p \geq \hbar/2$, by noting that the photon momentum $p = 2\pi\hbar/\lambda$ and $\Delta p \approx p\Delta\theta$ so $\Delta x\Delta\theta \geq \lambda/4\pi$. The underlying equivalence of these two approaches is discussed by Richard Feynman in *The Feynman Lectures on Physics* (Feynman et al., 1964).

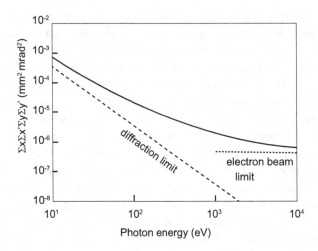

Figure 2.25 Comparison of the product of the horizontal and vertical effective total photon emittances for the case of ELETTRA with the values determined by the diffraction limit and the electron emittance. After Walker (1998), with permission under CC-BY-3.0 licence

$$\Sigma_x = \sqrt{\sigma_x^2 + \sigma_R^2}, \tag{2.21}$$

where σ_R is the value obtained from the diffraction limit outlined above. Similar expressions define the angular deviation in the horizontal direction and both spatial and angular deviations in the vertical direction. The product of the horizontal and vertical effective total photon emittances, $\Sigma_x \Sigma_{x'} \Sigma_y \Sigma_{y'}$ (the factor that appears in the denominator of the spectral brightness) is shown as a function of the emitted photon energy, while superimposed on the figure are dashed and dotted lines showing the diffraction-limited component alone and the electron emittance limit alone. At low photon energies the emittance product is close to the diffraction limit, but at high energies it is limited by the electron emittance and is not diffraction limited.

Notice that even if the electron emittance is sufficiently small to satisfy the diffraction-limited criterion, there may still be a limitation in matching the electron beam emittance to the diffraction-limited photon emittance in phase space. This is illustrated schematically in Fig. 2.26. In the case of a bending magnet source, L is only a few mm, so σ'_{ph} is large while σ_{ph} is small (~1 μm), but the electron beam size σ_e is much larger (~100 μm). The match in phase space is thus poor and the combined emittance is much larger than either component. By contrast, in an undulator, L is much larger (~a few m), so σ'_{ph} is small, while σ_{ph} is larger, leading to a significantly better match to the electron emittance at appropriate values of β.

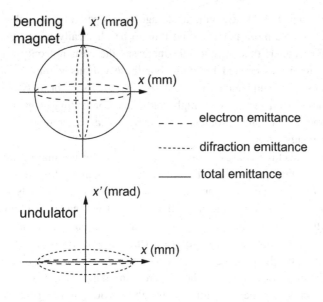

Figure 2.26 Highly schematic phase-space diagrams of electron and photon emittance from a bending magnet and an undulator. After Eriksson (1997). Reproduced with permission of the International Union of Crystallography

2.6 Practical Constraints: What Determines the Source Characteristics?

The preceding sections have provided a description of the basic theory of synchrotron radiation sources, including bending magnets, wigglers and undulators. Much of the theory has related to the properties of a single circulating electron in a storage ring with the underlying assumption that if there are N circulating electrons one obtains N times the output with no other change. The essential features of the accelerator physics (and funding!) that underpin the construction and operation of these sources has been largely ignored, apart from the issue of electron and photon emittance. It is instructive, however, to briefly describe some of the constraints that define the design and properties of these sources.

Even before considering the very important issue of electron emittance, the two obvious parameters that define the utility of a source are the machine energy and the circulating current. If only bending magnet radiation is to be used the choice of machine energy is largely determined by the highest photon energy of interest; over most of the spectral range the output is relatively insensitive to the electron energy, but the steep fall-off on the spectral output of bending magnets at high photon energies (Fig. 1.8) means that a good flux of high photon energies implies the need for higher electron energies. Of course, this comment assumes that the field strength of the bending magnets is limited to ~1 T. Much higher field strengths can be obtained using superconducting electromagnets, but with large associated running costs for liquid helium, higher capital costs and somewhat lower reliability. Also assuming the same bending magnet field strength, a storage ring operating at higher electron energies needs to have more magnets and thus to be physically larger, because the bend angles are smaller; this significantly increases the cost, not only for the ring itself, but also for its associated building. For these reasons it may be more cost effective, if accessing broad band radiation at high energies is the primary objective, to use a lower energy storage ring with high-field (superconducting magnet) multipole wigglers. In practice, of course, sources must serve a wide community of users with varying requirements.

The choice of machine energy is rather more critical for undulator radiation. As described above, there are practical limits on the minimum and maximum periods of undulators that can be used, and as a consequence undulators typically can deliver harmonic radiation over no more than about two decades of photon energy for any specific electron energy. Too low a machine energy means that high photon energy undulators are not possible (the required period would be too short). Too high a machine energy and it becomes impractical to deliver harmonic undulator radiation at low energies; the required undulator period is too long to fit a reasonable number of periods into a straight section of the ring. It may be possible to access somewhat lower energy by using a high-K undulator, but most of the radiation emitted would then be at much higher energies, causing severe problems in the beamline optics (Chapter 3). For these reasons users requiring photon energies in the UV and VUV (~5–200 eV) are best served by a storage ring energy of ~0.5–1.5 GeV, while those mainly interested in hard X-rays (~10–50 keV) ideally require ring energies of 6 GeV or more. In practice a ring energy ~3 GeV represents a compromise that attempts to serve both communities adequately at moderate cost at several national facilities

Of course, most users are keen to have the highest possible photon flux, so one might imagine one way to enhance the flux would simply be to increase the circulating current. In practice, typical values for current storage ring sources are ~100–300 mA. Why not increase this to 1 A or more? A higher current (and more synchrotron radiation energy loss) would require more RF power, which is expensive but not the critical limitation. Instead, this is one of several parameters that is limited by the accelerator physics. Evidently a storage ring requires a stable circulating current of electrons in a narrowly-defined orbit. Stability implies that if the beam is perturbed in some way the forces acting on it are such as to suppress the perturbation. In practice instabilities are a major issue; these arise from conditions under which a perturbation is actually amplified rather than suppressed. Instabilities can lead to catastrophic loss of the circulating beam. They can arise from interactions between the electron beam, its resulting magnetic fields and the coupling of these fields with induced fields in conducting components of the ring structure, including the vacuum vessel walls but also other structures within this vessel that are closer to the beam. Some of these interactions are found to be greatly increased at high circulating currents, so beam instabilities are a key issue that restricts the practical circulating current that can be achieved. These interactions can also cause disturbances of the beam that do not lead to instability, but do increase the electron beam emittance.

The general issues concerning electron beam emittance have already been outlined above, although the details of the associated accelerator physics fall outside the scope of this book. Some comments on the importance of the electron beam emittance are in order, however. As shown in Section 2.5, the photon emittance, the quantity that actually determines the kind of user experiments that can be performed, is not limited by the electron emittance at low photon energies in the UV, VUV and even SXR spectral regions. For hard X-rays, however, this is no longer the case, and it is satisfying the most demanding users of this community that is driving the search for ways of achieving diffraction-limited emittance in this energy range. Ultimately, the size of the photon emittance limits the size to which the photon beam can be focused onto a sample (because the photon beam emittance is conserved in the focusing optics). The driver for such small photon spots on a sample is the investigation (e.g. X-ray diffraction) of very small samples, and the achievable spatial resolution in certain methods of X-ray imaging (see Chapter 8). Notice, though, that not all users want the very high flux density that can be achieved in this way: the rate of radiation damage, a substantial problem in the study of soft matter (including biological material), is determined not by the total flux, but by the flux density, so the problem is greatly enhanced by a highly-focused beam of radiation.

2.7   Coherence

A property of radiation from storage rings and free electron lasers that has become of increasing interest is coherence. Ultimately, coherence is the property of a light source that ensures two or more waves can interfere such that the resulting intensity is obtained from the square of the sum of the amplitudes rather than the sum of the intensities of the contributing waves. To achieve this one must have a well-defined phase relationship between the two waves in time and space. The benchmark for coherent light sources is set by conventional lasers. In these devices light is generated by an optical transition within a lasing gas or solid, and mirrors at the ends of the device (one of which is less than 100% reflecting) establish an optical standing wave which triggers stimulated emission in the lasing materials. As a consequence separate photon emission events not only occur at the same wavelength, but have a well-defined phase relationship to the stimulating photons. Individual photons are thus coherent with one another.

By contrast, synchrotron radiation, be it from bending magnets or insertion devices, involves photon emission from individual electrons that have no knowledge of the emission by other electrons, so there is no well-defined phase relationship between individual photon emission, even when of the same energy[2].

---

[2] Strictly this is only true when the wavelength of the radiation is much shorter than the electron bunch length. However, as bunch lengths in synchrotron radiation sources are typically a few mm, truly coherent radiation is produced only at very much longer wavelengths than are typically exploited by users of such sources, although truly coherent synchrotron radiation has been observed in the far infrared/terahertz range with short bunch lengths. This issue is described more fully in the context of free electron lasers later in this chapter.

Within the definition described above for conventional laser emission, synchrotron radiation is thus not coherent. However, as mentioned above, a coherent light source may be defined as one that can give rise to interference patterns, the reference experiment to judge this being (in practice or as a *gedankenexperiment*) the classic Young's slits experiment of light optics. This experiment involves the interference of two components of the same wavefield passing through a pair of slits; in effect, a photon interferes with itself, so the issue of the relative phase of separate photons does not arise. However, this is only true for a point source, such that successive photons originate from exactly the same point in space relative to the two slits, ensuring that the two components of their wavefront at the slits have exactly the same relative phase.

Fig. 2.27 shows the basic Young's slits experiment, but displays the two interference patterns generated by two different point sources separated by a distance s. Considering first only one of these sources, the condition for the first minimum in the interference pattern produced by interference of the secondary waves passing through the two slits, separation d, is given by an angular separation at the screen of $\Delta\theta$ when

$$d \sin \Delta\theta = \lambda/2. \tag{2.22}$$

Consider, now, an extended source. The interference patterns from different parts of the extended source will be displaced in angle, but if this angular separation is sufficiently small, a single interference pattern will still be visible but somewhat 'blurred'. However, if the angular separation of these two parts of the source corresponds to that of the central maximum and the first minimum in the interference pattern from each part, as shown in Fig. 2.27, then an interference pattern will no longer be visible. This occurs when the angle subtended by the source (here shown as two separate sources to represent the edges of an extended source) is equal to the value of $\Delta\theta$ in Equation (2.22), namely, when

$$\sin \Delta\theta = \lambda/2d = s/D. \tag{2.23}$$

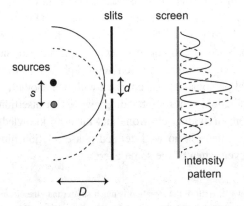

Figure 2.27 Schematic diagram of the Young's slits experiment showing the case in which two separate point sources give rise to interference patterns separated in angle by the angle between the central maximum and first minimum of each pattern.

From the point of view of the ability of an extended light source to allow the observation of such a two-slit interference pattern, the slit separation d in Equation (2.23) thus defines the limit of the effective transverse coherence length of the source; namely, the separation over which the source appears to be (or just not to be) coherent. This 'transverse coherence length', ξ_T is thus given by

$$\xi_T = \lambda D / 2s. \tag{2.24}$$

Notice that some authors define the angle $\Delta\theta$ as the angle between the interference patterns of the two sources being out of phase on one side, and on the other side, of the pattern formed by a central source. In effect, this defines a 'full-width', while Equation (2.23) defines a half-width. The resulting alternative definition of the transverse coherence length (Equation (2.24)) lacks the factor of 2 in the denominator.

For example, at an X-ray wavelength of 1 Å, a horizontal source size of 80 μm, and with an experiment placed 10 m from the source, the transverse coherence length is ~60 μm. The vertical source size is typically much smaller, leading to a larger transverse coherence length.

Fig. 2.27 is drawn with the source very close to the slits, with spherical wavefronts arriving at the slit. In reality, the source of a synchrotron radiation beam is very far from the diffracting object and the wavefronts will be essentially planar, but will arrive at slightly different angles, leading to the phase differences at the slits that cause a similar lateral shift of the diffraction patterns. In practice these different directions of arrival are determined both by the source size and the angular divergence of the beam (the two factors that define the emittance), so the transverse coherence length is actually determined by this angular range of the beam, which can replace the factor (s/D) in Equation (2.24). The transverse coherence length is therefore inversely proportional to the degree of collimation of the beam. For this reason the coherence length can be increased by placing a 'pinhole' aperture in front of the sample, as discussed in Section 8.6. Evidently, the fact that the degree of collimation determines the transverse coherence length means that this parameter is largest in low emittance sources. Indeed, a source that is 'diffraction-limited', as defined in Section 2.5, is deemed to be fully coherent. A measure of the transverse coherence of a source, referred to as the 'coherent fraction', F, is simply the ratio of the combined horizontal and vertical emittance of this ideal source and the actual total emittance of any given source,

$$F = \frac{(\lambda/4\pi)^2}{\Sigma_x \Sigma_{x'} \Sigma_y \Sigma_{y'}}. \tag{2.25}$$

In a somewhat similar way to the discussion of transverse coherence, one can define a longitudinal or temporal coherence parameter. In this case, too, the fact that the photons emitted from a synchrotron radiation source have no well-defined phase relationship at the same energy means that there is no longitudinal coherence in the way that a conventional laser is coherent, but the ability to detect interference effects is influenced by how monochromatic the source is. Specifically, if two photons have slightly different wavelengths, say λ and $\lambda - \Delta\lambda$, then the two waves go from being in phase, to being out of phase, over a distance $n\lambda = (n + 0.5)(\lambda - \Delta\lambda)$, so $\Delta\lambda \approx \lambda/2n$, so this the longitudinal coherence length is given by

$$\xi_L = n\lambda = \lambda^2/2\Delta\lambda. \tag{2.26}$$

In this case, too, a frequently-used alternative definition leads to the loss of the factor of 2 in the denominator of Equation (2.25), as remarked above.

Evidently, this longitudinal coherence parameter is not a function of the source, but of the monochromator placed after the source, which defines the emerging photon energy (or wavelength) resolution. Taking again the example of an X-ray wavelength of 1 Å, a typical crystal monochromator for use at this energy (based on two Bragg reflections from Si(111) – see Chapter 3) has $(\lambda/\Delta\lambda) \sim 10^{-4}$. So the longitudinal coherence length is ~0.5 μm.

The significance of these numbers will be discussed further in the context of experiments exploiting this type of coherence in Chapter 8.

2.8   Free Electron Lasers

As remarked above, modern storage rings that are built as dedicated synchrotron radiation sources, are designed particularly to exploit the radiation provided by insertion devices, and are generally referred to as 'third generation' sources. Storage rings with diffraction-limited emittances at short wavelengths (typically taken to mean 1 Å), still no more than a stated objective for the future, are often described as 'fourth generation' sources. However, this name is also often assigned to free electron lasers (FELs), which are capable of delivering soft and hard X-rays and are already operating at several sites worldwide. Although these devices are based on undulator radiation, they have a time structure very different from that of storage ring sources of synchrotron radiation and are, at least in this respect (and thus in some of their scientific applications), more akin to conventional fast lasers than conventional synchrotron radiation sources. Nevertheless, storage ring undulators and FELs are essentially complementary radiation sources that share many of their underlying physical principles, so a brief outline of the principles and properties of FELs is presented here.

As discussed in Section 2.7, the radiation emitted by different electrons passing through an undulator installed in an electron storage ring have randomly-related phases, even at the same wavelength, so the intensity seen from the many electrons, say N_e, in an electron bunch is simply $N_e I_e$, where I_e is the intensity produced by a single electron. By contrast, if the radiation from these different electrons were to be fully coherent, with exactly the same phase at the same wavelength, then the total intensity would be $N_e^2 I_e$, and as N_e is a very large number (~10^9 or more) there would be a huge intensity enhancement. In reality, such a massive enhancement is not achieved, but this underlying idea is what is exploited in a free electron laser. The key piece of physics that makes this possible, but which has not so far been discussed in the context of undulator radiation, is the effect of the interaction of the emitted radiation with the electrons as they pass through an undulator. Specifically, the

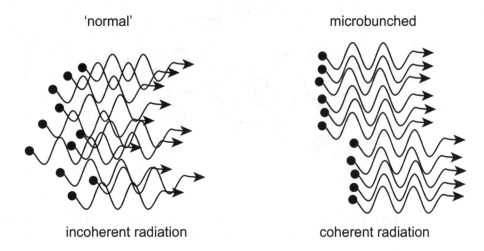

Figure 2.28 Schematic diagram showing electrons in part of an electron bunch emitting undulator radiation, without, and with, 'microbunching'.

transverse electric field of the radiation can interact with the transverse electric current induced by the lateral undulations of the electrons such that, depending on their exact relative phase, some electrons are accelerated and some are decelerated. As a result, the electrons can become 'microbunched', each microbunch being separated from the next by one wavelength of the emitted radiation. This effect is shown schematically in Fig. 2.28. Within the microbunches the relative phase differences of emitting electrons is small (the phase space available to photons is reduced) and the radiation is coherent. This coherence and the resulting huge enhancement in peak photon intensity are key features of FELs.

Of course, the extent to which this occurs depends on the intensity of the electromagnetic (EM) radiation field; in an undulator in a storage ring the effect is normally too weak to be very important. It can, however, be exploited in one of two different ways. In the spectral range from the infrared to the near ultraviolet, a range for which it is relatively easy to fabricate highly-reflecting mirrors at normal incidence, the EM field can be enhanced by creating an optical cavity around the undulator in essentially the same way as that in a conventional 'table top' laser. The radiation field, 'captured' in this cavity (but with partial escape through one of the mirrors), is thus enhanced by successive electron bunches passing through the undulator until significant microbunching occurs and the intensity of the radiation field becomes increasingly amplified.

Fig. 2.29 shows the key ingredients of such a device. The invention of the FEL is generally attributed to John Madey at Stanford University who, as a student, first presented the underlying theory (Madey, 1971) and, with his colleagues, reported the first successful operation of such a device (Elias et al., 1976; Deacon et al., 1977). In this first demonstration the undulator was not planar, as illustrated in Fig. 2.29, but helical, using a superconducting helical winding with a period of 3.2 cm. Electrons were injected at 43 MeV and the emitted radiation was detected as gain in an injected

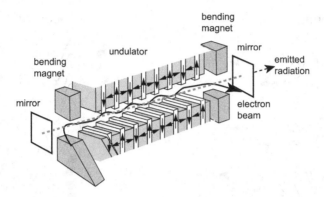

Figure 2.29 Schematic diagram of a FEL using an optical cavity for operation in the infrared and visible range. Bending magnets are used to guide the electron beam into and out of the undulator

laser signal at a wavelength of 10.6 μm. Oscillation was subsequently shown to be achievable at a shorter wavelength of 3.4 μm. Notice that the early theoretical descriptions of FEL operation were based on non-classical explanations and the concept of stimulated emission that underpins conventional laser operation, but it now seems more helpful for most purposes to focus on the purely classical physics of the Larmorian force exerted by the electric field vector of the EM field on the emitting electron and its micro-bunching consequences. Indeed the FEL can be thought of as an electron decelerator, with an appropriate choice of phase reversing the normal EM radiation flow.

The basic approach, illustrated in Fig. 2.29, has proved very successful for the production of infrared FELs, using a linac to inject electrons into the device, and has been the basis of a number of successful user facilities. A prime example is FELIX (Free Electron Laser for Infrared eXperiments), now based at Radboud University Nijmegen in the Netherlands; this user facility comprises four FELs injected by two different linacs, providing radiation over the wavelength range 2.7–1,500 μm (from the mid- to far-infrared/terahertz). Of course, the schematic layout of Fig. 2.29 could be installed in an electron storage ring, operating for conventional synchrotron radiation. Indeed, a successful demonstration of this capability on the ACO storage ring near Paris in 1983 (Billardon et al., 1983; Ellaume et al., 1984) was effectively the world's second FEL. In this case the storage ring was operated at the rather low energy of 166 MeV and laser radiation was observed in the visible at a wavelength of ~6,500 Å from an undulator with a period of 7.8 cm. Of course, the higher electron energy of most storage rings used for synchrotron radiation leads to higher energy photons (the photon energy scaling as γ^2 – see Equation (2.20)), but going to significantly higher photon energies precludes the use of a simple optical cavity because of the difficulty of making highly reflecting normal incident mirrors much beyond the near ultraviolet. Nevertheless, a multi-institutional development of a higher energy FEL installed on the ELETTRA storage ring in Trieste (Walker et al., 2001), using special oxide multilayer mirrors, did achieve lasing in the far ultraviolet down to a wavelength of 1,764 Å (corresponding to a photon energy of more than 7 eV) with the storage ring operating at an electron energy of 0.75 GeV (Curbis et al., 2005), significantly lower than its usual operational energy for synchrotron radiation users.

Of course, an important ultimate goal is to extend the range of FELs to much higher photon energies, from a few eV to a few keV, into the X-ray range. Clearly, this is an energy range in which there are no strongly reflecting mirrors near normal incidence and a rather different approach must be adopted. However, the basic idea of exploiting the interaction of the emitted undulator radiation with the electrons passing through it, to achieve microbunching, remains. Instead of increasing the radiation field intensity by creating an optical cavity, however, an alternative approach is to increase the length of the undulator sufficiently that the radiated intensity far along the undulator is sufficient to produce microbunching in a single pass. This is the underlying principle of the so called SASE (self-amplified spontaneous emission) FEL. The first demonstrations of SASE FELs were performed at microwave frequencies (cm wavelengths) by Gold et al. (1984) and Orzechowski et al. (1985) although the effect had been observed in this frequency range in a device known as the ubitron in 1957 (before the name free electron laser was coined) as described by Phillips (1988). More recently SASE FEL operation was demonstrated at a photon energy of 11 eV using the TESLA Test Facility (TTF) at the DESY laboratory in Hamburg (Andruszkow et al., 2000), but subsequently later facilities have demonstrated this mode of operation at much higher energies. For example, the first hard X-ray FEL (covering the 1 Å wavelength range), the LCLS (Linear Coherent Light Source) at Stanford in the USA (Emma et al., 2010) has been operating since 2009, while the European XFEL near Hamburg in Germany started user operation in 2017. As microbunching starts to occur, the radiation intensity increases sharply; this further enhances the microbunching instability, leading to even further enhancement of the radiated power, which grows exponentially until saturation, according to:

$$P = \alpha P_n \exp\left(s/L_g\right), \tag{2.27}$$

where P_n is the initial power that initiates the instability, α is a coupling coefficient dependent on the excited mode and L_g is the power gain length.

Achieving this mode of operation requires overcoming some important challenges in the properties of the electron beam that must be delivered into the long undulator. Specifically, it is necessary to significantly increase the interaction of the electrons with the radiation that produces the microbunching, and thus the gain, relative to that achieved in an electron storage ring. FELs based on optical cavities are low-gain devices with a gain per pass of only a few percent (but sufficiently greater than its losses to ensure overall gain), but a SASE FEL must operate at high gain, typically defined as a gain greater than unity, although in practice significantly higher values are required. The basic theory of the SASE FEL under these conditions was first described by Bonifacio, Pellegrini & Narducci (1984). This gain can be written as

$$G = 4\pi\rho N_u \tag{2.28}$$

where N_u is the number of periods in the undulator and ρ is known as the universal FEL parameter or Pierce parameter, which for a helical undulator is given by

$$\rho = \left(\frac{K^2 e^2 \lambda_0^2 n_e}{64\varepsilon_0 \pi^2 \gamma^3 m_0 c^2}\right)^{1/3}. \tag{2.29}$$

For a linear undulator there is a further scaling factor that depends on the harmonic. Here n_e is the electron density in the bunch and ε_o is the permittivity of free space. The gain is thus proportional to the cube root of the electron density and inversely dependent on the electron energy defined by γ. A consequence of this is that in order to achieve high gain, and thus to render a SASE FEL viable, one needs extremely high electron densities in the beam, implying low electron emittances and high values (in the kA range) of the instantaneous current. Achieving a small energy spread is also crucial. A high electron density also implies short bunch lengths, although a further requirement influences this parameter. Specifically, although the electrons passing through the undulator are travelling at almost exactly the speed of light, the transverse oscillations mean that the average speed at which they travel along the axis is lower, as discussed earlier when introducing the general properties of undulators in Section 2.3.2. This means that the emitted radiation moves ahead of the emitting electrons in a time Δt by a length ΔL

$$\Delta L = (c - v_z)\Delta t = c\Delta t\left(1 - \hat{\beta}_z\right). \tag{2.30}$$

So, taking Δt to be the time the bunch spends in the full length of the undulator,

$$\Delta L = \lambda_0 N_u\left(1 - \hat{\beta}_z\right) = N_u \lambda. \tag{2.31}$$

To ensure the radiation field and electrons interact over the full length of the undulator the bunch length should be longer than this value. Of course, for a hard X-ray undulator, with a wavelength λ of ~1 Å this is not a very stringent requirement, even with values of N_u of several thousand (corresponding to total undulator lengths of hundreds of metres). For example, a bunch length of 3,000 Å, corresponding to a pulse duration of 1 fs, is well below anything so far achievable, although simulations suggest future progress is possible.

Obviously the overall length of the undulator in a SASE FEL must also be chosen to ensure that sufficient gain is achieved within its length to reach saturation; the intensity grows exponentially with the distance along the undulator as $\exp(z/L_g)$ where L_g is the power gain length,

$$L_g = \frac{\lambda_0}{4\sqrt{3}\pi\rho}. \tag{2.32}$$

Typically the distance to achieve saturation is ~20 L_g, while a representative value of ρ is ~4 × 10^{-4} for hard X-ray FELs (the value is a factor of ~10 larger for soft X-ray FELs). Taking this hard X-ray value, the number of undulator periods required (~$(20/4\sqrt{3}\pi\rho)$) is ~2,300 and if λ_o is a few cm this leads to the requirement for a minimum total undulator length of ~120 m. The radiated energy at saturation is approximately

$$E_{sat} = \rho N_e E,$$ (2.3)

where E is the energy of the electrons and N_e the number of electrons, but this can be re-expressed as the number of photons, of energy E_{ph} per electron as

$$N_{sat} = \rho E / E_{ph}.$$ (2.34)

So if one takes $E = 17.5$ GeV, $E_{ph} = 10$ keV $\left(\lambda \sim 1.2 \text{ Å}\right)$ and $\rho = 4 \times 10^{-4}$ then $N_{sat} \approx 700$. This is to be compared with the number of photons emitted per electron in an undulator without SASE amplification, which is ~0.01–0.1. This implies a gain in photon emission of ~10^4–10^5 relative to that achieved with conventional undulator radiation.

Fig. 2.30 shows a highly-simplified schematic diagram of the key components of a SASE FEL. As remarked above, the undulator(s) required can be very long; the total length allocated to undulators at the European XFEL facility at Hamburg is 1.6 km, although the individual undulators are shorter than this. For this reason alone, it would clearly not be practical to incorporate such an undulator into a storage ring, but in fact the choice of a linear accelerator to provide the high-energy electron beam is based on far more significant technical reasons. To achieve SASE lasing one requires an extremely low emittance and short bunch length not achievable in a storage ring, due to the need to maintain stable bunches that can circulate in the ring for many hours. Electron bunches that pass only once along a linac-based FEL device retain the basic character of the initially injected bunches, including any bunch compression, but in a storage ring these characteristics are lost after travelling around the ring for only ~10–100 ms. Nevertheless, the technical challenges of a SASE FEL are considerable, and Fig. 2.30 omits many components that are essential for effective operation, such as magnetic focusing of the beam in between multiple sections that make up the complete undulator. The initial electron gun is based on photoelectron emission from a photocathode excited by a pulsed laser, intrinsically producing rather short bunches. The electrons are then accelerated through a long series of high-power RF cavities with bunch compression at various points. To overcome the potential problem of excessive heating due to the current induced in RF cavity walls at high power, these are constructed from superconducting materials and must be operated at liquid helium

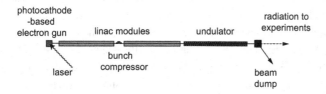

Figure 2.30 Highly-simplified schematic diagram showing the layout of the main components of a SASE FEL. The bunch compressor shown is based on a magnetic chicane (c.f. Fig. 2.32).

temperatures. A magnetic deflector is installed after the SASE undulator(s) to dump the electron beam while the emitted radiation passes on to the user experiments. At the European XFEL facility, the total length of the laboratory is 3.4 km, operating at electron energies up to 17.5 GeV with a bunch duration of 100 fs.

As remarked above, a key requirement for SASE operation is a very small electron emittance and a short bunch length. In a storage ring, which operates at a fixed energy, the electron emittance is conserved, and it is this quantity that defines the beam properties. In the case of a SASE FEL it is usual to quote instead the *normalised electron emittance*, the product of the electron emittance and the energy expressed as the dimensionless quantity γ; it is this quantity that is conserved as the electrons are accelerated along a linac. At the European XFEL the normalised electron emittance is 1.4 μm rad so at the energy of 17.5 GeV the emittance is ~40 pm rad, very much smaller than the horizontal emittance in a storage ring (~3 nm rad) though similar to the vertical emittance in a storage ring. Indeed, this electron emittance is ~50% of the diffraction limit for the photon emittance ($\lambda/4\pi$), so this source achieves the goal aimed at for future storage rings of having an emittance that is diffraction limited at a wavelength of 1 Å; the resulting photon emittance is ~70 pm rad. The bunch length of 100 fs is also much shorter than typical values for a storage ring of ~20 ps, greatly increasing the electron density in the individual bunches.

Clearly, the combination of the SASE amplification, the reduced electron emittance and the short bunch length means that the *peak* brightness achieved in such a source is hugely increased relative to that of even the most modern electron storage rings. For example, the European XFEL at a photon energy of 1 Å delivers a peak brightness of ~5 × 10^{33} photons/0.1% bw/s/mm^2/mrad^2. Of course, the *average* brightness is much lower; each RF pulse delivers 3,000 bunches but the repetition rate is only 10 Hz, so the average brightness is lower by a factor of 3 × 10^{-9} giving ~1.6 × 10^{25} photons/0.1% bw/s/mm^2/mrad^2 (other current X-ray FEL facilities have overall bunch rates at least one order of magnitude lower). This may be compared with the average brightness available at the ESRF of ~10^{21} photons/0.1% bw/s/mm^2/mrad^2. Of course, the short bunches and high peak brightness make the FEL particularly attractive for challenging time-resolved experiments; in this regard the source is more comparable to conventional high-power fast lasers, although the accessible wavelength range is very different.

One important disadvantage with the SASE amplification scheme, however, is that it does not lead to truly monochromatic radiation nor do the individual flashes of radiation have the same intensity. This arises because the initial radiation that is amplified has the character of shot noise leading to pulse-to-pulse intensity variations of several percent as well as an overall spectral wavelength variation of tenths of one percent.

Fig. 2.31 shows an example of the spectral output from the LCLS (Linear Coherent Light Source) X-ray FEL facility at Stanford in the USA, the first such facility to become operational. When operated in the normal SASE mode the output radiation spectrum from a single electron bunch (a 'single shot' measurement) comprises many narrow peaks covering a range of 20 eV or more around the nominal energy of 8.3 keV (Fig. 2.31(a)), while averaged over many such radiation pulses one obtains a relatively smooth but quite broad spectrum (Fig. 2.31(b)).

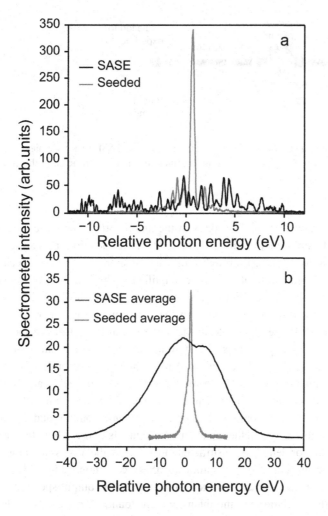

Figure 2.31 Spectral output from the LCLS FEL operating at a nominal photon energy of 8.3 keV (wavelength 1.5 Å). (a) shows the output from a single shot (a single bunch), (b) the average of many bunches. The black lines show the output with the FEL operating in SASE mode, while the grey lines show the output using a self-seeded mode as described in the text. Reprinted by permission from Macmillan Publishers Ltd: Nature Photonics, Amann et al. (2012). Copyright (2012)

Fig. 2.31 also shows the results of the first experiments to operate the FEL in a different mode which overcomes the problem of the broad and noisy spectral output. The basic idea is to 'seed' the FEL with radiation of the desired energy in order that this single wavelength is the one that is amplified in the undulator. At significantly lower photon energies this seeding can be achieved using a high harmonic of a conventional laser, but clearly this is not possible in the hard X-ray energy range. The solution is 'self-seeding', namely to extract photons of the desired energy from early in the SASE process along the FEL's undulators, and use this to seed the

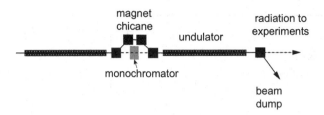

Figure 2.32 Simplified schematic diagram of the layout in a SASE FEL to achieve self-seeding at a specific photon energy. The lateral displacement of the electron beam in the chicane is greatly exaggerated in this diagram.

amplification achieved in the later parts of the undulator(s). The basic idea is shown (also in a highly-simplified fashion) in Fig. 2.32. SASE radiation from an early part of the undulator array is passed through a monochromator to select the specific wavelength of interest and continues on to be amplified in the later part of the undulator array. However, one implication of introducing such a monochromator is that the radiation suffers a time delay relative to the electrons, so in order to ensure that the radiation is still able to couple to the electron bunch to achieve the necessary amplification, the electrons are passed through a magnetic chicane to introduce a compensating time delay to the electrons. A novel monochromator design was used in this case by Amman et al. (2012) to minimise the radiation delay and thus also to ensure a magnetic chicane with the required delay could be accommodated. A further key requirement is the appropriate location of the monochromator and chicane in the undulator array; too early and the monochromated intensity is too low, too late and the remaining undulator length is insufficient to achieve the necessary amplification. Despite significant challenges in optimising these conditions, Fig. 2.31 shows the clear benefit of this self-seeding in the resulting radiation output spectrum, although large shot-to-shot variations in the intensity were found. A review by Seddon et al. (2017) summarises this use of self-seeding and other developments at various operating FEL facilities, in both the hard and soft X-ray energy ranges, including the application of these FELs to a range of scientific areas.

2.9   The Energy Recovery Linac

It should be clear from the foregoing discussion that current electron storage ring undulators and FELs are very different sources of essentially the same synchrotron radiation, namely, radiation arising from the passage of relativistic electrons through magnetic deflection structures. In a storage ring, electrons in a large number of relatively long bunches continually circulate the storage ring and produce radiation from bending magnets, undulators and wigglers. This radiation can often be regarded by its users as quasi-continuous. For example in the Diamond Light Source in the UK there are commonly 936 electron bunches, each of length ~20 ps circulating with a frequency of 534 kHz leading to a time between bunches of 1.9 μs, so in fact the source is pulsed with a mark-space ratio of about 1%. Although the source is therefore pulsed, the mark-space ratio is clearly ill-suited to most time-resolved experiments, and indeed the pulse (bunch) length is rather long for many time-resolved processes of interest. Reduced 'bucket-filling' modes of operation can certainly decrease the mark-space ratio significantly, but with much reduced average radiation intensity – not an attractive prospect for most users who are not interested in time resolution but typically require the maximum average flux. The source appears quasi-continuous to most users because the relatively long bunches with the total number of electrons distributed between many of them means that the peak radiation intensity is generally low enough to avoid saturation of any of the detectors used by experimenters that rely on counting of scattered or emitted individual photons or electrons, excited by the incident photons.

By contrast, FELs produce much shorter pulses of fully coherent radiation, with very much higher peak intensities and even higher brightness, which are, on average, far more widely spaced in time. Of course, the extreme peak intensity output arises to a significant degree because of the FEL amplification process. However, the ability to deliver both much shorter pulses (electron bunches) and lower electron emittance than in a storage ring stems from the fact that the electron beam passes only once along the linear accelerator and undulators, and the electron bunches do not have to be maintained in a constant recirculation orbit. Evidently if one were to pass electrons through the radiation-producing magnetic structures around a storage ring only once, the resulting radiation could have the short pulses and low emittance characteristic of the FEL. Unfortunately, delivering new electron bunches of this type at the near MHz rate of the bunches in a storage ring would involve unrealistically huge amounts of power input. A way to solve this problem is the energy recovery linac (ERL). The basic idea is that if one installs a linac in a closed loop (ring) with electron circulation, then it is possible, when the electron bunch has completed one orbit, for it to re-enter the linac with the opposite RF phase, causing the electrons to decelerate, thereby recovering most of their energy in the form of RF power. Moreover, this RF power can be stored for a sufficiently long time to accelerate a new electron bunch around the orbit. The basic idea is not new, although the first experimental demonstration of the idea seems to have been implemented on a linac-based infrared FEL at the Jefferson Laboratory in the USA (Neil et al., 2000). More recently a number of accelerator physics projects around the world are under way to explore the potential applications further, including for higher photon energies, as described by Bilderback et al. (2010). Indeed, a detailed design study was completed in 2006 for an ERL-based facility for the VUV and XUV spectral range, named 4GLS (4th Generation Light Source: www.4gls.ac.uk/) to be built in the UK (Flavell et al., 2005), but construction was never funded. An upgrade to the existing Cornell High Energy Synchrotron Source (CHESS) to create an ERL X-ray source was proposed in 2013 (www.classe.cornell.edu/Research/ERL/PDDR.html). A detailed review of ERLs is to be found in Merminga (2020).

Fig. 2.33 shows the schematic layout of the 4GLS proposal that illustrates both the basic design of an ERL for this spectral energy range but also possible add-on

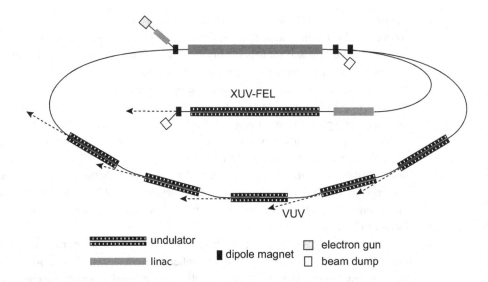

Figure 2.33 Simplified schematic diagram of the 4GLS ERL-based proposal for a VUV/XUV facility.

facilities exploiting this design. The outer ring shows the basic ERL concept comprising a linac injected by an electron gun based on a laser-excited photocathode (as in the linac-based FELs described in Section 2.8) and an electron circuit (achieved with bending magnets and appropriate focusing elements – not shown) that takes the electron beam through a series of conventional undulators delivering radiation to a range of different experiments, but then returns the beam to give up its energy in the main linac before being dumped (at low energy). Of course, the main linac could also be used to deliver electrons to a FEL undulator (without energy recovery) and the 4GLS design mentioned above proposed to include such a bypass, including additional linac acceleration to produce higher energy radiation. The separate injector for this component, and the laser-induced seeding of the FEL are omitted from the diagram for simplicity.

There is no doubt that a hybrid design of this type has a number of advantages over a conventional storage ring, providing both conventional storage ring undulators sources and the extreme intensity and brightness of a linac-based FEL. The improved emittance and bunch length possible in such a machine could provide one way to achieve a diffraction-limited 'ring source' at higher photon energies. Whether this concept will be fully exploited in the future remains to be seen. As yet, however, funding for no major new user facilities based on this concept have been approved.

2.10   'Compact' Sources

As should be clear from the preceding sections, synchrotron radiation and FEL sources are large-scale facilities with total costs of hundreds of millions of dollars, euros, pounds or their equivalent. In most cases they are national or international facilities dependent on government funding. Nevertheless, various ideas for much simpler and less expensive facilities have been explored; despite the fact that such designs are clearly not capable of competing with the latest state-of-the-art large-scale facilities over their broad range of capabilities, they may provide important niche capabilities. A particular motivation for a 'compact' machine designed to deliver soft X-rays at high flux was the potential for its use in X-ray lithography for the manufacture of semiconductor devices. The one such machine that came to fruition, the Helios-1 storage ring facility with a circumference of just 9.6 m, was developed jointly by accelerator scientists at the UK's Daresbury Laboratory and a manufacturer, Oxford Instruments (Wilson et al., 1993). It was purchased by the IBM corporation and installed in their Advanced Semiconductor Technology Center in East Fishgill in New York state in 1991 for X-ray lithography and micromachining (Morris et al., 1995), where it operated reliably for most of a decade. The second such ring (Helios-2) was established as the Singapore Synchrotron Radiation Facility in the National University of Singapore, where it is used not only for lithography, but also for a range of imaging, scattering and spectroscopy studies from the infrared to the soft X-ray energy range. These rings operate at an energy of 700 MeV with a critical photon energy of 1.47 keV (critical wavelength 8.45 Å). The extremely compact design is achieved by having only two bending magnets, each leading to a 'bend' of 180°, and each being based on superconducting magnets with a field of 4.5 T to achieve these bends in a relatively small space (bending radius of 0.519 m). A fan of bending magnet beamlines emerges from each of these magnets. The arrangement is shown in the highly simplified schematic of Fig. 2.34 which omits some of the focusing components and the details of the injection kicker magnets. While the

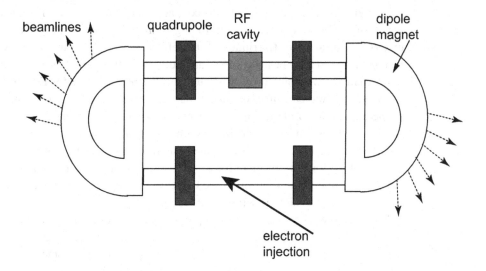

Figure 2.34 Simplified schematic diagram showing the general design of the Helios storage ring but omitting many of the detailed components.

machine is certainly not a 'table top' instrument, it is remarkably compact (circumference 9.6 m) and could actually be shipped and installed intact.

A very different approach, using an even smaller electron storage ring (circumference 4.6 m), operating at much lower energies below 50 MeV, was not built to generate conventional synchrotron radiation, but has been shown to be capable to delivering hard X-rays (up to 35 keV) through a process known as inverse Compton scattering. Compton scattering is an energy- and momentum-conserving ('billiard ball') collision between a photon and a nominally stationary electron; the electron recoil causes the photon energy to decrease and thus to increase the associated wavelength. The very small change in the wavelength means that the effect is only detectable if the photon wavelength is small (i.e. the energy is large), but an additional Doppler-like wavelength shift due to the initial motion of the scattering electron means that the phenomenon can be exploited to investigate the momentum distribution of electrons in materials, as described in Section 6.3.3. However, if the photon is involved in a collision with an electron approaching at relativistic speeds, the scattered photon can show a huge *upward* shift in energy (i.e. the inverse of the usual Compton scattering behaviour). Specifically, if the incident photons from a conventional laser have an energy E_1 and are directed for head-on collisions with relativistic electrons whose energy is given by γ, then the scattered photon energy, E_2 is

$$E_2 = \frac{4E_1\gamma^2}{1 + \gamma^2\theta^2} \tag{2.35}$$

where θ is the angle of emission relative to the direction of the electron propagation. For on-axis emission ($\theta = 0$) this simplifies to $E_2 = 4E_1\gamma^2$, so for $\gamma = 100$ (i.e. an electron energy of 51 MeV) the scattering photon energy is 4×10^4 times larger than the incident photon energy; 1 eV becomes 40 keV! Notice that the scattered photon energy falls quite sharply off-axis; at an angle $\theta = 1/\gamma$, E_2 is reduced by a factor of 2. This very large shift to high emitted photon energy can also be understood in a wave description of the effect, the electron interacting with the incident infrared laser beam acting as an undulator with a periodicity corresponding to one half the wavelength of this light. Recall that the undulator Equation (2.20) shows that the wavelength of emitted radiation $\lambda \propto \lambda_0/\gamma^2$, where λ_0 is the periodicity of the undulator, so the factor of ~10^5 reduction in λ_0 (e.g. from 5 cm to 0.5 μm) can be compensated by a reduction in γ by a factor of ~300 (e.g. from 3 GeV to 10 MeV) to achieve the same emitted photon energy. The main challenge in successful exploitation of this process is ensuring spatial and temporal overlap of the electron and photon pulses. A useful review of the technique and its development has been provided by Krafft & Priebe (2010).

The idea of exploiting this effect to produce a source of hard X-rays was explored in many laboratories worldwide, but it is only recently that a commercially available machine that appears to be capable of routine operation has become available. The design and performance of this facility, manufactured by Lycean Technologies in the USA and installed at the Technical University of Munich in Germany, was published by Eggl et al. (2016). The key components of this machine are shown in a highly

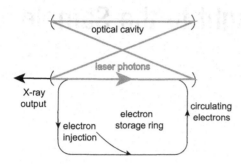

Figure 2.35 Highly simplified representation of the key components of a compact light source based on the inverse Compton effect.

simplified form in Fig. 2.35. It comprises a racetrack electron storage ring, operated at energies of 25–45 MeV, with an overlapping optical cavity in one straight section into which two Nd:YAG laser pulses (wavelength 1,063 nm, corresponding energy 1.166 eV) are stored. While the dependence on θ in Equation (2.35) shows clearly that the output is not strictly monochromatic, these authors showed that using a defining aperture to limit the off-axis radiation to an accepted 4 mrad cone led to a bandwidth ($\Delta E/E$) of 3.6% at an emitted photon energy of ~25 keV. This instrument design has been shown to be capable of enabling exploitation in a number of synchrotron radiation applications including an example of phase contrast tomography (see Chapter 8) of a mouse (Eggl et al., 2015) and, by Huang et al. (2020), dispersive X-ray absorption (both XANES and EXAFS – see Section 5.6). Of course, this device delivers a single beam of X-rays at a single energy, which is determined by the electron energy in the storage ring. It is therefore very different in concept from a conventional true synchrotron radiation source that is capable of simultaneously delivering photons of widely different energies to a range of different users.

3 Getting the Light to the Sample
Beamlines

3.1 Introduction

Chapter 2 describes the way that the sources of radiation, be they bending magnets, undulators, wigglers or FELs, produce a beam of radiation with certain characteristics in terms of lateral dimensions, divergence and spectral character. In the case of bending magnet and wiggler radiation the spectral output is 'white' (i.e. it comprises a broad continuum of wavelengths). The radiation output from undulators is much more concentrated in specific wavelength regions that are harmonically related, while the FEL delivers radiation centred around a single harmonic. However, very few experimental techniques that use this radiation can exploit a 'white' spectrum; more commonly, they require monochromatic radiation with a significantly narrower bandwidth than is delivered directly, even by the harmonic sources. A key requirement for most users is therefore some kind of monochromator that can be used to select the wavelength(s) required for the users' experiments. This is a key feature of almost all 'beamlines'.

There are, however, a range of other key instruments that need to be included in the 'beamline' between the source and the experimental end-station, components that address safety issues and other optimisations of the radiation arriving at the end-station.

Fig. 3.1 shows a highly simplified schematic of a beamline defining some of the associated vocabulary. Notice that the storage ring is enclosed in a concrete shield wall to ensure that stray radiation is not emitted into the outer hall, which is occupied by users. Synchrotron radiation itself is not the only source of this hazard; additional radiation arises both from the energetic electron beam striking molecules in the residual pressure of the ultra-high vacuum (UHV) of the storage ring (gas bremmstrahlung) and from striking the walls of, or components in, the storage ring vacuum vessel. Recall that some electrons are being constantly lost from the beam during operation of the storage ring due to high-angle electron-electron scattering and scattering by residual molecules in the vacuum; these electrons must then strike the walls or other solid components in the storage ring. A further radiation hazard can arise from unintended beam 'dumps' due to sudden problems in the electron beam control system, such as the power supplies of the magnet structures and the RF cavity; these may cause the whole electron beam to collide with the walls of the containment vessel or 'masks' or 'skimmers' within this vessel. All of these high energy electron

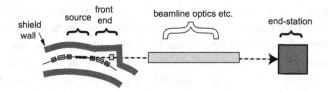

Figure 3.1 Simplified schematic layout showing the main components of a synchrotron radiation beamline. On the left is shown the source, here an undulator or wiggler, located in the storage ring between bending and focusing magnets within a concrete radiation shield wall. Additional 'front end' components are located within this shield wall. Further down the beamline and outside the shield wall are the various optical components followed by the user experiment end station. Not shown are the radiation-protection enclosures ('hutches') that generally surround the beamline optics and the end-station instrumentation.

collisions can lead to the emission of neutrons as well as high energy photons. Concrete walls with thicknesses up to 1 m or more can absorb most of this radiation, as can much thinner lead sheet shielding; high atomic-number materials like lead are more effective absorbers than concrete for the X-rays, but are less effective than concrete for neutrons.

As indicated by Fig. 3.1, the key parts of a beamline are a front end, with components fitted within the main shield wall of the storage ring and under the control of the machine operators, a second set of components, mainly optics for focusing and monochromating the radiation, and the end-station which is specific to the experimental techniques to be exploited by the users. Key components within the front end are concerned with radiation safety, but also remove some of the unwanted radiation that can otherwise pass on down the beamline. Collimators in the form of long narrow channels bored to blocks of high atomic number metals (typically tungsten) not only prevent most stray radiation from travelling down the beamline (most sources of this radiation are not aligned with the beamline axis) but can also select out the central cone of undulator radiation. For beamlines designed to exploit 'hard' X-radiation (photon energies above ~8 keV), it is also possible to install filters that can absorb much lower energy radiation and prevent this passing down the beamline. Another key component of the front end is a beam stop consisting of a block of tungsten or similar material, which can be moved into the beam to ensure no radiation passes down the beamline when it is not required.

One further important component of the front end is one or more vacuum valves to protect the integrity of the UHV environment of the storage ring vessel. Some experimental end-stations, particularly those devoted to investigations of the physics and chemistry of solid surfaces, must also operate under ultra-high vacuum, and on these beamlines it is common for the whole beamline (from the radiation source to the sample in the end-station including its optical components) to be contained in UHV vessels with no separation of the vacuum of these components under user operation. Clearly under these circumstances it is essential that the storage ring vacuum is protected from vacuum failures in the beamline, with fast-acting valves triggered by any sudden pressure rise in the beamline components. At least one valve must also be under the control of the experimental user group so that this can be closed for added security when the radiation is not being used, but also to allow maintenance, repair or modification work on beamline components that may necessitate some sections being brought up to atmospheric pressure. A second category of beamlines that require a common storage ring/beamline vacuum are those exploiting low photon energies below ~1–2 keV in the vacuum ultra-violet and soft X-ray energy range, because at these energies the radiation is strongly absorbed in air and in most available window materials. While many experiments in this photon energy range also use UHV surface science instrumentation, another important category is low photon energy experiments investigating the properties of atoms and molecules in the gas phase. In this case the analysis chamber of the end-station is necessarily operated at a higher pressure of the gas being investigated (to provide a target of sufficient density), and differential pumping is required to ensure UHV pressures can be retained in the section of the beamline closest to the storage ring.

Of course, at much higher photon energies in the hard X-ray range, absorption due to passage of the beam through air is much less significant, and indeed in early beamlines designed to operate in this energy range the optics were often operated in a helium atmosphere contained in relatively simple enclosures and even plastic pipework. Nowadays such beamlines are generally also operated under UHV conditions to overcome not only gas-phase absorption but also problems arising from gas-phase ionisation and contamination of optical components due to cracking of residual gases in the extremely bright radiation beams. This contamination of optics is particularly troublesome on beamlines operating in the soft X-ray energy range just above the C K-edge (~280 eV), as carbon build-up due to gas-phase molecular cracking can lead to strong (wavelength-dependent) attenuation of the beam intensity over a period of a few months of operation, even under UHV conditions, necessitating the implementation of methods of *in situ* cleaning of the optics.

The main optical components of beamlines play two important roles, focusing and monochromating. A key underlying issue influencing both of these functions is the properties of mirrors operating at different photon energies, so this general property will be described first before describing the instrumentation in which they play a part.

3.2　　Mirrors

In the visible range of the electromagnetic spectrum we are used to domestic mirrors having a high reflectivity even at normal incidence, and indeed even the surface of a sheet of glass, with a refractive index of ~1.5, is quite strongly reflecting. In the far ultraviolet and X-ray energy range this is far from true. In this case the refractive index of the mirror material is actually very slightly less than unity, usually written as

$$n = 1 - \delta + i\beta, \tag{3.1}$$

δ is the deviation in the real part of the refractive index from unity, which may be only $\sim10^{-6}$ at X-ray energies; and β, the imaginary component, accounts for absorption. This refractive index can be related to the forward (zero angle) atomic scattering factor of the radiation of the atoms comprising the solid, as a function of angular frequency, ω, $f_r^0(\omega) - if_i^0(\omega)$, where the suffices r and i denote the real and imaginary components,

$$n(\omega) = 1 - \frac{n_a r_e \lambda^2}{2\pi} \left[f_r^0(\omega) - if_i^0(\omega) \right], \tag{3.2}$$

where n_a is the atomic density, λ is the wavelength and r_e is the classical electron radius

$$r_e = \frac{e^2}{4\pi\varepsilon_0 mc^2}, \tag{3.3}$$

with e being the electron charge, ε_0 the permittivity of free space, m the electron mass and c the speed of light.

Under these conditions high reflectivity is only achieved at or close to the critical angle for total reflectivity. The condition for this can be readily derived from the usual Snell's law equation, which relates the angle of refraction ϕ_r to the angle of incidence, ϕ_i and the refractive index, n (Fig. 3.2), these angles being defined relative to the surface normal

$$\sin\phi_r = \sin\phi_i/n \tag{3.4}$$

In the case of a refractive index n that is less than unity, the refraction angle is larger than the incidence angle, so when the incidence angle reaches the critical angle of incidence, ϕ_c, the refracted wave does not penetrate the surface but runs parallel to the surface such that $\theta_r = 90°$. If we can assume there is no absorption ($\beta = 0$) then

$$\sin\phi_c = 1 - \delta. \tag{3.5}$$

Evidently this critical value of the angle of incidence is quite close to 90° (i.e. the incidence is grazing to the surface). It is therefore more useful to discuss the critical reflectivity conditions in terms of the *grazing* angle of incidence, θ_i, which is the complement of ϕ_i (see Fig. 3.2), so

$$\cos\theta_c = 1 - \delta$$

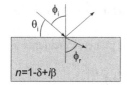

Figure 3.2 Refraction of radiation incident from air or vacuum (refractive index unity) into a material with a refractive index less than unity.

and if this angle is small, expressing $\cos \theta_c$ by its series expansion and taking only the leading terms leads to

$$\theta_c \simeq \sqrt{2\delta}. \tag{3.6}$$

The amplitude of the reflectivity can be determined using the usual Fresnel equations, describing the incident, reflected and refracted waves and applying the necessary matching conditions at the interface (a particularly clear derivation of the results that follow using this approach may be found in Attwood (1999). Two sets of such equations must be solved corresponding to the two perpendicular polarisation components, those with the polarisation vector in the plane of incidence (p-polarisation) and those for the polarisation vector perpendicular to the plane of incidence (s-polarisation). For s-polarisation one obtains

$$R_s = \frac{\left| \cos \phi_i - (n^2 - \sin^2 \phi_i)^{1/2} \right|^2}{\left| \cos \phi_i + (n^2 - \sin^2 \phi_i)^{1/2} \right|^2} \tag{3.7}$$

Notice that if one takes $\phi_I = 0$ and assume that both δ and β are very small, then at normal incidence one obtains

$$R_{s,0} \simeq (\delta^2 + \beta^2)/4, \tag{3.8}$$

confirming the earlier assertion that the normal incidence reflectivity is very small (notice that at normal incidence R_s and R_p must be equal as there is no longer any distinction between 'in-plane' and 'out-of-plane' of incidence). If, instead, one takes $\phi_I = \phi_c$, then at the critical angle

$$R_{s,\theta_c} = \frac{1 - \dfrac{\sqrt{2\delta\beta}}{\delta + \beta}}{1 + \dfrac{\sqrt{2\delta\beta}}{\delta + \beta}}. \tag{3.9}$$

Notice that if $\beta = 0$ (i.e. one has no absorption), then the reflectivity at the critical angle is unity, as was implicit in the earlier discussion. On the other hand, if $\beta = \delta$, then the reflectivity at the critical angle is only 0.17. At the critical angle, there is no refracted beam passing into the solid, but there is an evanescent component to this wave decaying into the solid, so if absorption is strong significant flux is removed from reflected wave. The decay length of this evanescent wave decreases as the grazing incidence angle decreases below the critical angle, so at more grazing incidence the reflectivity is increased.

The reflectivity with the polarisation vector in the plane of incidence is given by

$$R_p = \frac{\left| n^2 \cos \phi_i - (n^2 - \sin^2 \phi_i)^{1/2} \right|^2}{\left| n^2 \cos \phi_i + (n^2 - \sin^2 \phi_i)^{1/2} \right|^2}. \tag{3.10}$$

Notice that this expression differs in one important way from that for R_s. Specifically, R_s increases monotonically as the incidence angle, ϕ_i, increases (as the grazing incidence angle decreases), but R_p has a sharp minimum (with a value of zero if n is real) when $\phi_I = \phi_B$ defined by

$$n^2 \cos \phi_B = \left(n^2 - \sin^2 \phi_B\right)^{1/2}, \tag{3.11}$$

where ϕ_B is the Brewster angle – the angle at which the reflected wave resulting from unpolarised incident radiation becomes polarised, whence

$$\tan \phi_B = n. \tag{3.12}$$

For a typical glass in the visible range of the spectrum n has a value of about 1.5 and may be regarded as real (very weakly absorbing), leading to the familiar value of the Brewster angle of 56°. If n is complex, the reflectivity does not go to zero, but has a sharp minimum at the Brewster angle. This angle corresponds to the condition at which the reflected and incident beams are separated by an angle of almost exactly $90°$ $\left(\theta_B \simeq \frac{\pi}{4} - \frac{\delta}{2}\right)$, an angle at which the transverse field modulations of the p-polarised incident wave cannot give rise to transverse modulations in this reflection direction.

Based on tabulations of the atomic scattering factors f, such as those by Henke, Gullikson & Davis (1993), calculations of the reflectivity of different mirror materials as a function of the incident photon energy can be readily performed. Indeed, these authors have performed such calculations over the photon energy range of 50 eV to 30 keV for several mirror materials of interest, and Fig. 3.3 shows such a set of results, for different grazing incidence angles, for gold. The grazing incidence angles are expressed in milliradians (mrad); 1 mrad = 0.0573°,

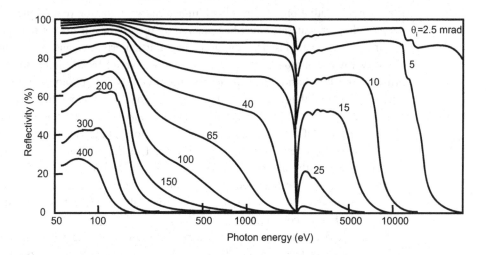

Figure 3.3 Calculated reflectivity of unpolarised radiation from gold as a function of photon energy at a series of different grazing incidence angles. Adapted from Henke, Gullikson & Davis (1993) with permission from Elsevier

so 2.5 mrad corresponds to a grazing angle of 0.14°, 400 mrad corresponds to a grazing angle of 23°. The sharp dip in the reflectivity around 2,200 eV is due to the M_4 and M_5 absorption edges in Au associated with excitation of electrons in the 3d states of the atoms into the continuum (photoionisation), while the weaker structure at slightly higher energies is associated with the other M edges (excitation of 3p and 3s electrons). Evidently one would not choose gold as the mirror coating of optics to be used for this energy range, substituting a different material (such as Pt) with absorption edges at quite different energies.

Apart from the specific problem of absorption edges, Fig. 3.3 shows the general trend for an effective cut-off in reflectivity at high energies, with the energy of this cut-off increasing as the grazing incidence angle is reduced. At a grazing incidence angle of $0°(\phi_I = 90°)$ the theoretical reflectivity is unity (see Equation 3.7), but of course, this is not useful in practice and indeed at very grazing angles the effect of surface roughness, ignored in the derivations above, plays an increasingly important role in degrading the reflectivity. What is clear from Fig. 3.3, however, is that in the X-ray energy range up to ~10 keV very significant intensity losses arise for grazing incidence angles greater than about 0.5° (~9 mrad) and even in the extreme ultra-violet (EUV) or soft X-ray energy range up to 1–2 keV, grazing incidence angles should preferably be no larger than about 1° (~17 mrad).

There are a number of further practical implications of these conclusions. Firstly, mirrors act as low-pass filters. For experiments that exploit the highest energy photons, the removal of the unwanted lower energy photons can be achieved by the use of conventional filters, specifically foils of suitable materials that absorb the low energy radiation, but transmit the higher energy radiation and thus act as high-pass devices; such filters may even be installed in the front end of some beamlines as mentioned in the previous section. By contrast, in beamlines designed to use lower-energy radiation, the unwanted higher energy radiation (e.g. higher harmonics of undulator radiation) may be removed by the use of mirrors set to appropriate grazing incidence angles. This requirement is particularly relevant when working at the very lowest photon energies that can be delivered by an undulator, because bringing the first harmonic radiation to a sufficiently low photon energy implies operating the undulator at high values of K, approaching the conditions associated with a wiggler (or at least a 'wiggulator'!) when *most* of the emitted radiation is at higher energies. Of course, one reason for trying to exclude unwanted radiation (be it at higher or lower energies than that of interest) is that this will reduce the power loading to which the beamline optics is subjected. One consequence of removing many photons outside the spectral range of interest by a filter or a mirror is that these components suffer a high power load, typically necessitating the installation of some method of cooling.

In this regard, another implication of the need to use very grazing angles on mirrors can be both a disadvantage and an advantage. At grazing incidence, the 'footprint' of a radiation beam on the mirror becomes greatly elongated (by a factor $(1/\cos\theta_i)$, necessitating large (long) mirrors. Consider, for example, bending magnet radiation from a 3 GeV storage ring. The angular divergence is $\sim 1/\gamma$ or about 0.2 mrad, so if a mirror is placed 10 m from the source in the storage ring the size of the beam at the mirror is dominated by the divergence (rather than the source size) and is ~2 mm. If this radiation beam strikes the mirror at a grazing incidence angle of 2.5 mrad then the length of the footprint is ~2 m. Of course, as discussed in Chapter 2, the divergence from an undulator is a factor of a few smaller than that of the bending magnet radiation, and such a small grazing angle may not be necessary, but mirrors with lengths of tens of cm or more are commonly required. The positive aspect of this is that the heat load on the mirror due to the intense radiation (including that at higher and lower energies than required for the experiment in the end-station) is distributed over a large area and thus less likely to lead to very high local temperatures, although it is still often necessary to cool the mirrors to minimise the heating and any resultant distortion of the mirror surface.

In practice (grazing incidence) mirrors perform one or more of three functions in beamline optics. Firstly, as described above, they can be used as low-pass filters, so on beamlines designed to serve experiments that exploit lower photon energies a cooled plane mirror may be the first optical element encountered by the 'white' or multi-harmonic radiation. Secondly, mirrors may be used to change the beam direction (through small angles), often to compensate for deflections produced by other optical elements. Thirdly, they may be used to focus the radiation beam. Key aspects of this focusing role can be best understood by a brief description of the properties of spherical mirrors when used at grazing incidence.

Fig. 3.4 shows a schematic diagram of the focusing of an object at O to an image at I by grazing incidence onto a spherical mirror. Notice that the angle of incidence at the curved mirror surface is necessarily different across the fan of radiation emerging from the object. Analysis of this simple geometrical optical problem (see, for example, Michette, 1993) leads to the conclusion that the effective focal length of the mirror in this tangential or meridianal plane is

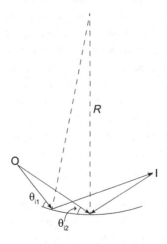

Figure 3.4 Image formation at I of an object at O, in the ('tangential') plane of incidence, by grazing incidence at a spherical mirror of radius of curvature R.

$$f_T = \frac{R \sin \theta_i}{2}. \tag{3.13}$$

As θ_i is necessarily small, this focal length is very much smaller than the usual value of $R/2$ for near-normal incidence to a spherical mirror. In effect the mirror is foreshortened in this plane. By contrast, the focusing in the perpendicular 'sagittal' plane is described by a focal length

$$f_S = \frac{R}{2 \sin \theta_i}. \tag{3.14}$$

These equations are often referred to as the Coddington Equations after Henry Coddington who published A Treatise on the Reflection and Refraction of Light in 1829, but it seems their original derivation was much earlier (Kingslake, 1994). At the small grazing incidence angles discussed above the resulting difference in the focal lengths in these two planes is very large, specifically

$$\frac{f_S}{f_T} = \frac{1}{\sin^2 \theta_i}. \tag{3.15}$$

So, for example, if the grazing incidence angle is 5 mrad this ratio is 40,000! This difference in the tangential and sagittal focal properties is known as astigmatism, and clearly at grazing incidence a spherical mirror is highly astigmatic.

Of course, Equations (3.13) and (3.14) are expressed in terms of a single incidence angle, whereas on the curved mirror surface the incidence angle differs across the incident fan of radiation. This clearly means that the effective focal length for different parts of this radiation fan is different; thus, this arrangement therefore suffers from spherical aberration, with different off-axis radiation focusing at different positions.

There are effectively two alternative ways of overcoming the strong astigmatism. One is to separate the roles of the tangential and sagittal focusing by using two separate focusing mirrors involving reflections in the two perpendicular planes (Fig. 3.5), an arrangement named after Kirkpatrick & Baez (1948) who first proposed this as a method of building an X-ray microscope. If these are cylindrical mirrors there is no focusing in one plane, making the analysis simpler, although in practice the focusing of a spherical mirror in the sagittal plane is so much weaker than in the tangential plane that the difference between optical properties of cylindrical and spherical mirrors becomes rather marginal. Indeed, by using a pair of mirrors with elliptical, rather than cylindrical or spherical surfaces, it is also possible to correct from the spherical aberration (Suzuki, Uchida and Hirai, 1989).

The alternative solution to overcoming the astigmatism of a single spherical mirror is to use an aspheric mirror with different radii of curvature in the two orthogonal directions. In particular, one can use a toroidal mirror whose reflecting surface has the shape of a section of the inside of a bicycle tyre (see Fig. 3.6).

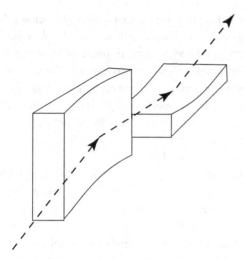

Figure 3.5 Kirkpatrick-Baez arrangement of a 'K-B' pair of cylindrical or spherical mirrors to produce stigmatic focusing.

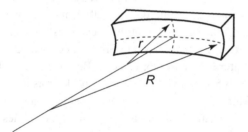

Figure 3.6 Toroidal mirror with tangential and sagittal curvatures of R and r providing stigmatic focusing at a specific grazing incidence angle.

3.3   Focusing Lenses

Of course, in the visible range of the spectrum, a far more common way of focusing radiation is through the use of refractive lenses. Such lenses are typically made from some form of glass and so are limited in range from the near infrared to the near ultraviolet; quartz glass extends the range further into the ultraviolet, and the use of materials such as LiF and CaF_2 can extend the range further to photon energies of ~10 eV. Beyond this essentially all available materials are too absorbing. Of course, lenses rely on refraction for their focusing properties, and in the near-visible region the materials used have refractive indices significantly greater than the value of unity that characterises vacuum and air (typically ~1.3–1.5), so a convex lens has a strong positive focusing effect. As remarked in Section 3.2, at X-ray energies materials have a refractive index of less than unity.

This raises an interesting possibility; if X-rays passing through such a solid meet a spherical hole in the solid, this hole has a higher refractive index and thus acts as a convex focusing lens. Similarly, an X-ray passing through vacuum can be focused by a concave lens made of a material with a refractive index of less than unity. Of course, the refractive index of the material is only very slightly less than unity, so the focusing effect is very weak, but a succession of such 'vacuum lenses' can produce the weak focusing necessary to bring about a focal spot over a sufficiently long distance. This is the principle on which *compound refractive lenses* (CRLs), for use on the X-ray energy range, is based.

Fig. 3.7 illustrates this idea with a diagram of a single concave cylindrical lens fabricated from a material with a refractive index $n = 1 - \delta$ and a long series of such lenses made by drilling a series of cylindrical holes into the solid. The focal length of just one lens is $R/2\delta$, where R is the radius of curvature of the lens (the radius of the drilled holes) while N such lenses leads to a focal length of $R/2N\delta$. Of course, these horizontally drilled holes only produce focusing in the vertical plane, but a similar set of vertically drilled holes produce horizontal focusing. An obvious problem with this approach is absorption in the lens material (only the real part of the refractive index is considered in the simplified description above). This absorption can be reduced by fabricating the device with a material having a low atomic number (so all absorption edges are at low energies) but can still be significant. For example, a device like that shown in Fig. 3.7. containing 160 biconcave lenses of radius of curvature 1.9 mm fabricated in Be gave a focal length of 93 cm at a photon energy of 6.5 keV, but only gave a peak transmission of 9% (Beguiristain et al., 2002). More sophisticated designs can ameliorate this problem and compound refractive lenses are now in use at several beamlines. Further discussion of these, and of other focusing schemes with short working distances based on capillaries and Fresnel zone plates, will be described more fully in the context of microscopies (Chapter 8).

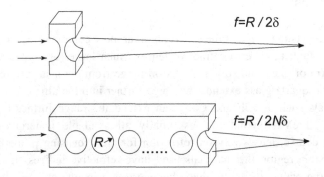

$$f=R/2\delta$$

$$f=R/2N\delta$$

Figure 3.7 Schematic diagram showing the basic principle of X-ray focusing with one and a succession (N) of concave cylindrical lenses fabricated from a material with a refractive index $n = 1 - \delta$.

3.4   Monochromators: From Visible to Soft X-Rays

The first report of the separation of 'white' light into its constituent colours seems to be that of Isaac Newton in 1671. Popular science books frequently show images recreating his experiment, showing him holding a prism in the beam of sunlight passing through a hole in a window blind and displaying the resulting rainbow-like spectrum on a screen. Fig. 3.8 shows Newton's own sketch of this experiment. Notice that this includes a second prism that could be used to investigate the refraction of light of a particular colour passing through a hole in the screen. Fuller details of this experiment and Newton's deduction may be found in the rather more recent paper by Fara (2015).

It is, of course, perfectly possible to make a spectrograph, and indeed a mono-chromator, using exactly the same effect of dispersive refraction in a suitable trans-parent material (typically a glass) for use in the visible range of the spectrum, and indeed somewhat beyond the visible depending on the absorption properties of the prism material. However, even in this spectral range, it is more common to use a transmission diffraction grating, typically on a transparent support), but designed to have the effect of a periodic array of 'slits'. This arrangement is shown on the left-hand side of Fig. 3.9. The standard description of how this device works is then to regard each slit as a source of 'Huygen's secondary wavelets' emanating from the slit; this is the process of diffraction at each slit. The condition for constructive interference between the waves emanating from successive slits is then that the path-length difference is an integral number (m) of wavelengths, namely

$$d \sin \theta = m\lambda. \tag{3.16}$$

If $m = 0$, one has zero order diffraction with white light continuing in the same direction (perpendicular to the grating) as the incident radiation. For larger values of m the direction of emission is dependent on the wavelength, λ, producing the required

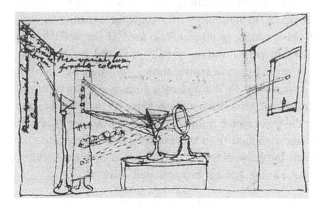

Figure 3.8 Isaac Newton's sketch of his light dispersion experiment. After Fara (2015). Reproduced under Creative Commons CC BY 4.0 licence

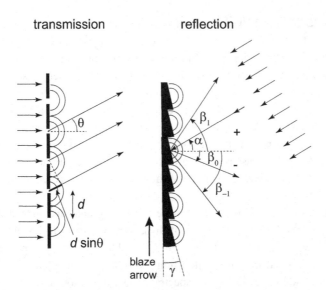

Figure 3.9 Schematic diagram showing transmission and reflection diffraction gratings and the definitions of the angles and distances that determine the diffraction conditions. The blaze angle, γ, of the reflection grating is exaggerated for clarity.

dispersion. Typically one exploits only the first order ($m = 1$) diffraction, but higher orders may be observed at larger angles.

For radiation with energies above the absorption cut-off for transmission through the support material that is present in a traditional transmission grating, this method cannot be used, but instead one can use reflection gratings. These are typically made by 'ruling' (effectively scratching) a periodic array of 'lines' on a suitable surface. By appropriate choice of the shape and orientation of the ruling device (typically a diamond tip) it is possible to produce a 'blazed' grating such as that shown in Fig. 3.9; the surface has a sawtooth profile with a well-defined blaze angle, γ, defined as in Fig. 3.9, designed to optimise performance at a particular wavelength (as described below). Other surface profiles are also used. Holographic gratings generally have an essentially sinusoidal surface profile produced by recording an optical interference pattern onto a photosensitive coating followed by appropriate etching. Laminar gratings have essentially flat-bottomed grooves and can be manufactured to a low surface roughness using ion-beam milling, although laminar-like profiles can also be obtained using the holographic technique. The condition for coherent interference between the scattered radiation from successive periods of the reflection grating is similarly determined by the path-length differences in both the incidence (not now assumed to be normal to the grating) and scattered waves:

$$d(\sin \alpha + \sin \beta) = m\lambda \tag{3.17}$$

where β is the angle between the diffracted beam and the normal to the average grating surface orientation (not the local surface of the blaze facets) and α is the incidence angle defined in the same way. The sign of the angle β is defined as positive if it lies on the same side of the grating surface normal as the incidence angle, and negative if it appears on the opposite side (see the $+$ and $-$ signs in Fig. 3.9). In Fig. 3.9 the diffraction angle β is shown distinguished by subscripts that define the diffraction order (the value of m) 0 (the zero order diffraction), and 1 and -1, first order of the diffraction in the direction of the blaze arrow and in the opposite direction. These are conventionally known as 'outside' and 'inside' orders, respectively. The angle β_0 is the zero-order diffraction angle and corresponds to the specular direction relative to the average grating surface.

Of course, this specular geometry relative to the grating surface does not correspond to the usual condition for reflection from a plane mirror because of the sawtooth surface profile. However, the condition from specular *reflection* from the blaze facets does correspond to a diffraction condition at which the grating efficiency (the intensity of the diffracted beam) is at its highest. The diffraction Equation (3.17) corresponds to the condition in which scattered radiation from one part of a blaze facet is in phase with scattering from the equivalent position on all the other facets. However, if this geometry *also* corresponds to that of specular reflection from the individual facets, then scattering from *all* positions on all the facets is in phase. This coherence of scattering from different parts of a planar surface is, of course, what defines the specular reflection properties of a plane mirror. This second condition requires that

$$\alpha + \beta = 2\gamma. \tag{3.18}$$

One further aspect of diffraction gratings that can become important in their use in monochromators on synchrotron radiation beamlines is the effect of overlapping orders. Equation (3.17) shows that a particular geometry – the angle of incidence α and the diffracted beam angle β defines the product $m\lambda$. Thus several different values of the wavelength can satisfy this condition with different values of m. For example, the $m = 1$ condition defines a wavelength, say λ_1, but the $m = 2$ condition defines a second wavelength $\lambda_2 = \lambda_1/2$, and so on. The presence of this higher order radiation (with photon energies higher by a factor of 2, 3, etc.) can clearly be a problem if true monochromatic radiation at a single energy is required. One solution, or at least a partial solution, is to ensure there is a reflecting mirror with an angle of incidence low enough to retain the first order radiation but large enough to largely exclude the higher orders, as mentioned in Section 3.3. Such a mirror, incorporated into a monochromator, may even be arranged to change its angle of incidence depending on the spectral range in which the monochromator is operated.

Detailed designs of grating monochromators vary, but Fig. 3.10(a) shows a particularly simple design, the Czerny-Turner monochromator, for use in the energy range from visible to near ultra-violet. Notice that a 'collimating' spherical mirror is used to ensure that, despite the divergent fan of radiation entering from the entrance slit, all the of this radiation strikes the plane grating at the same incidence angle, α; a range of different incidence angles would lead to a range of different wavelengths emerging at any specific emission angle, β, as is shown by Equation (3.17). A second spherical mirror then brings all the radiation emitted at the same angle to a single focus

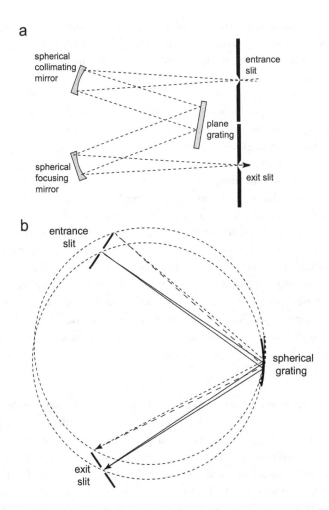

Figure 3.10 Optical layout of two designs of normal incidence grating monochromators, namely the (a) Czerny-Turner and (b) Seya-Namioka designs using plane and spherical gratings respectively. (b) shows the Rowland circle and grating position for two different wavelengths demonstrating how the instrument fails to maintain focus as the wavelength is scanned.

at the exit slit. Scanning the selected photon energy requires only rotation of the plane grating. A key aspect of this instrument is that it involves three reflections, two from spherical mirrors and a third from a planar reflection grating, that are all quite close to normal incidence. This clearly limits the useful photon energy range to low energies. The basic design predates extensive use of synchrotron radiation and may be used with conventional laboratory radiation sources such as a hydrogen discharge lamp. The useful photon energy range is strongly limited by the presence of the three large angle reflections. One way to overcome this is to use a spherical grating, combining both the focusing and the dispersion into a single optical element. Such a design is illustrated in Fig. 3.10(b) which shows the layout of a Seya-Namioka monochromator.

The object and image of the spherical focusing must both lie on the 'Rowland circle' which has a diameter equal to the radius of curvature of the grating, and is tangential to its surface. The problem with this general arrangement is that rotating the grating, in order to change of output wavelength, causes the Rowland circle to rotate, so that fixed entrance and exit slits no longer satisfy this focusing condition. For experiments on a synchrotron radiation source there are clear advantage to having fixed entrance and exit slits as the storage ring is not moveable and the experiment located behind the exit slit also often involves large and heavy equipment. It transpires, however, that if the angle between the incident and diffracted radiation is 70.25°, the resulting defocusing is relatively small and may be tolerated (Namioka, 1959). A range of designs of these so-called normal incidence monochromators (NIMs), with a large scattering angle only off the grating, can be used up to photon energies of ~40 eV.

Of course, for use on a synchrotron radiation beamline the geometry of Fig. 3.10(a), in particular, with the monochromatic radiation heading back towards the source, is not very convenient, and alternative designs that involve much smaller total angular deflections such that the monochromatic radiation emerges at a much smaller angle relative to the incident radiation are often more convenient. Monochromators for use at higher photon energies, from the vacuum ultraviolet to soft X-rays with energies up to ~1 keV or more, clearly require all reflections to be at grazing incidence and this intrinsically leads to a smaller total angular deflection. These grazing incidence monochromators (GIMs) typically involve rather more complex designs although there are also clear benefits to keeping the number of such reflections to a minimum, as losses occur at each one. A further requirement is generally that the direction and position of the monochromatic beam should be constant as the wavelength is varied in order that it is not necessary to move the sample and associated instrumental chamber that must receive the radiation, the latter often comprising a large and heavy ultra-high vacuum chamber. A range of different designs have emerged using plane gratings, spherical gratings and even toroidal gratings.

Fig. 3.11(a) shows the optical layout of a particularly simple design based on a toroidal grating, which was implemented on the Berlin BESSY storage ring in the 1980s (Dietz et al., 1985). Like the Seya-Namioka NIM, the only optical component is the grating and the wavelength is adjusted only by rotating this grating, achieving imperfect but satisfactory focus over the operating range. Toroidal grating monochromators (TGMs) became quite popular for operation for photon energies up to ~200 eV, but this instrument was used successfully up to ~1,000 eV. This monochromator was designed to achieve a high throughput of flux from a bending magnet source at modest energy resolution. Notice that there is no physical entrance slit; the electron beam source acts as a virtual entrance slit. Of course, this slit-less mode of operation relies on a small source (small electron beam) in the (vertical) plane of dispersion, which is very stable in position. One feature of many monochromator designs in use in the 1980s and 1990s, all installed on bending magnets of storage rings, was an attempt to capture as large as possible an angular range of the radiation in the horizontal plane in order to maximise the output flux. This is, of course, particularly difficult to achieve using only grazing-incidence optics. Nowadays, with the dominance of undulator beamlines that emit an intrinsically narrow fan of radiation, this criterion is no longer so important.

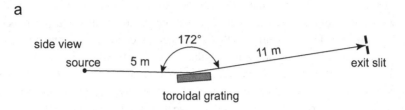

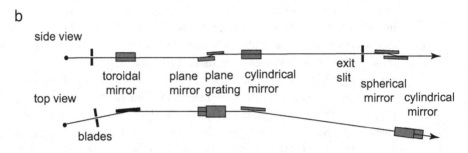

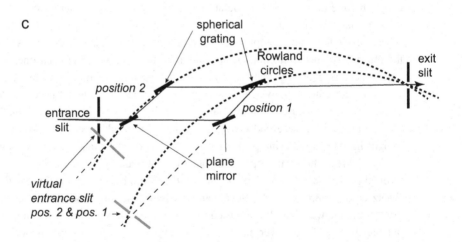

Figure 3.11 Three different optical layouts for grazing incidence grating monochromators for the SXR photon energy range. (a) shows a toroidal grating monochromator with the grating as the only optical component as described by Dietz et al. (1985). (b) shows a plane grating monochromator described by Follath (2001). (c) shows how the use of a spherical grating as the only focusing component, moving and rotating together with a plane mirror, can keep the fixed entrance and exit slits close to the Rowland circle (described by Senf, Eggenstein & Peatman, 1992).

Much more common than TGMs now, are GIMs based on either spherical gratings or plane gratings (SGMs and PGMs). The general layout of a PGM design installed on several undulator beamlines on the BESSY II storage ring in the 2000s is shown in Fig. 3.11(b) (Follath, 2001). Essentially the same design is the basis of instruments on newer beamlines at, for example, the MAXIV facility in Sweden. The overall length of this beamline is more than twice that of the high-energy TGM of Fig. 3.11(a), in part because the larger higher-energy BESSY II ring and its associated shield wall means that the first optical element is necessarily further from the source due to physical constraints. Notice that in this design the monochromatic beam that emerges is horizontal and at the same height as the original undulator beam. The wavelength is varied by rotating the plane grating, but also rotating the plane mirror in front of the grating. The initial toroidal mirror provides sagittal collimation to ensure a parallel beam of radiation hits the grating (as in the case of the Czerny-Turner NIM in Fig. 3.10(a)), improving the energy resolution. The final spherical and cylindrical mirrors provide focusing onto the sample.

The problem of varying the emitted wavelength and keeping focus with a fixed entrance and exit slit, using a spherical grating as the only focusing element, has been described above in the context of the Seya-Namioka NIM. It is, however, possible to design a grazing-incidence grating monochromator of this type that also retains near-focus condition by the use of additional (plane) mirrors and more complex movements of the optical components, as described by Senf, Eggenstein & Peatman (1992) and illustrated by the optical layout shown in Fig. 3.11(c). The diagram shows two different positions (and rotation angles) of the spherical grating and an associated plane mirror, corresponding to the conditions for the monochromator to output two different photon energies. As the grating is rotated it is also displaced such that the Rowland circle passes through the exit slit, while the virtual entrance slit produced in the plane mirror is very close to the Rowland circle. Coordinated motions of several components are a common feature of many GIMs, and indeed an early Rowland circle instrument used at the Stanford Synchrotron Radiation Laboratory in the late 1970s has some similarities to the design shown in Fig. 3.11(c), but also involved movement of the entrance slit located at a reflecting mirror; the presence of a large lever arm used to coordinate these movements mechanically led to this being known as the 'Grasshopper' monochromator (Brown, Bachrach and Lien, 1978) (albeit a grasshopper with only one leg!), although later modifications of this design used individually-controlled digital movement mechanisms (Brown, Stott and Hulbert, 1986). More common designs of SGMs nowadays have extra focusing optics leading to an arrangement quite similar to that of Fig. 3.11(b), but with a spherical grating instead of the plane grating; this has also been shown to be an effective practical design (Senf et al., 2001).

While the curved surface of a spherical surface produces focusing in both mirrors and gratings, it is also possible to produce a focusing effect in a plane grating if the line spacing is varied across the grating in an appropriate manner. This causes the angle β of the diffracted radiation at a particular wavelength to vary across the gratings, compensating, for example, for different angle of incidence on the grating from a divergent source. An early example of a GIM covering the soft X-ray energy range based on such a grating was reported by Itou, Harada & Kita (1989).

While the detailed design of individual monochromators lies outside the scope of this book, it is appropriate to comment on the main criteria that apply. These are the photon energy range, the energy resolution and the transmittance or efficiency. One parameter that influences all of these factors is the chosen density of lines on the dispersing grating. In this regard, it is important to recognise an important feature of the behaviour of gratings when used at grazing incidence. Typical grating densities used in these monochromators fall in the range ~500–2,000 lines/mm. 2,000 lines/mm corresponds to a line spacing of 0.5 μm, and it is difficult to rule gratings with spacings much smaller than this. This line spacing (0.5 μm = 5,000 Å) is very much larger than the wavelength of the radiation being diffracted (e.g. photon energies of 120 eV to 1,200 eV correspond to wavelengths of 100 Å and 10 Å, respectively), which might lead one to doubt that such gratings can be used at these photon energies. Notice, though, that if the incident and exit radiation falls on opposite sides of the normal to the grating surface, Equation (3.17) effectively becomes

$$d(\sin|\alpha| - \sin|\beta|) = m\lambda. \tag{3.19}$$

So if one has grazing incidence and grazing exit both of the sine terms are large (close to unity) but almost cancel, so the quantity on the left hand side is significantly smaller than d; in effect the grating spacing is foreshortened in this grazing geometry. Of course, the two sine terms cancel if $\alpha = \beta$, but this is the zero order condition (specular geometry from the grating surface) with $m = 0$ and no dispersion. However, it is instructive to determine the degree of angular dispersion, which is one factor that determines the resolution that can be obtained in a monochromator. If the angular dispersion is large, two similar wavelengths are separated in angle, β, by a relatively large angle, and so may be separated by passage through an appropriate width of exit slit. The size of this angular dispersion can be simply determined by differentiation of Equation (3.17), leading to

$$D = \frac{d\beta}{d\lambda} = \frac{m}{d\cos\beta}. \tag{3.20}$$

In a GIM, using both grazing incidence and grazing emission, the angle β is large – only slightly less than 90°, so $\cos\beta$ is small, for example, if $\beta = 86°$ (4° grazing exit) then $\cos\beta = 0.07$, and as this factor appears in the denominator of Equation (3.20), grazing exit greatly increases the angular dispersion. This dispersion, however, is only one factor determining the ability of the monochromator to separate two similar wavelengths. If the monochromator were to be illuminated by a line source comprising two similar wavelengths, say λ_1 and λ_2, then stronger dispersion separates them by a larger angle, but the ability to distinguish them also depends on the angular width that each line presents in the exiting beams. This width is determined by the total number of grating lines that are illuminated, N. This is the same result as one obtains when extending of the usual Young's double slit interference experiment to multiple slits. With only two slits the width of the interference maxima is just half of their spacing, but as additional slits are added at the same spacing, intermediate (much weaker) interference peaks appear, the intermediate interference minima forcing each peak to become narrower.

Resolving power is usually written as

$$R = \frac{\lambda}{\Delta\lambda} = \frac{h\nu}{\Delta(h\nu)} \qquad (3.21)$$

and for a grating

$$R = mN. \qquad (3.22)$$

The implication is that, for example, using a grating with 1,000 lines/mm and a total illuminated width of 100 mm one could achieve a resolving power of 10^5. In practice imperfections and aberrations in the optical components lower this expectation, but resolving powers of ~10^4 or more are certainly achievable. A common feature of many monochromators is to construct them such that two or three different gratings can be used, mounted side by side; they can be changed by a lateral shift of the grating mount; indeed, sometimes alterative mirrors are mounted in a similar fashion. This arrangement is particularly relevant when the instrument is designed to cover a large energy range of one decade or more in energy. As remarked above, undesirable higher order radiation, when the monochromator is set to deliver first-order radiation at lower energies, can often be suppressed by using larger grazing angles in these optical components at these lower energies.

The third significant parameter that is sought to be optimised in a monochromator design is the overall transmittance or efficiency: what fraction of the incident radiation at a given photon energy emerges from the monochromator? In the higher photon energy, grazing incidence range, key factors are the grating efficiency but also losses in the mirror reflections. As remarked above, blazed gratings are particularly efficient when operating close to the condition for specular reflection from the blazed faces, so the blaze angle, combined with the associated groove spacing, can be chosen to optimise the transmission in particular wavelength ranges. This is a further factor favouring the use of instrument designs that allow the grating to be changed to allow efficient operation over a wide energy range. A combination of these factors typically leads to an upper useful energy limit for grating monochromators in the 1–2 keV range. Two key factors limit the use of gratings to higher photon energies, namely the dispersion and the impact of surface roughness. For wavelengths of only a few ångström units (a photon energy of 2 keV corresponds to a wavelength of ~6 Å), surface roughness on little more than the atomic scale can lead to significant losses. Moreover, a blazed grating at these energies would need a blaze angle of no more than about 0.25°, a requirement that is particularly difficult to achieve while maintaining low surface roughness. Nevertheless, a laminar grating with 600 lines/mm has been shown to be capable of operating at a grazing incidence angle of 0.5°, with an efficiency of several percent, up to photon energies of 8 keV (Heinmann, Koike and Padmore, 2005), although the energy resolution achievable at such high energies is limited due to the weak dispersion. Nevertheless a blazed grating (blaze angle 0.4°) with 600 lines/ mm has been shown to yield efficiencies of 16.5% at 1.8 keV and even 5% at 6 keV (Cocco et al., 2007), capable of being used in a 'soft-to-hard' PGM installed on the

Elettra storage ring in Italy to achieve a resolving power of more than 2,500 over the energy range to 2.2 keV (Gianoncelli et al., 2016).

3.5   X-Ray Monochromators

Despite these achievements in using manufactured gratings, with periodicities of $1-2$ μm, the fact that hard X-rays have wavelengths of ~1 Å or less clearly suggests that diffracting structures with a periodicity of a few Å would be more appropriate; such structures exist naturally in the form of crystalline solids. Monochromators for use in this hard X-ray energy range are thus almost invariably based on diffraction from crystals rather than from manufactured gratings. The standard equation determining the conditions for diffraction maxima in this case is Bragg's law

$$2d \sin \theta = m\lambda, \tag{3.23}$$

where d in this case is the interlayer spacing of the crystalline Bragg scattering planes and θ is the grazing incidence angle relative to these planes. While superficially similar to Equation (3.16), there are, however, important differences in what the two equations actually describe. Fig. 3.12 shows the scattering conditions from which Bragg's law is derived. The X-ray scattering is treated simplistically as arising from planes of scattering atoms in the crystal in a specular geometry. The pathlength difference, $2d \sin \theta$, is that of scattering paths between successive planes of atoms; when this corresponds to an integral number of wavelengths constructive interference gives to an intense diffracted beam. This is thus a condition for constructive interference between scattering from successive layers of atoms; in the Fig. 3.12 these layers are implied to be parallel to the surface (although this is not necessary). This interference phenomenon thus depends on the one-dimensional periodicity perpendicular to the surface, whereas the diffraction grating Equation (3.16) describes the interference between scattering from equivalent positions in a laterally periodic system. In reality of course, a crystal is not one-dimensionally periodic, but three-dimensionally periodic, so the general condition for coherent interference requires satisfying three different conditions simultaneously. Bragg's Law actually ensures that

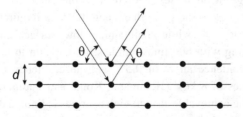

Figure 3.12 Schematic diagram showing the scattering paths that are compared in deriving Bragg's law.

all three conditions are satisfied by assuming that the reflection from the scattering planes must be specular.

More generally, evaluating the interference between scattering from the individual atomic scatterers within the three-dimensionally periodic crystal leads to the three Laue conditions. The three-dimensional periodicity of a crystalline solid can be described by three primitive translation vectors, **a**, **b** and **c** that define the primitive unit cell of the crystal. All equivalent positions in the crystal are then related by the set of vectors $\mathbf{r} = u\mathbf{a} + v\mathbf{b} + w\mathbf{c}$ where u, v and w are integers. The three Laue conditions for constructive interference of scattering (i.e. for pathlength differences between scattering from symmetrically equivalent sites to be an integral multiple of the X-ray wavelength, $\lambda = 2\pi/|\mathbf{k}|$, where **k** is the wavevector[1]) from equivalent points in the three-dimensionally periodic structure are then

$$\mathbf{a}.\Delta\mathbf{k} = 2\pi h \quad \mathbf{b}.\Delta\mathbf{k} = 2\pi k \quad \mathbf{c}.\Delta\mathbf{k} = 2\pi l, \tag{3.24}$$

where h, k, and l are integers and

$$\Delta\mathbf{k} = \mathbf{k}_{out} - \mathbf{k}_{in} \tag{3.25}$$

is the change in the wavevector between the incident and outgoing X-rays, commonly known as the scattering vector. A simple connection with the representation of Bragg's law of Fig. 3.12 can be made by taking the 'reflecting' atomic layers to be the (001) planes of a simple cubic structure. In this case the interlayer spacing, d, is simply $|\mathbf{c}|$ while $|\Delta\mathbf{k}| = 2d \sin\theta \frac{2\pi}{\lambda}$ so the third Laue condition becomes

$$\mathbf{c}.\Delta\mathbf{k} = 2d \sin\theta \frac{2\pi}{\lambda} = 2\pi l, \tag{3.26}$$

which gives Bragg's law, Equation (3.23) (with m being replaced by l – just different notation for the integral order of the diffraction). Notice that this specular reflection from these scattering planes is actually zero order diffraction with respect to the lateral periodicity of the crystal. The diffraction grating is used in first or higher order diffraction relative to its one-dimensional lateral periodicity (zero order would not disperse the radiation) whereas the analogy using Bragg's law to describe the behaviour of the crystal diffraction corresponds to zero order with respect to the lateral periodicity but first or higher order relative to the periodicity perpendicular to the surface of the crystal.

The three-dimensional periodicity of a crystal leads to a significant practical difference between linear grating monochromators and crystal monochromators. As discussed in Section 3.4. A grating disperses the incident radiation so that a broad fan of radiation emerges from the grating with different wavelengths appearing at different exit angles. By contrast, Bragg's law tells us that polychromatic radiation incident on a crystal leads to specular reflection of a single wavelength of radiation (or, more

---

[1] Notice that here a factor of 2π is used in the relationship between the wavevector and the wavelength, as is normal in discussing electron wavevectors in physics. The alternative definition, common in X-ray crystallography, is to omit the factor of 2π.

exactly, a harmonic series of wavelengths) but *no* radiation at other exit angles. The specular reflection from a linear grating is polychromatic because it is zero order with respect to the lateral periodicity, but from a crystal this refection is not zero order in diffraction from the periodicity perpendicular to the surface, but picks out only discrete harmonically-related wavelengths. Of course, because the crystal is three-dimensionally periodic there may be higher-order (non-zero order) interference associated with the lateral periodicities at other angles, but these also occur only at very specific directions and wavelengths. The broad fan of radiation from a linear grating is replaced by a set of discrete diffracted beams from a crystal, corresponding to directions and wavelengths that match all three Laue conditions simultaneously.

One practical implication of this difference is that while it is possible to vary the wavelength that is scattered into a fixed direction from a diffraction grating by simply rotating the grating, the same is not true for diffraction from a crystal. Using the geometry of Fig. 3.12 rotating the crystal does change the wavelength of the diffracted radiation, but the direction of this beam changes; indeed, for this specular geometry the direction of the exiting monochromatic beam changes by twice the rotation angle of the crystal. Clearly it is necessary to use a second optical element to restore the direction of the incident radiation that must be delivered to a fixed sample. The large deflection angles involved at high photon energies clearly precludes the use of mirrors for this purpose, but a second identical crystal arranged at the same incident Bragg angle provides a conceptually simple solution. This is shown schematically in Fig. 3.13. Much the most common crystals used are Si set to the (111) reflection. Large Si crystals can be grown to exceptionally high levels of perfection (dislocation-free) and thus have the potential to perform to their theoretical resolution limits. Indeed, a common practice in the early use of such instruments was to use a single block of silicon with a central section removed such that the two crystals were actually part of the same single crystal; this is known as a 'channel-cut' crystal. This arrangement does have two disadvantages, however. First, as the whole channel-cut crystal is rotated, and the incidence angle on each crystal changes, the height difference between the incident and exiting beams also varies, so the exit beam shifts on a fixed sample located further down the beamline. Secondly, an increasing problem with higher intensity (undulator and wiggler) sources is that the first crystal, which receives a 'white' beam, becomes much hotter than the second crystal. This changes the lattice parameter of this crystal due to thermal expansion, and thus changes the diffracted wavelength. By using two separately-mounted crystals, not only can the crystal

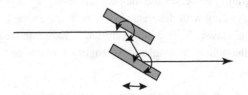

Figure 3.13 Basic arrangement of a double-crystal hard X-ray monochromator.

temperature difference can be reduced by appropriate cooling, but also small rotation angle differences of the two crystals can compensate for any lattice parameter difference. Moreover, the relative heights of the two separate crystals can be adjusted as the crystals rotate, to keep the height of the exiting beam fixed. Of course, maintaining the exact rotations of the two crystals to keep both on the Bragg condition for a given wavelength is a demanding mechanical problem, but achievable particularly using modern digital stepper motors.

Another key difference between grating and crystal diffraction monochromators concerns the factors determining the energy resolution. A grating produces a broad fan of energy-dispersed exiting radiation, but if a parallel beam of 'white' radiation is incident on a plane grating, and the diffracted radiation is then focused, one factor determining the energy resolution is the width of the slit through which the radiation is passed, as this then selects a particular narrow angular range of the dispersed radiation. By contrast, because crystal diffraction only produces discrete beams of monochromatic radiation (and not a wide dispersed fan of radiation) there is no need for an exit slit to define the exact energy range of the output radiation if the incident radiation is beam is parallel. Of course, if the radiation incident on the crystal(s) is not parallel, a narrow fan of dispersed radiation *is* produced, the wavelength range being related to the range of incident (Bragg) angles. In addition to these considerations of the degree of dispersion and exit slit sizes, there are also fundamental limits to the resolution of both types of monochromator. In the case of a grating, as explained earlier, this is determined by the total number of illuminated grating lines and thus by the size of the 'footprint' that the incident radiation makes on the grating. By contrast, because the specular Bragg 'reflection' in a crystal monochromator functions in zero order with respect to the lateral periodicity of the crystal, the size of this footprint is not relevant to the achievable resolution. In this case it is interference between scattering from successive atomic layers of the crystal that achieve the energy selection so, in exact analogy to the grating case, it is the number of these layers that are illuminated that determine the resolution. This number is a property of the crystal material and the overall photon energy, but not of the instrument design.

This number of contributing layers is determined by the extinction length of the X-rays in the crystal material. Specifically, because each atomic layer scatters some of the incidence X-ray back out of the crystal, the amplitude of this incidence radiation is attenuated as it propagates deeper into the crystal. Of course, elastic scattering is a conservative process, so inside an infinite crystal there could be no resulting attenuation, but in a real (finite-sized) crystal, some X-rays are scattered out of the material, so elastic scattering can lead to attenuation. One consequence of this finite depth of scattering is that the total reflectivity of the nominally exact Bragg condition (effectively implying that the reflectivity would appear as a delta-function as one varies the incidence angle or the wavelength) must actually have a finite width. Specifically, taking account of the multiple scattering of X-rays in the crystal (the so-called dynamical theory of X-ray diffraction) leads to the conclusion that, for a non-absorbing crystal, this reflectivity curve has a flat top – the 'Darwin plateau' (Fig. 3.14).

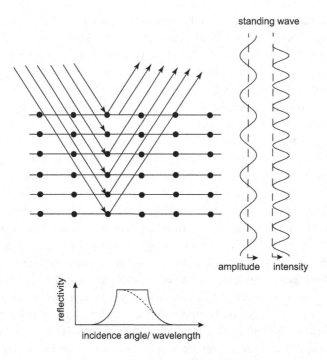

Figure 3.14 Schematic diagram showing the properties of the X-ray standing wave created under the conditions of a Bragg reflection together with the reflectivity as a function of incidence angle (the 'rocking curve') at fixed wavelength, or of wavelength at fixed incidence angle. The dashed curve includes the effect of absorption, the continuous curve shows the reflectivity with no absorption. On the right is sketched the X-ray standing wave that results from the interference of the incident and 'reflected' X-rays.

The width of this plateau in the incidence angle (the 'rocking curve' width) is found to be

$$\Delta\theta = 2(|P|\Gamma(F_H F_{\bar{H}})^{1/2}/(\sin 2\theta_B), \tag{3.27}$$

where θ_B is the Bragg angle, P is a polarisation factor equal to unity for σ-polarisation (X-ray **A**-vector perpendicular to the scattering plane) and $\cos(2\theta_B)$ for π-polarisation (X-ray **A**-vector in the scattering plane), and F_H and $F_{\bar{H}}$ are the structure factors for the $H(hkl)$ and $-H(\overline{hkl})$ scattering, while

$$\Gamma = (e^2/4\pi\varepsilon_0 mc^2)\lambda^2/\pi V, \tag{3.28}$$

V being the volume of the crystallographic unit cell and other parameters having their usual meaning. Sweeping through the Bragg reflection at fixed incidence angle while varying the photon energy, E, leads to a width of

$$\Delta E = E|P|\Gamma(F_H F_{\bar{H}})^{1/2}/(\sin^2\theta_B). \tag{3.29}$$

As may be expected, stronger scattering (larger values of F_H and $F_{\bar{H}}$) leads to a wider peak because the extinction length is shorter; the X-rays penetrate less far into the crystal, and thus the number of atomic layers contributing to the Bragg diffraction is reduced.

While the width of this reflectivity curve is determined by this elastic scattering, the shape is also influence by inelastic scattering (i.e. absorption). An important phenomenon influencing this shape is the fact that interference between the incident and reflected waves gives rise to an X-ray standing wave, as shown schematically in Fig. (3.14). In the simple case illustrated, in which the crystal has just one atom per unit cell, the intensity of this standing wave has the same periodicity as the atomic scattering planes and at the low angle or energy side of the reflectivity curve the nodal planes of the standing wave coincide with the atomic planes. At this condition, therefore, there is no absorption due to photoelectron emission from the atomic cores. However, as one sweeps through the reflectivity curve, the phase of the standing wavefield shifts, so that at the high angle or energy side, the antinodes of the standing wave lie on the atomic planes, leading to maximum photoelectron emission at the atomic cores. This effect forms the basis of the X-ray standing wave (XSW) technique for structure determination described more fully in Section 4.6, but from the point of view of the properties of a crystal monochromator the important conclusion is that the reflectivity curve becomes asymmetric, with high reflectivity on one side and reduced reflectivity, due to enhanced absorption, at the other (high angle, high energy) side. Evidently in a double-crystal monochromator the spectral shape of the emerging radiation corresponds to a self-convolution of these individual reflectivity curves for the two reflections.

As may be inferred from Equation (3.29) to achieve high resolution (a low value of ΔE) one requires weak backscattering (low values of F_H), favouring higher energies and materials comprising atoms with low atomic numbers. Silicon is not only a good choice in terms of the ability to grow large single crystals of exceptionally high perfection, but also meets this requirement for a relatively low atomic number. By far the most numerous double-crystal monochromators are based on Si (111) reflections. Typical resolving powers $(E/\Delta E)$ obtained with these instruments are $\sim 10^4$, which corresponds to a resolution of ~ 1 eV at a photon energy of 10 keV. One way to achieve much higher resolution (inevitably with a loss of intensity) is to use much higher-order reflections, which correspond to larger values of $\sin \theta_B$ for any chosen wavelength. One example of this is a monochromator installed at the ESRF in Grenoble (Verbeni et al., 1996) used for extremely high resolution inelastic scattering experiments (see Chapter 7). This instrument uses higher order (nnn) reflections from Si crystals but with a Bragg angle of almost exactly 90°; this results in achieved resolutions of 150 meV at a photon energy of 13.84 keV using (777) reflections, and 0.50 meV at 25.70 keV using (13 13 13) reflections. Notice that the use of such a large Bragg angle not only increases the value of $\sin \theta_B$ in the denominator of Equation (3.29), but also renders the instrument able to accept a larger acceptance angle of radiation, avoiding the need for pre-focusing, because the rocking curve width becomes much broader at near-normal incidence to the scattering planes. According to Equation (3.27) $\Delta\theta$ scales as $1/\sin^2\theta_B$ which would imply an infinite rocking curve with when θ_B is exactly 90°; in reality the approximations used to derive this equation break down at such large values of the Bragg angle, but the implication that there is a very large increase in $\Delta\theta$ as one approaches these angles is correct. Notice that this extremely high resolution monochromator at the ESRF is only required to operate over very narrow energy ranges; this is achieved by varying the temperature (and thus the lattice parameter) of the Si crystal. A key limitation of the use of higher order reflections at Bragg angles very close to 90° is that only very specific photon energies can be accessed. An alternative approach is to use multiple 'bounces' from several different Si crystals, generally using asymmetric reflections; this provides more flexibility in the choice of wavelength, but loses the high efficiency of the strong backscattering method with its wider acceptance angle. An example of a four-bounce monochromator is that reported by Yabashi et al. (2001), which achieved a resolution of 120 µeV at a photon energy of 14.41 keV.

One further factor relevant to the choice of crystal material for X-ray monochromators is the lowest accessible energy, which is determined by the largest atomic interlayer spacing d. In the case of silicon the largest value of d corresponds to the (111) orientation for which the value is 3.14 Å, so the lowest energy Bragg scattering occurs at a wavelength of 6.28 Å, corresponding to a photon energy of 1,974 eV. However, this corresponds to a Bragg angle of 90° which is not achievable with a conventional double-crystal monochromator arrangement such as that shown in Fig. 3.13; the practical lower-energy limit for such instruments based on Si (111) reflections is ~2,100 eV. Lower energies can be achieved, however, with other crystal reflections such as InSb (111) ($d = 3.74$ Å leading to a practical low energy limit of ~1,750 eV) and beryl $(10\bar{1}0)(d = 7.97$ Å giving a practical low energy limit of ~820 eV). However, these materials are far more susceptible to radiation damage, requiring further precautions to reduce the flux density of the incident radiation.

One rather different type of material that can be used in X-ray monochromators is a fabricated multi-layer structure, such as that illustrated schematically in Fig. 3.15. These structures typically involve 50–100 or more alternating layers of low and high atomic number materials such as W and C, or W and B_4C, each of a few atomic layers, producing a structure with a period, Λ, of a few tens of ångström units. In effect, these are artificial crystals with large $d(\Lambda)$ spacings. However, while in a real crystal the diffraction condition corresponding to Bragg's law arising from scattering of individual atoms within an atomic layer, in these multilayer structures it is optical reflections from successive layers that produce the interfering emitted wave components. The amplitude of reflection from each layer is determined by the same equations governing mirror reflectivity described in Section 3.2. However, in discussing mirror reflectivity it has been implicitly assumed that the incident and scattered radiation is travelling through air or vacuum with a refractive index of unity. In the case of the multilayer film, all reflections but the outermost one involve the interface between two materials, and it is the difference in the two refractive indices of these two materials that determines the reflectivity. Optimum criteria for suitable pairs of materials are thus

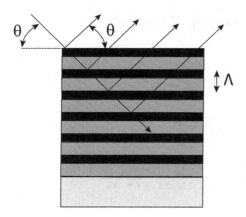

Figure 3.15 Schematic side view of a multilayer film showing the Bragg-like specular scattering geometry.

large differences in δ (the variable component of the real part of the refractive index), which favours high atomic number/low atomic number combinations, but also small values of β, the parameter that determines the degree of absorption as the radiation penetrates the multilayer.

The analogy with crystal monochromators might lead one to expect that the large layer spacings of multilayer films (or order ten times larger than the d spacings of the most commonly used crystals, such as Si) would render them most useful for use with soft X-rays, but in fact they are widely used for certain applications with hard X-rays. To achieve the required high reflectivity for hard X-rays, one must operate at grazing incidence angles, but this is precisely the condition required by Bragg's law if the interlayer spacings are large. Of course, compared to a crystalline material, the number of layers contributing to the diffraction in a multilayer is small, and this inevitably leads to much lower achievable resolving power which scales as this number of contributing layers. As such multilayer monochromators may be used for experiments that only require a 'pink' beam, with a relatively wide band pass. They may also be used a pre-monochromators to reduce the heat load on high resolution crystal monochromators. Fig. 3.16 shows the behaviour of such a multilayer having a particularly large number (500) of double layers, consisting of $Ni_{93}V_7$ and B_4C, with a periodicity of 20 Å, used on the ID15 beamline at the ESRF (Morawe, Carau and Peffen, 2017). Specifically, Fig. 3.16 shows the reflectivity as a function of incidence angle at a single photon energy of 8,048 eV. Two peaks are shown at angles of approximately 2.22° and 4.45° corresponding to the 1st and 2nd order Bragg conditions. At an angle of 2.22° (~40 mrad) the reflectivity of a single layer at this energy may be expected to be very low (c.f. Fig. 3.3, although this relates to reflection from gold) but the coherent interference of reflections from some 500 layers leads to a total reflectivity of ~60%. By contrast, when the angle is doubled the total reflectivity falls by a factor of 1,000. Coincidentally, if the photon energy were to be doubled, this would also be reflected at 2.22° in second order, but at this higher energy the reflectivity of each layer would be

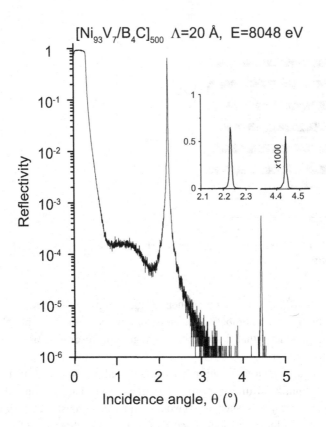

Figure 3.16 Reflectivity as a function of incidence angle at a photon energy of 8,048 eV of a multilayer comprising 500 double layers of $Ni_{93}V_7$ and B_4C with a periodicity of 20 Å. Data courtesy of Christian Morawe, ESRF

greatly reduced; a multilayer film thus has good higher-order suppression. Notice that the main graph in Fig. 3.16 has a logarithmic reflectivity scale; the insets show the two main peaks on linear scales. As expected, the peak in the 2nd order is roughly a factor of 2 narrower than that in the 1st order.

Despite the effectiveness of multilayers at hard X-ray energies they can, indeed, also prove valuable for soft X-rays. The criteria for the component materials are similar to those at higher energies, although for energies ~1 keV Si proves to be a competitive material for the low atomic number component because this energy falls well below the Si K-absorption edge but far above the L-edges, leading to particularly weak absorption. Of course, in general absorption is a more-strongly limiting factor for soft X-rays than for hard X-rays, and has a very significant influence on the monochromating performance. In particular, the depth to which the incident radiation can penetrate (and be reflected out of the multilayer) is typically constrained not by the number of layers that can be fabricated, but by the absorption. This reduced number of contributing reflecting layers has the consequence of reducing the spectral resolving power (which scales as the inverse of this number). One way of ameliorating this

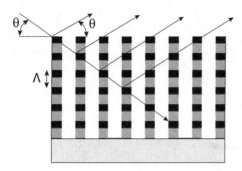

Figure 3.17 Schematic side view of a lamellar grating multilayer. As shown, this is operated in specular reflection, which corresponds to zero order with respect to the grating structure.

problem is to etch regular grooves in the multilayer to produce a lamellar grating multilayer, as shown schematically in Fig. 3.17. The objective of these devices is not to exploit the dispersive properties of the lateral periodicity of the grating. Instead, operated in zero order with respect to the grating periodicity, a significant fraction of the radiation path through the multilayer structure passes through vacuum, so the depth of penetration, and with it the resolving power, is enhanced.

3.6   Incident Beam Flux Monitors

The intensity of the beam being delivered to an experiment after passing through all the beamline components depends on the photon energy delivered by the source (insertion device or bending magnet) and selected by the monochromator, but also by the size and positioning of various apertures (most notably entrance and exit slits of monochromators), losses in the optics and on the circulating electron current in the storage ring, which is time-dependent due to the loss of electrons, but also due to periodic top-ups of this current. In many experiments it is important to have some way of monitoring these variations, albeit generally as a relative change of intensity rather than an absolute measurement. The signal from these monitors can then be used to normalise the intensity of some property being measured in the experimental end-station. For hard X-ray beamlines the most common such device is a gas-filled ionisation chamber. Two alternative geometries of such devices are shown schematically in Fig. 3.18.

As implied by the name, these devices function by detecting the ionisation of the gas atoms (typically inert gas atoms) by the incoming X-rays, the photons creating a photoemitted electron and a positively-charged ion. Bias voltages of typically a few hundred volts cause the electrons and ions to be collected at the two electrodes, and the resulting current is detected. In fact this general type of device has several different operating modes depending on the applied voltage and gas pressure. At relatively low voltages the charge collected is simply proportional to the number of electron-ion pairs created by the photoionisation events, and thus to the number

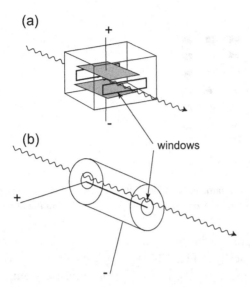

Figure 3.18 Schematic diagrams of two different designs of gas ionisation chambers based on (a) parallel plate electrodes and (b) a central wire anode in a cylindrical conducting tube that acts as the cathode.

of incident X-ray photons. This is the mode of operation relevant to a measurement of the incident flux on a beamline.

At higher voltages, and using a wire anode, the high electric field close to the wire (the field scales as the inverse of the radial distance from the centre of the wire) causes the electrons to be accelerated, resulting in further inelastic atomic collisions and the production of an increasing number of electron-ion pairs. This cascade or 'Townsend avalanche' leads to an amplified pulse of charge that allows the incident flux to be determined by photon counting. This process, operated under conditions of saturation of the amplification process, is the operating principle of the well-known Geiger counter or Geiger-Müller tube widely used for detecting of alpha, beta and gamma radiation arising from radioactivity. At intermediate gain, however, this becomes a 'proportional counter', a device that enables single photon counting, but in which the amount of charge in each pulse is proportional to the energy of the incident X-rays photons. One can therefore not only count the number of incident photons, but also determine the energy of each one. This is possible because the high energy photoelectron created in the initial photoionisation event can undergo a series of inelastic collisions with other atoms, ionising each of them. As the ionisation energy loss is typically only 20–40 eV, the total number of electrons released (before further amplification in the Townsend avalanches) from an X-ray photon of energy, say, 10 keV can be large (in this example, of the order of 500). This number is proportional to the energy of the incident X-ray. Proportional counters can be valuable detectors for X-ray scattering and emission experiments, but the pulse counting modes are limited by the dead time to count rates of no more than

~1 MHz, leading to saturation at the very high X-ray fluxes generally associated with the incident radiation. Notice that while most X-rays may pass through the gas chamber without energy loss, the mean-free-path of the energetic photoelectrons in the gas is very much shorter, and these electrons can be expected to deposit all of their energy in inelastic collisions within the detector.

Of course, when used to monitor the incident radiation flux at a beamline (and indeed in any other measurement) an ionisation chamber does absorb some of this flux through the gas photoionisation events. However, a gas (at ~ atmospheric pressure) has a very low atomic density and this absorption is typically less than 1% for hard X-rays. There is also some loss in transmission through the windows required to maintain the pressure difference inside and outside the detector. These problems become far more severe, however, at much lower photon energies for which no suitable window materials are available. For experiments in the VUV and soft X-ray energy range, with all experiments and beamline necessarily operating under ultrahigh vacuum conditions, one solution it to interpose a fine high-transparency conducting mesh in the beam and either measure the drain current at this mesh (a result of absorption in the mesh material and consequent emission of photoelectrons and low energy secondary electrons), or to measure this total electron emission from the mesh using a channel electron multiplier positioned to the side of the mesh and biased to accelerate these electrons into its front-cone. Of course, the use of a mesh is not ideal with a very narrowly focused beam, as small beam movements can lead to large variations in the effective transmission if the beam size is of the same order as the grid spacing. If a thin foil of a conducting material can be identified that is not strongly absorbing at the photon energy of interest, replacing the mesh by this foil overcomes this problem. An alternative approach is to use the drain current (or electron emission) at the conducting coating of the last focusing mirror in the beamline.

4   Crystalline Structural Techniques

4.1   X-Ray Diffraction Basics

X-ray diffraction is much the most widely used method for determining the structure of crystalline solids and is widely exploited utilising conventional laboratory instrumentation. The phenomenon of X-ray diffraction was first discovered by Laue, while the two Braggs (William Henry Bragg and his son William Lawrence Bragg) provided a simple mathematical description of the wave phenomenon and exploited the effect further (publishing a small book on the subject – Bragg & Bragg (1914)). These three scientists shared the 1914 Nobel Prize in Physics for their discovery. As the basic technique forms part of the standard armoury of any crystallography laboratory, one might ask why one would choose to leave the convenience and security of one's home laboratory to use a synchrotron radiation source. The answer lies in the very considerable difference in the character of the available sources.

Laboratory X-ray sources are based on high energy ~20–150 keV electron bombardment of metal target material, typically Cu, Mo or W. This leads to X-rays being emitted over a range of 2π steradians (the vast majority of which are unused due to collimation) from a relatively large area (typically $\sim mm^2$) of the target. Evidently this leads to only a very small fraction of the emerging X-rays reaching a sample at a distance of at least a few cm from the source, and the effective photon emittance (the product of the lateral dimensions and the divergence of the source), even after collimation, is very large, making it extremely difficult to focus much of the total X-ray flux onto a very small area of a sample (recall that emittance is conserved in focusing optics). The available flux for X-ray diffraction, particularly from small samples, is thus very low when compared with that provided by the intrinsically narrow (and more intense) beam of X-rays produced by a synchrotron radiation source.

A second limitation of such sources is the spectral output. Fig. 4.1 shows a simplified representation of the output from a typical source using a Cu target material. It comprises narrow peaks superimposed on a continuum background. These peaks are often referred to as 'characteristic radiation', because their wavelengths are characteristic of the target material and its core level binding energies. Specifically, they are a result of X-ray emission from Cu atoms that have been core-ionised by the incident electrons, removing a 1s electron; the core-hole is then refilled by an electron from a more shallowly bound state of the atom, the emitted X-ray energy corresponding to the energy released in this refilling. The strongest peak, at the longest wavelength,

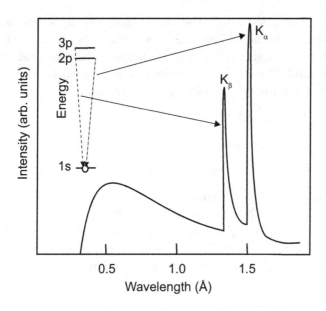

Figure 4.1 Schematic graph showing the spectral output from a conventional X-ray source with a Cu target. Also shown is a representation of the electronic transitions giving rise to the main peaks.

corresponds to electrons falling down from the 2p levels into the 1s hole. The peaks are labelled according to the X-ray notation of the principle quantum number of the core hole ($n = 1,2,3\ldots$ correspond to the K, L, M $\ldots$ shells) with a suffix defining the sub-shell from which electrons have fallen. Strictly this strongest K_α peak is a doublet ($K_{\alpha1}$ and $K_{\alpha2}$), corresponding to electrons falling from the $2p_{1/2}$ and $2p_{3/2}$ states, and the associated wavelengths are 1.5406 Å and 1.5444 Å. Other weaker peaks correspond to electron decay from other more shallowly bound states. The background on which these peaks sit is due to bremsstrahlung ('braking radiation'), arising from less specific energy loss processes, and has a maximum photon energy (and thus a minimum wavelength) equal to the energy of the energetic electrons that bombard the Cu target. This continuum background is very much weaker than the peaks and thus very much less intense than the continuum offered by synchrotron radiation. As a consequence, synchrotron radiation is the source of choice if one has very small or weakly-scattering samples, if one needs a very narrow incident beam to distinguish closely-separated diffracted beams, if one needs to collect data rapidly (to study time-dependent processes) or if one needs to be able to freely vary (choose) the photon energy/X-ray wavelength.

The basic theory of X-ray diffraction is covered in many undergraduate textbooks in physics, chemistry, crystallography and materials science, but for completeness a brief description is presented here. The condition for coherent interference between scattering from equivalent sites in the three-dimensionally periodic crystalline solid have already been given in Section 3.5. Specifically, taking the three primitive translation vectors, **a**, **b** and **c** to define the primitive unit cell of the crystal, all equivalent positions in the crystal are then related by the set of vectors $\mathbf{r}_c = n_1\mathbf{a} + n_2\mathbf{b} + n_3\mathbf{c}$ where n_1, n_2, and n_3 are integers. The three Laue conditions that must all be satisfied for constructive interference of scattering from all equivalent points in the three-dimensionally periodic structure are then

$$\mathbf{a}.\Delta\mathbf{k} = 2\pi h \quad \mathbf{b}.\Delta\mathbf{k} = 2\pi k \quad \mathbf{c}.\Delta\mathbf{k} = 2\pi l, \tag{3.24}$$

where h, k, and l are integers[1] and

$$\Delta\mathbf{k} = \mathbf{k}_{out} - \mathbf{k}_{in} \tag{3.25}$$

is the 'scattering vector', also often denoted by the symbol **q**. As remarked upon in Section 3.5, the factor of 2π is included here in the relationship of $|\mathbf{k}|$ and the wavelength, λ. In the simple single scattering ('kinematical') theory, one assumes that a parallel beam of monochromatic X-rays (represented by a plane wave) is incident on a perfectly-ordered crystal and takes the sum of the amplitudes of the scattering from all the constituent atoms

$$A_{\Delta k} = \sum_{n_1=1}^{N_1}\sum_{n_2=1}^{N_2}\sum_{n_3=1}^{N_3} \exp\left(i\,\Delta\mathbf{k}\cdot\mathbf{r}_c\right) \sum_j f_j \exp\left(i\,\Delta\mathbf{k}\cdot\mathbf{r}_j\right). \tag{4.1}$$

Here the first three summations are over all the unit cells in a crystal of some finite size, while the final summation is over the atoms within the unit cell, with locations defined within the unit cell by the vector $\mathbf{r}_j$ and their atomic structure factors (also known as form factors) f_j (this is the same quantity, written with its real and imaginary parts separately, $(f_r - if_i)$, that appears in Equation (3.2)). For an infinite crystal this amplitude is only non-zero when the three Laue conditions are satisfied, namely, when

$$\Delta\mathbf{k} = \mathbf{G}_{hkl} = h\mathbf{a}^* + k\mathbf{b}^* + l\mathbf{c}^*, \tag{4.2}$$

where $\mathbf{a}^*$, $\mathbf{b}^*$ and $\mathbf{c}^*$ are the primitive translation vectors of the reciprocal lattice, defined as:

$$\mathbf{a}^* = \frac{2\pi\mathbf{b}\times\mathbf{c}}{V} \quad \mathbf{b}^* = \frac{2\pi\mathbf{c}\times\mathbf{a}}{V} \quad \mathbf{c}^* = \frac{2\pi\mathbf{a}\times\mathbf{b}}{V} \quad \text{and} \quad V = \mathbf{a}.\mathbf{b}\times\mathbf{c}, \tag{4.3}$$

V being the volume of the primitive unit cell. At this condition all values of $\exp\left(i\,\Delta\mathbf{k}\cdot\mathbf{r}_c\right)$ are unity, so Equation (4.1) becomes

$$A_{hkl} = N_1N_2N_3 \sum_j f_j \exp\left(i\,\mathbf{G}_{hkl}\cdot\mathbf{r}_j\right). \tag{4.4}$$

This equation is often written as

---

[1] The use of the letter k to describe both the integer in the Laue condition (which becomes a label of the reciprocal lattice and one of the Miller indices of a lattice plane – see below) but also the X-ray wave-vector is perhaps superficially confusing, but the wave-vector is always a vector **k** whereas the number k is a scalar.

$$A_{hkl} = N_1 N_2 N_3 F_{hkl}, \tag{4.5}$$

where

$$F_{hkl} = \sum_j f_j \exp\left(\mathrm{i}\, \mathbf{G}_{hkl} \cdot \mathbf{r}_j\right) = \sum_j f_j \exp\left(2\pi i\left(hx_j + ky_j + lz_j\right)\right) \tag{4.6}$$

is known as the geometrical structure factor, x_j, y_j and z_j being the coordinates of the jth atom within the unit cell. Notice that as the X-rays actually scatter off the electrons surrounding the atoms one can also write the geometrical structure factor in terms of the scattering length density which (but for a scaling factor equal to the classical electron radius) equals the electron density at a position defined by the vector $\mathbf{r}$ within the unit cell, $\rho_e(\mathbf{r})$

$$F_{hkl} = \int_{unit\ cell} \rho_e(\mathbf{r}) \exp\left(2\pi i(\mathbf{G}_{hkl}.\mathbf{r})\right) dv. \tag{4.7}$$

The integral being over volume elements dv within the unit cell.

An extremely useful graphical representation of the consequences of the three Laue conditions is provided by the Ewald sphere construction in reciprocal space, an example of which is shown in Fig. 4.2(a). The reciprocal lattice is denoted by filled black circles at the lattice points. The incident X-ray wavevector, $\mathbf{k}$, is drawn to end at the origin of the reciprocal lattice (point 000) and a sphere is drawn of radius $|\mathbf{k}|$ about the origin of the vector $\mathbf{k}$. Vectors drawn from this point to all reciprocal lattice points that are intersected by this sphere correspond to diffracted beams consistent with the three Laue conditions. Notice that all these radial vectors of the Ewald sphere have the same length, $|\mathbf{k}|$, and thus correspond to elastic scattering (energy conservation), while the fact that they end at reciprocal lattice points means that the corresponding scattering vector, $\Delta\mathbf{k}$, is a reciprocal lattice vector as required by Equation (4.2). One example, $\Delta\mathbf{k} = \mathbf{G}_{202}$, is highlighted in Fig. 4.2(a). Fig. 4.2(b) shows how the 006 and 202 diffracted beam conditions shown in (a) are related to (specular 'reflection') Bragg's Law conditions (see Section 3.5, Equation (3.23)) for 6th order scattering from (001) planes and 2nd order scattering from (101) planes, respectively. The labels of the lattice points of the reciprocal lattice correspond to the Miller indices of sets of scattering planes in real space.

One striking feature of the diagram of Fig. 4.2 is that for a specific incidence direction and a specific X-ray wavelength (i.e. for a specific value of $\mathbf{k}$), very few of the reciprocal lattice points intersect the Ewald sphere and thus correspond to diffracted beams. Indeed, formally, the reciprocal lattice points are mathematical points that are infinitesimally sized, so the Ewald sphere may not intersect any of them. This is only true, however, for an infinitely periodic crystal; for a finite-sized crystal, the exact diffraction conditions of Equation (4.2) are relaxed, and can be represented as finite-sized reciprocal lattice 'points' within the Ewald sphere construction, the size of these 'points' being inversely related to the physical size of the crystal. However, to sample a large number of reciprocal lattice points (i.e. to obtain a large number of

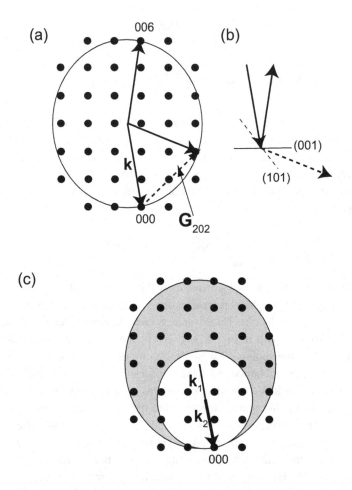

Figure 4.2 (a) an example of the Ewald sphere construction for a simple cubic lattice showing two diffracted beams; the reciprocal lattice vector, **G**, associated with one of these is labelled. (b) shows how these two beams are related to (specular) Bragg reflections from the associated scattering planes. (c) shows the same Ewald sphere construction as in (a) but with a range of different incident wavelengths such that all reciprocal lattice points in the shaded area give rise to diffracted beams.

diffracted beams) one must either use 'white' incident radiation covering a range of wavelengths (a range of Ewald sphere radii – see the shaded area of Fig. 4.2(c)) or use a range of different crystal orientations. These two conditions correspond to the standard laboratory X-ray diffraction conditions of Laue diffraction from a single crystal using the continuum bremmstrahlung radiation, or powder or rotation crystal diffraction using the characteristic line emission of standard laboratory sources. Using synchrotron radiation the 'white' beam of the full range of continuum radiation is normally replaced by a 'pink' beam with a limited range of wavelengths, removing radiation with wavelengths much shorter or longer than those of potential value in the diffraction experiment. Some examples of the applications of some of these different methods are given in Section 4.4.

4.2   X-Ray Diffraction Detectors

The development of synchrotron radiation sources over the last few decades has meant that the intensity of the incident radiation arriving at a sample has typically increased by many orders of magnitude, while the fact that the beams can be much narrower means that it is possible to detect much smaller angular separations of diffracted beams. To reap the benefit of these advances, however, it has been essential to also make major advances in detectors that can handle higher incident fluxes without saturating, that can detect the very small angular differences in diffracted beams, and can cope with the huge dynamic range of the signals to be detected from the very intense to the very weak. Another key advance has been in parallel detection of different diffracted beam directions and different scattered (and emitted) photon energies. Some of the advances in detectors have had comparable impact in synchrotron radiation studies to the improvements in the sources themselves (in many cases at significantly lower costs!).

Historically, X-ray diffraction patterns were recorded on photographic plates or films. The obvious disadvantage with this approach is that it was necessary to develop the films in order to see that data, a process with a potential timescale of hours. Photographic materials also have limited resolution and a non-linear response that makes it difficult to extract reliable diffracted beam intensities. Fortunately, far more rapid digital methods of recording, in some cases with higher spatial resolution, are now available, which also offer the advantage of direct interfacing to computers for direct read-out and processing of beam locations and intensities. Many of these developments have not only been implemented at synchrotron radiation sources but also in X-ray diffraction facilities in users' own laboratories, and in medical applications of X-rays, although synchrotron X-ray diffraction experiments present some most severe detector challenges.

Even now some types of diffraction experiments still use 'single point' detectors – detectors with a relatively narrow acceptance angle – that are scanned physically on diffractometer arms to measure individual diffracted beams. However, it is now far more common for the majority of experiments to use detectors with one- and two-dimensional (1D and 2D) position sensitivity, although single-point detectors still play an important role in some experiments. 2D detectors with high spatial resolution, initially developed to tackle the challenges presented by diffraction from the very large unit cells of macromolecular crystals (including protein crystals), are increasingly being adopted by a much wider range of diffraction experiments.

Conceptually, perhaps the simplest type of 1D position sensitive detectors involved a minor modification of the type of gas ionisation chamber shown in Fig. 3.18(b), which has a central wire in a tubular conducting tube. By feeding the high voltage to both ends of the wire, and detecting the time difference between the arrival of a charge pulse at each end, the location along the wire of the exciting radiation can be determined. Indeed, this type of 1-D gas ionisation detector can be constructed in a curved geometry with the centre of its radius of curvature located at the diffracting crystal, thereby directly (and linearly) relating the angle of scattering to the location of the initiating charge pulse on the wire. Ultimately, the maximum count rate that such a device can detect is limited by the dead time: the time the state of the gas in the detector takes to recover from one ionisation event to be ready to detect a further ionisation event. Modifications to the gas mixture by the addition of a quenching agent – typically methane – to the dominant inert gas, can reduce this dead time. As remarked in Section 3.6, when operated as a proportional counter, these devices do provide energy resolution in the X-ray detection. As diffraction involves only elastic scattering, this capacity might appear to be irrelevant for diffraction experiments, but inelastic scattering (notably core level ionisation of atoms in the scattering material) can also occur, leading to the emission of fluorescent X-rays (equivalent to the 'characteristic' radiation of a conventional X-ray source – Fig. 4.1) at lower energies. This fluorescence leads to a detected background under the diffracted intensity if energy resolution is not available, and in some experiments this becomes a major limitation, especially for the detection of weak diffracted beams.

The need to replace photographic films and plates for medical X-ray imaging led to the development, initially by the Fuji company, of 'image plates', which were first introduced commercially in the early 1980s and also became adopted at many synchrotron radiation facilities for recording X-ray diffraction patterns. Image plates are effectively storage phosphor screens in which an initial electronic excitation induced by X-ray impact remains unrelaxed until it is 'read out' using Photon-Stimulated Luminescence (PSL). Specifically, they are typically based on barium halides with Eu^{2+}-doping of Ba sites, BaFX where X is commonly Br but may also be Cl or I or some mixture of these. Incident X-rays photoionise Eu^{2+} to local Eu^{3+} species, leading to the creation of electrons in the conduction band that can become trapped at X^- vacancies, thereby producing colour centres. These can then be caused to emit blue light (wavelength ~3,900 Å), which can be detected using a photomultiplier tube, by illuminating the image plate with focused red laser light (typically from a HeNe laser) that is rastered over the device to give a sequential read-out of the initial excitations. The colour centres can then be erased by brief illumination to a bright white light source, allowing the plate to be reused many times. The read-out time can be as little as a few seconds, but other advantages over photographic emulsions are a high dynamic range in which the response is linear, while spatial resolution is typically ~100–200 μm. However, while this technology is still widely used in medical applications, other more advanced detectors are now more commonly exploited at synchrotron radiation facilities.

A traditional alternative to the use of gas-filled detectors are scintillation detectors, which are used in a wide range of radiation detection applications. Both types of device are capable of handling count rates in excess of 1 MHz, but gas-filled proportional counters do suffer from aging effects that limit their useful lifetimes. The underlying principle in the scintillation detector is the conversion of the incident

X-ray energy into visible radiation that can be detected by a photomultiplier tube. As in a gas-filled proportional counter, the incident X-radiation photoionises a core state of an atom in the detector, and the resulting photoelectron then undergoes a series of inelastic collisions, each involving energy losses of a few eV. In the case of a scintillation detector these inelastic collisions ultimately lead to electrons in excited states that decay with the emission of visible light photons, the number of these photons being proportional to the energy of the incident X-rays. Like the gas-phase proportional counter, the detected signal (in this case the intensity of the visible light emitted) therefore provides information on the energy of the detected X-ray photon. Notice that in both gas-phase and solid-state detectors this proportionality is a result of multiple inelastic scattering events of the photoelectron emitted by the X-ray photon, not of multiple inelastic scattering events of the X-ray photon itself. In fact there are only three possible types of scattering event that can lead to energy loss of a photon, namely pair production (for which the photon energy must be more than 1 MeV), Compton scattering, which has a low cross-section at typical X-ray energies, and photoionisation, in which *all* the energy of the photon is lost in a single event. By contrast electrons can undergo partial energy loss in collisions with other electrons (bound or otherwise), the initial energy being shared between the two electrons that emerge from the collision. Notice that in a scintillator the average value of these energy losses is significantly larger than the energy of the visible photons that emerge; initial excitations are to a range of different excited states, with subsequent decay by non-radiative processes before reaching the energy level from with the radiative decay occurs. Although both organic and inorganic scintillator materials have been used in a range of radiation detector experiments, it has been inorganic materials, traditionally alkali halides, and thallium-doped NaI in particular (with a scintillation emission peaking at a photon energy of ~3 eV), which have been used in synchrotron radiation diffraction studies. The dead time of a Tl-doped scintillator – the time it takes the material to recover from one scintillation event before a second one can be detected – is ~250 nS, limiting the detectable count rate with linear performance to ~1 MHz. Some newer materials, such a $YAlO_3$ activated by Ce, have dead times that are an order of magnitude smaller. Higher count rates can thus be achieved with suitable matching electronics.

Fig. 4.3 shows a schematic diagram of a photomultiplier linked to a scintillator. The photomultiplier consists of an evacuated glass tube with a photocathode coating on the inside of the front face and containing a series of electrodes, known as dynodes, at increasingly large positive voltages, terminating in a final anode. The visible light passes through the glass envelope to the photocathode where it creates low energy photoelectrons that are then accelerated to the first dynode of an electron multiplier. The collision with this dynode leads to the production of several low energy secondary electrons that are then accelerated to the second dynode. Repetition of this process leads to an increasing cascade of secondary electrons that are collected at the anode. Evidently this overall sequential conversion process from an X-ray photon to a multiple electrons, then to visible photons, and then to low energy photoelectrons that are then multiplied to a finally detected electron charge pulse, is less than 100%

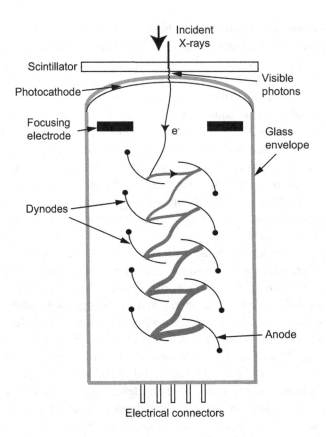

Figure 4.3 Schematic diagram of a scintillator combined with a photomultiplier, showing the main components of the photomultiplier.

efficient, yet is nevertheless effective. An important area of potential losses lies in the photocathode. If this is too thin an insufficient fraction of the incident visible photons give rise to photoelectrons (instead simply passing straight through), but if it is too thick, some of the photoelectrons fail to escape into the vacuum. Note that most materials have a work function in the range 4–5 eV, meaning that photoemission cannot occur with incident photon energies less than this value; in order to detect the lower energy photons of the visible spectrum the photocathode is thus based on alkali metals with unusually low work functions, but the kinetic energies of the photoelectrons that escape are nevertheless generally less than 1 eV. Similar concerns regarding the capture of the X-ray photons could potentially be relevant to the scintillator, which needs to be thick enough to ensure than almost none of the X-ray photons pass through without producing a photoionisation event, although if the scintillator material is transparent to its emitted visible photons there is no equivalent problem of their absorption in the material.

An alternative way to detect the visible light emission from a scintillator or a phosphor is with a photodiode. Omitting the photoemission stage of the photomultiplier means that this is more efficient, a semiconductor band gap of 1–2 eV allowing each incident photon to create an electron-hole pair, but the total charge to be detected is small without the amplification that occurs in a photomultiplier tube. An important advantage of moving to some type of semiconductor-based detector, however, is the possibility of using a 2D array of small detectors, such as photodiodes, laid out on an integrated single device. Another realisation of this idea is to use a CCD (charge-coupled device) such as those used in digital cameras (including those fitted to mobile phones). In effect a CCD comprises an array of switchable photosensitive capacitors that accumulate the charge produced by the incident photons, thus making it an analogue time-integrating detector. The 'CCD' name originates from the way the device is read out, by moving the charges sequentially between a series of capacitive bins.

More recently, however, there have been major advances in the use of hybrid photon counting pixel detectors for the high-speed 1D and 2D recording of powder and single crystal diffraction patterns at synchrotron radiation beamlines. Of course, insofar as any digital image comprises an array of pixels, this terminology could be applied to any detector capable of producing such an image, but the terminology comes from the particle physics community, which developed these detectors in vast numbers for tracking energetic particles resulting from high-energy physics particle collisions. An interesting review of the early background to the development of these detectors for X-ray diffraction was presented by Hall (1995). A key difference of these devices from other 2D imaging detectors, such as CCDs, is that each pixel has its own integrated electronics for processing the signal arriving at its particle detector. In each pixel a silicon-based semiconductor sensor is used to detect the incoming X-rays directly via the photoelectric effect, giving rise, as described in the context of both scintillators and gas-filled proportional counters, to lower-energy (electron-hole) excitations, the number of which is proportional to the energy of the incident X-ray. These sensors are each bonded directly to integrated read-out chips capable of processing the resulting charge pulse and selecting the threshold (and thus the incident X-ray energy) above which the signal is fed via a digital-to-analogue converter to a counter that registers the number of detected X-ray photons with energies above the value determined by the in-built discriminator. The result is an energy-selected 2D digital photon-counting image of the diffraction pattern. A widely-used example of this technology is the PILATUS (e.g. Fig. 4.4) and EIGER 2D detectors and the 1-D MYTHEN detector developed at the Swiss Light Source (e.g. Kraft et al., 2009; Bergamashi et al., 2010), now commercialised by DECTRIS and adopted by many facilities around the world. This type of detector is largely replacing other types of X-ray detector on many synchrotron radiation beamlines due to their low noise, high count-rate potential and ability to collect large amounts of spatially-resolved data in short times.

As remarked at the beginning of this section, two particular challenges for X-ray diffraction detectors at a synchrotron radiation source are firstly the need to handle very high count rates, but also to have high angular (and thus spatial) resolution to detect closely spaced diffracted beams or fine features in images. In the case of 1- and 2D pixel-based detectors, this spatial resolution is ultimately determined by the pixel

Figure 4.4 Photograph of a Pilatus detector installed on the I07 beamline at the Diamond Light Source.

dimensions and their spacing in the device, although an effect known as 'blooming' due, for example in a CCD device to charge leakage between adjacent charge-collecting 'bins' as they saturate, can degrade the resolution. For a single-point detector this resolution is typically defined by the entrance slits or apertures in front of the detector. The angular resolution is also influenced by the size of the illuminated area of the diffracting sample by the incident X-radiation, leading to the definition of a 'point-spread function'.

Fig. 4.5(a) shows the effect of using an aperture of size A in front of a detector collecting scattered X-ray from a region of width S on a sample. The angular range accepted in this arrangement is given by $\Delta\theta = (S + A)/D$ so if, for example, both S and A are 1 mm and D is 1 m the accepted angular range (and thus the angular resolution of the measurement) would be 2 mrad or $0.12°$. In practice on a synchrotron radiation beamline S is typically significantly less than 1 mm, but if the two contributing widths, S and A, differ significantly, the angular resolution is dominated by the larger of these two quantities. Reducing the detector aperture may improve the angular resolution but reduces the detected signal proportionally.

Fig. 4.5(b) shows an alternative arrangement in which a diffracting crystal (typically Si(111)) is interposed between the sample and the detector. In effect this crystal becomes a monochromator in reverse, or a simple spectrograph. However, because the diffracted X-rays being detected have a discrete wavelength that is defined by the monochromatic incident radiation prior to the elastic scattering in the sample, it only scatters X-rays into the detector that arrive at the crystal in the very narrow range of incidence angles, θ_a, that correspond to the Bragg condition for scattering at this particular wavelength. The angular resolution in this arrangement is thus defined by the rocking curve width of the analysing crystal. While this angular range is very well defined, X-rays scattered from the sample at this angle are detected independent of the lateral position from which they emerge from the sample. This is clearly a more efficient method of achieving high angular

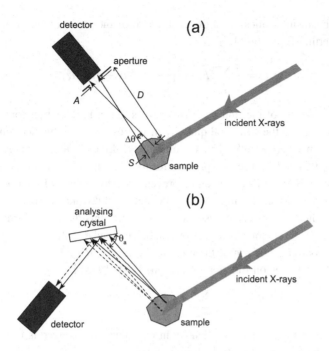

detector   (a)

aperture

A   D

Δθ   incident X-rays

S   sample

analysing   (b)
crystal

θ_a

incident X-rays

detector   sample

Figure 4.5 Schematic diagram showing in (a) the accepted angular range of a point detector fitted with an aperture, and in (b) the impact of using an analysing crystal before the detector on the accepted angular range.

resolution than reducing the size of the detector aperture of Fig. 4.5(a). Of course, the small angular width of the analyser rocking curve inevitably leads to a weak signal; this is a necessary consequence of the high resolution. To compensate for this one can use multiple crystals and associated detectors for parallel detection of multiple scattering angles. Multiple analysing crystal (MAC) detectors are used, in particular, for powder diffraction (see Section 4.4), generally combined with 1D or 2D array detectors.

4.3   X-Ray Diffraction Structure Determination

Notice that the Ewald sphere construction simply shows the consequences of scattering from a crystal of a particular periodic lattice, so identifying the directions of the detected diffracted beams allows one to determine this periodicity – specifically the size and shape of the unit cell defined by the primitive translation vectors. The information regarding the location of the atoms within the unit cell is contained in the diffracted beam intensities. These intensities are simply the square of the scattered amplitudes, A_{hkl} (Equation (4.5)), or more precisely, the modulus squared of A_{hkl}, because A_{hkl} is a complex number, representing an amplitude and a phase; the phase is lost in the intensity. As A_{hkl} is simply proportional to the structure factor F_{hkl}, the diffracted beam intensities are a direct measure of the relative values of $|F_{hkl}|^2$. Strikingly, inverting Equation (4.7)

$$\rho(x, y, z) = \frac{1}{V} \sum_{hkl} F_{hkl} \exp\left(-2\pi i(hx + ky + lz)\right) \tag{4.8}$$

where V is the volume of the unit cell. This shows that a Fourier transform of F_{hkl} actually corresponds to the structural quantity of interest, namely the electron charge distribution within the unit cell, $\rho(\mathbf{r})$. Unfortunately, the quantity measured experimentally is $|F_{hkl}|^2$, not F_{hkl}, (i.e. the phase information is lost). This is known as the 'phase problem' in X-ray diffraction, and it prevents the direct use of Equation (4.8) to obtain the real-space structure directly from the measured diffracted beam intensities. A number of methods have been devised to try to recover this missing information, as outlined below. However, some information can be obtained directly by taking what is effectively the Fourier transform of the measured quantity, $|F_{hkl}|^2$.

Specifically, the Patterson function (Patterson, 1934, 1935) is

$$P(u, v, w) = \sum_{hkl} |F_{hkl}|^2 e^{-2\pi i(hu + kv + lw)} \tag{4.9}$$

This corresponds not to the electron density in real space but to what is formally the convolution of the electron density with its inverse

$$P(u, v, w) = \rho_e(\mathbf{r}) \otimes \rho_e(-\mathbf{r}), \tag{4.10}$$

where $\otimes$ denotes a convolution. However, as $\rho(\mathbf{r})$ is real, it is equal to its complex conjugate, so P is a self-convolution of $\rho_e(\mathbf{r})$. A convolution is an integral over all possible values of some parameter x (e.g. a position or a time) of the product of one function at the parameter value x and a second function at a parameter value of $u - x$. Mathematically, the convolution $C(u)$ may be written as

$$C(u) = f(x) \otimes g(x) = \int f(x)g(u - x)dx. \tag{4.11}$$

Perhaps the most common way that a convolution is encountered in experimental science is in making any measurement with an instrument that (inevitably) has some kind of instrument function that limits the quality of the measurement. A typical example would be a spectral measurement using an instrument with only a finite spectral resolution. What is actually measured in this case is the convolution of the true property with the instrument function, causing the spectral peaks to be broadened. In the case of the Patterson function, which represents a map in real space, the quantity that is mapped is not $\rho_e(\mathbf{r})$, the electron density, which would have peaks (only) at the locations of the atoms in the unit cell defined by the vectors, $\mathbf{r}$, relative to its origin (recall that most of the electrons are localised at the atoms in strongly bound core levels). Instead it is a map of the self-convolution of $\rho_e(\mathbf{r})$, so all the *interatomic* vectors within the unit cell appear in $P(u, v, w)$ as vectors with the same length and direction but drawn from the origin. For simple structures a Patterson map can give very strong clues to the true structure.

To solve more complex structures it is necessary to address the phase problem more directly. In attempting to exploit the potential of Fourier transforms of the scattered amplitudes and phases it is important to recognise that the phases are typically more important than the amplitudes. This is not entirely surprising, as the relative positions of scatterers in the unit cell appear in the phase of the structure factor. Fig. 4.6 shows a graphic illustration of this importance of phase; taking the Fourier transform of two simple images, and then taking the inverse transform after switching the amplitude and phase components of the two images, leads to reconstructions that are clearly dominated by the phase components. Despite this paramount importance of phases, it transpires that reasonable 'guesses' at the phases of a set of X-ray diffraction structure factors, for which only the moduli of the amplitudes can be measured, can often form the basis of an effective iterative optimisation routine exploiting very general constraints, such as the fact that the electron density can never be negative. One method of gaining a first step in determining the phases is through variations of the 'heavy atom method', which exploits the fact that if a crystal contains predominantly low atomic number atoms (C, N, O etc.), as is true for many organic and biological molecules, but also contains a small number of atoms of much higher atomic number, then the scattering of these

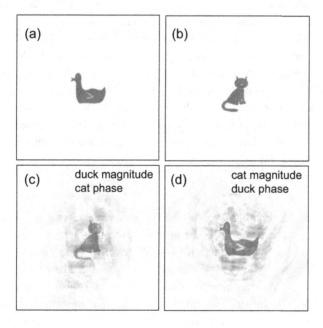

Figure 4.6 A simple illustration of the importance of phase in image recovery via Fourier transforms. (a) and (b) show two objects, drawings of a duck and a cat. After taking Fourier transforms of these objects, reconstructions are created by switching the amplitudes and phases derived from these objects. (c) and (d) show the results. Clearly it is the phases, not the amplitudes, which dominate the appearance of the reconstructed images. Copyright Dr. Kevin Cowtan, University of York, reproduced with permission

'heavy' atoms tends to dominate the diffracted intensities, and thus also to dominate Patterson function maps. This is because the atomic structure factor is proportional to the atomic number Z, which determines the number of associated electrons. Of course, biologically-related molecules are generally large (with very high molecular weights), with correspondingly large unit cells; this leads to a very small spacings in the reciprocal lattice, and thus to a vast number of potential diffracted beams, giving a large information base with which to address the great complexity of determining the structure. Nevertheless, it is often possible to arrive at a good structural model for the location of the heavy atoms, from which one can calculate their phases; by initially using these phases for the whole structure, one can calculate a predicted set of diffracted beam intensities. This forms the basis for an iterative computational process in which improved 'guesses' of the true structure are tested for their ability to reproduce the measured diffracted beam intensities.

A second method that builds on the heavy atom method is that of isomorphous replacement: replacing specific atoms in the unknown structure by different, generally much heavier, atoms. Of course, this approach relies on the new crystal having the same general structure (i.e. it is isomorphous). What is in some senses a variant of these methods, but which exploits a particular strength of synchrotron radiation, is the use of anomalous scattering. Changes in the incident photon energy around the absorption edge associated with particular atoms can cause strong variations in the scattering properties of these atoms, so this can provide similar additional information to isomorphous replacement, but without the need to grow new crystals. As described in Section 3.2 in the context of the optical properties of mirrors in the X-ray energy range, the atomic form factor changes sharply when the X-ray photon energy is close to that of an absorption edge of the atom (i.e. the energy required to photoionise one of the atomic core levels). In Section 3.2 the impact of this change is discussed in the context of mirror reflectivity (see Fig. 3.3), but here the interest is in the form factor itself. Revisiting this in terms of its impact of X-ray diffraction, the atomic form factor was expressed in Equation (3.2) as

$$f(\omega) = f_r(\omega) - if_i(\omega), \tag{4.12}$$

namely, it has real and imaginary parts, the imaginary part accounting for absorption. Indeed $f_i = \sigma/2r_e\lambda$, where σ is the atomic adsorption cross-section (r_e is the classical electron radius – see Equation (3.3)). Evidently, when the photon energy increases through a core level binding energy, there is a sharp rise in absorption and thus in the value of f_i. A consequence of this is that there is also a sharp change in the real part of the form factor, because the real and imaginary parts are related by the Kramers-Kronig relation. Specifically, writing the atomic form factor in a slightly different way as

$$f = f_0 + f' + if'', \tag{4.13}$$

where f_0 is the usual Thomson scattering form factor (equal to the atomic number Z but for a relativistic correction that is very small for low atomic number atoms), the relationship between the real and imaginary parts of the form factor that are influenced by the absorption edge is given, in terms of the photon energy, E, as:

$$f'(E) = \frac{1}{\pi r_e hc} \int_0^\infty \frac{\varepsilon^2 \sigma(\varepsilon)}{E^2 - \varepsilon^2} d\varepsilon = \frac{2}{\pi} \int_0^\infty \frac{\varepsilon f''(\varepsilon)}{E^2 - \varepsilon^2} d\varepsilon \qquad (4.14)$$

An example of this change in the form factor close to an absorption edge is shown in Fig. 4.7. These data were obtained as a part of a study of the crystal structure of human chorionic gonadotropin (hCG), a hormone associated with the early weeks of pregnancy, by Wu et al. (1994). The selenium-containing proteins in the material investigated provided the basis of the anomalous scattering structural determination, so while the data shown in Fig. 4.7 were derived from absorption measurements of the complete hCG crystal, they are clearly dominated by the scattering from the 'heavy' Se atoms.

The changes in both the real and imaginary parts of the form factor close to an absorption edge evidently mean that recording diffraction data at different photon energies in this range is somewhat equivalent to isomorphous replacement in that the form factor of an atom in the crystal changes, but there is no need to grow new crystals with different constituent atoms, and no associated risk that the new crystal may not be truly isomorphous. Notice that these changes in the real and imaginary part of the form

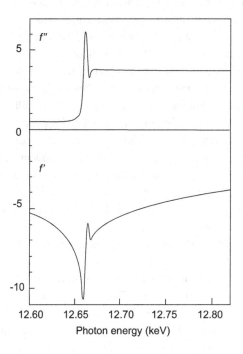

Figure 4.7 Form factor components for Se scattering around the Se K-edge (corresponding to photoionisation of the 1s state) obtained from absorption measurements on hCG crystal described in the text. After Wu et al. (1994) with permission from Elsevier

factor mean that the phase of the scattering by the atom, ϕ, also changes because one can write $f = f_m e^{i\phi}$, where

$$f_m = \left| f_r^2 + f_i^2 \right|^{1/2} = \left| (f + f')^2 + f''^2 \right|^{1/2},$$

$$\phi = \tan^{-1}(f_i/f_r) = \tan^{-1}(f''/(f_0 + f'))$$

$$(4.15)$$

Evidently, the easy tunability of the photon energy of synchrotron radiation delivered to an experiment is ideal for exploiting anomalous scattering, and indeed has led to the development of the Multiple-Wavelength Anomalous Diffraction (MAD) technique that has proved invaluable for structural investigations of proteins, and in macromolecular crystallography in general. The form factor measurements shown in Fig. 4.7 were part of a MAD investigation of the structure of hCG described above. MAD is in many ways the natural successor of Multiple Isomorphous Replacements (MIR) that had proved very successful in the past in solving complex structures.

4.4   Standard X-Ray Diffraction Methods

As described in Section 4.1, if a static single crystal is exposed to a monochromatic X-ray beam, rather few diffracted beams will be observed; the requirements of energy and momentum conservation, illustrated in the Ewald sphere construction, are too stringent. To observe many diffracted beams one must either widen the range of incident wavelengths, or move the crystal to a range of different orientations. These two options lead to two fundamentally different experimental techniques for investigating single crystals. Laue diffraction uses 'white' incident radiation (at a synchrotron radiation source the wavelength range is generally limited by the pre-optics or by the use of un-monochromated undulator radiation, the resulting radiation being referred to as 'pink'), containing a continuum of wavelengths over some range. Alternatively, using a monochromatic beam of X-rays to record rotation patterns, the single crystal is rotated or oscillated in angle in front of the incident beam.

Laue diffraction can be recorded in both backscattering (sometimes referred to as a Bragg geometry) and forward-scattering (transmission) geometry, but in practice backscattering is primarily used for the purposes of determining the orientation of highly-absorbing crystals of known materials conducted using the bremsstrahlung radiation of laboratory-based sources. The use of 'white' or 'pink' radiation leads to a large number of diffracted beams being recorded for a fixed geometry, so data can clearly be collected much faster than if the crystal has to be scanned through a range of different orientations. However, if the main interest is in obtaining accurate diffracted beam intensities at a precisely known wavelength, one must use the rotation or oscillation technique. A Laue pattern necessarily has a higher background, while diffracted beams may also correspond to several different orders of diffraction at different (multiply-related) wavelengths. An energy-selective detector can help to separate these orders, but the energy resolution achievable is much inferior to that of a standard X-ray monochromator used for monochromatic X-ray illumination.

The particular strength of the Laue method lies in time-resolved studies, following rapid changes in the detailed structure as a result of some change in imposed conditions, particularly in macromolecular X-ray diffraction (MX) studies.

An example of this is in the investigation of photo-activated protein dynamics based on a 'pump-probe' approach, also used in a wide range of both laboratory- and synchrotron radiation-based experiments. In the specific case described here a short laser pulse is used to produce the initial excitation (the pump) while a short exposure to the synchrotron X-radiation (probe) collects the (Laue) X-ray diffraction data. Different time delays between the pump and probe allow the dynamics to be investigated.

Fig. 4.8 shows the significant difference in the recorded Laue diffraction patterns, in this case from a crystal of myoglobin (a much-studied model system with unit cell dimensions of many tens of ångström units), resulting from the use of 'white' wiggler radiation and 'pink' undulator radiation. The undulator radiation used to produce this figure covered a wavelength range of only ~3% of the mean value, insufficient to intersect all potentially accessible reciprocal lattice points, and thus producing a smaller number of diffracted beams than the broad spectrum of the wiggler radiation, but the recorded pattern has a much lower background. The much-reduced number of incident photons from the undulator also reduces the problem of radiation damage to fragile biological molecules. The problem of the smaller number of recorded beams can be overcome by recording additional patterns of different central wavelengths of the undulator radiation. Of course, the key advantage of using the Laue technique rather than single-wavelength rotation patterns is its ability to capture the many diffracted beams in a very short time, allowing dynamics on sub-ns timescales to be investigated. Bourgeois et al. (2007) provide a fuller description of this approach.

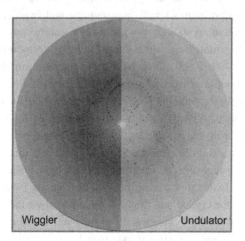

Figure 4.8 Laue diffraction from myoglobin; two half-patterns are shown, the left-hand half being recorded with 'white' wiggler radiation while the right-hand half was recorded using 'pink' undulator radiation. After Bourgeois et al. (2007) with permission of The Royal Society of Chemistry (RSC) on behalf of the Centre National de la Recherche Scientifique (CNRS) and the RSC

Indeed, using the more intense and much shorter pulses of radiation delivered by an X-ray free electron laser it is possible to obtain similar pump-probe MX structural data on the fs timescale (e.g. Aquila et al., 2012).

Of course, in view of problem of radiation damage mentioned above, one might expect that a crystal of this type of molecule would be completely destroyed by the pulse of radiation from an X-ray free-electron laser. This is, indeed, the case. However, if the duration of this radiation pulse ~70 fs is short compared with the timescale of atomic motions in the crystal (recall typical pre-factors describing vibrational properties on molecules are $\sim 10^{13} s^{-1}$) then the diffraction data may be recorded before the crystal structure has time to be significantly altered. Of course, these short pulses are not available in conventional synchrotron radiation, in which bunch lengths are some three orders of magnitude longer. Under these conditions radiation damage in organic molecules in general is an important problem, and is particularly significant if the incident radiation beam is focused to a very small spot to allow study of small crystals, because the radiation damage scales as the incident flux density, not the total flux. One important way of ameliorating this problem is by lowering the temperature of the sample. This approach led to the development of cryo-electron microscopy which led to the award of the Nobel prize in Chemistry in 2017 to Joachim Frank, Richard Henderson and Jacques Dubochet, but the underlying principles are equally relevant for X-ray crystallography of these fragile biological crystals. Indeed, the primary event giving rise to most radiation damage from X-rays is photoionisation, generating energetic photoelectrons that in turn generate a cascade of lower energy electrons. This effect, leading to chemical bond breaking, is clearly not temperature dependent. What a reduction in temperature can influence, however, is the diffusion of molecular fragments, and notable free radicals, through the crystal, leading to further damage. The problem of radiation damage had been recognized early in the development of protein crystallography, leading to careful monitoring of the effects. However, it is rather more recent that the benefits of cryo-cooling have been appreciated (Henderson, 1990, 1995), albeit with some attendant problems, and have become commonly exploited (e.g. Burmeister, 2000; Garman, 2010).

Fig. 4.9 illustrates the underlying principle of the rotation method for recording single crystal diffraction patterns using monochromatic incident radiation. In such an experiment the crystal is rotated with a fixed incident direction so the reciprocal lattice sweeps through the Ewald sphere. A more convenient way to represent this in a figure is to rotate the Ewald sphere through a fixed reciprocal lattice. Notice that there are 'blind' regions of the reciprocal lattice that are 'missed' by the rotating Ewald sphere. Depending on the symmetry of the crystal and the rotation axis chosen the reciprocal lattice points that are missed may be symmetrically equivalent to others that are captured; if not, partial rotation about a different axis may be necessary. Fig. 4.9 is drawn with a spacing of the reciprocal lattice that is only about a factor of 2–3 times smaller than the X-ray wavelength, implying that if the wavelength is, for example, 1 Å, the real space lattice parameter would be only ~2–3 Å. In reality, structure determination using the rotation method at a synchrotron radiation beamline is applied to far more complex structures, typically involving protein or other macromolecular

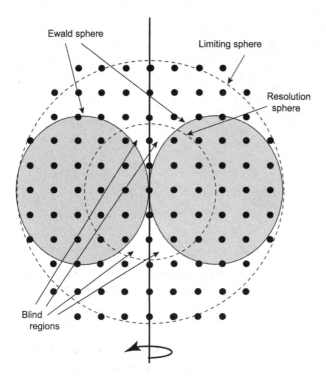

Figure 4.9 Schematic illustration of how the reciprocal lattice is sampled by a rotation or oscillation diffraction experiment.

crystals, with lattice parameters of hundreds of ångström units or more, so the spacing of the reciprocal lattice points is a tiny fraction of the value shown in the diagram. In such a situation there are two important consequences. First, even a rotation of a few degrees will cause the Ewald sphere to pass though many reciprocal lattice points, thus producing many diffracted beams. If the rotation were over a much wider range it is likely that many overlapping unresolved diffracted beams would arise. For this reason such experiments are performed by recording sequentially data from a series of experiments, which each cover only a small rotational angular range, in what is referred to as an oscillation, rather than rotation, experiment.

Also shown in Fig. 4.9 is the 'limiting sphere', a sphere centred on the origin of the reciprocal lattice with a radius of $2\,|\mathbf{k}| = 4\pi/\lambda$. This surrounds all reciprocal lattice points that are accessible with the value of the wavelength used in the experiment (i.e. all reciprocal lattice points that could be swept by the Ewald sphere with rotation about more than a single axis). The significance of this is that it defines the largest reciprocal lattice vector that can be accessed, thus defining the smallest real-space separation that can be measured to be $\sim\lambda/2$. Notice that this limit is related very simply to Bragg's law, $2d\sin\theta = n\lambda$, because the smallest value of n is 1 and the largest value of $\sin\theta$ is 1, so the smallest value of d is $\lambda/2$. In effect the limiting sphere defines the limit of the spatial resolution. However, in macromolecular X-ray crystallography

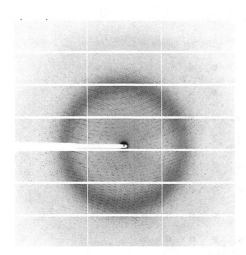

Figure 4.10 A snapshot of an X-ray diffraction pattern taken on a Pilatus 6M detector from *A. thaliana* Rubisco crystals. Courtesy Laura Gunn, Uppsala University

(MX), the true resolution is often limited by the quality of the crystals, so a smaller 'resolution sphere' may show the true range of usefully accessible reciprocal lattice points.

There is no doubt that MX studies form one of the most important parts of the range of applications of synchrotron radiation X-ray diffraction, and while the underlying scientific motivation and understanding falls outside the scope of this book, with its focus on applications in the physical sciences, it is illuminating to recognize the complexity of such investigations. Fig. 4.10 shows just one of the diffraction patterns recorded in a rotation method investigations of crystals of *Arabidopsis thaliana* Rubisco (RuBisCo is Rubulose-1,5-biphosphate carboxylase/oxygenase) taken on a Pilatus 6M detector at beamline ID-29 of the European Synchrotron Radiation Facility (ESRF) at a wavelength of 0.978 Å in the investigation reported by Valegård et al. (2018). This is just one of the 1,860 diffraction patterns, each recorded over a rotation angle range of 0.05°, that were used to solve the structure of this crystal with unit cell dimensions a = b = 111.9 Å and c = 197.7 Å; a total of 192,721 unique diffracted beams were measured in the analysis. Notice that the four-fold symmetry means the total angular range sampled could be restricted to only slightly more than 90°. Rubisco (Rubulose-1,5-biphosphate carboxylase/oxygenase) is an enzyme involved in the photosynthesis of plants, specifically in the first major step in carbon fixation.

The molecular structure derived from this study, to a resolution of 1.5 Å, is shown in Fig. 4.11. Fig. 4.11(a) shows the asymmetric unit of *A. thaliana* Rubisco, while Fig. 4.11(b) shows top and side views of the complete Rubisco molecule comprising four of the asymmetric units. One asymmetric unit is shown in grey, with the rest of the enzyme in black. In Fig. 4.11(c) one molecule of a ligand, 2-carboxy-D-arabinitol 1,5-bisphosphate (2-CABP), is shown bound at the catalytically active site within the

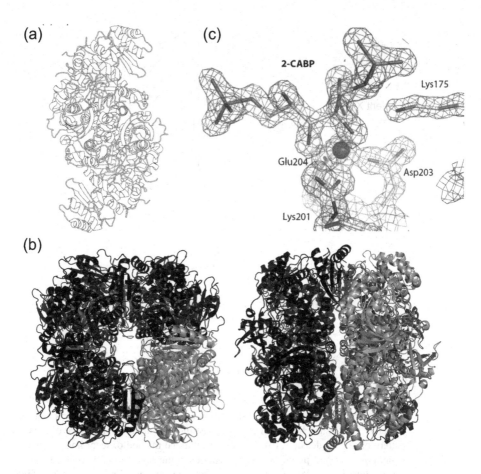

Figure 4.11 The molecular structure derived of *A. thaliana* Rubisco, derived from a set of 1860 diffraction patterns, one of which is shown in Fig. 4.9, as described in the text. (a) shows the asymmetric unit of the molecule, (b) the complete molecule as at top and side view. (c) shows at much higher magnification the region around the catalytically active site. After Valegård et al. (2018). Reproduced with the permission of the International Union of Crystallography

calculated electron density map. The Mg^{2+} ion that stabilizes Rubisco's catalytic residue (lysine 201) is shown as a sphere. Notice that Figs. 4.11(a)–(b) are ribbon drawings widely used in protein structure visualization in which smooth curves are interpolated through the polypeptide backbones, with α-helices shown as coiled ribbons (e.g. Richardson, 2000).

A rather different standard X-ray diffraction technique that can be used for detailed structure determination of materials for which sufficiently large crystals cannot be grown, but is also often used for materials characterisation, such as identifying the coexistence of multiple structural phases, is powder diffraction. In this case the sample is in the form of a fine powder, so tiny crystals in all possible orientations are represented in the sample. The scattering of the monochromatic incident X-rays then leads to cones

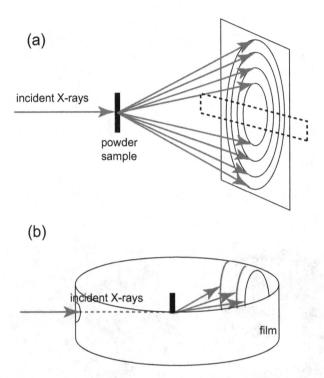

Figure 4.12 Schematic diagram of the arrangement for a powder diffraction pattern. (a) shows the basic geometry and the rings of diffracted beams projected onto a flat screen or other planar detector. The dashed line outlines a planar strip, showing the appearance of the diffraction pattern on such a detector. (b) shows the traditional method of recording a powder diffraction pattern on a strip of photographic film encircling the sample that detects the full 360° pattern (apart from close to 180° scattering where a hole is punched in the film to allow in the incident beam to enter the 'camera').

('Debye-Scherrer cones') of diffracted beams corresponding to the 2θ deflection, about the incident direction, by individual crystals in the powder that are oriented at a Bragg angle of θ relative to the incident beam, as shown in Fig. 4.12(a). Historically, in the traditional laboratory-based powder diffraction technique, the diffracted beams were detected on a strip of photographic film wrapped around the sample on the inside of a circular powder diffraction 'camera', as shown in Fig. 4.12(b).

At powder diffraction beamlines of synchrotron radiation sources a range of different detectors may be used. Fig. 4.13 Shows an example of the arrangement of at the I11 powder diffraction station at the Diamond Light Source. The diffractometer is fitted with a curved one-dimensional linear array position-sensitive detector (PSD), which captures part of the range of the traditional film, but also groups of multiple analysing crystal (MAC) single point detectors. Notice that the photograph also shows a carousel in which multiple samples can be stored, a robotic arm performing automatic changing of these samples. Robotics of this kind are also extensively used in macromolecular crystallography.

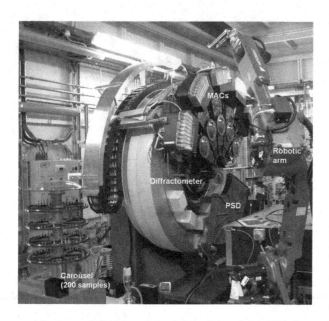

Figure 4.13 Photograph of the main end-station of the I11 powder diffraction beamline at Diamond Light Source. Courtesy Chiu C. Tang

Powder diffraction is still widely used as a standard laboratory tool for materials characterisation, but the high intensity narrow monochromatic beams of synchrotron radiation allow higher resolution data to be collected much more quickly, thus enabling investigations that follow the influence of external conditions, such as temperature and pressure, that may lead to subtle structural changes or even phase transitions.

Of course, following changes of this kind are commonly constrained by the fact that most synchrotron radiation experiments are scheduled by the facility for very short periods of time (at most a few days, and often significantly less), precluding time-resolved studies on a longer time scale. A novel branch line at I11 has been implemented to address this problem, allowing a series of several different samples to be mounted and conditioned in different ways simultaneously, over extended periods of time, automatically switching them into the beam at intervals to record the progress of the processes being investigated (Murray et al., 2017). In this case a two-dimensional array planar detector can provide adequate resolution, and is installed on this branch line. Fig. 4.14 shows the output from such a detector recorded at this beamline on a reference sample. The inset pattern provides a very direct link to the schematic visualization of Fig. 4.12.

Also shown in Fig. 4.14 is a plot of the extracted diffracted beam intensities as a function of the total scattering angle 2θ. In otherwise well-characterised systems identification of the crystals in the sample can be achieved by comparing these data with databases of solved structures, the structure then being refined to optimise the agreement between theory and experiment by an iterative technique known as Rietveld (1969) refinement.

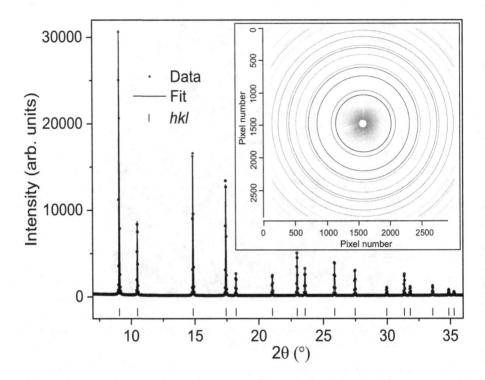

Figure 4.14 The inset shows a powder X-ray diffraction pattern diffraction pattern of a CeO_2 powder standard sample recorded using an incident X-ray energy of 25 keV. The main graph shows the extracted experimental diffracted beam intensities ('data') and the computed fit following Rietveld refinement, as a function of the scattering angle 2θ. The lines at the bottom show the angles of the expected *hkl* scattering planes. After Murray et al. (2017) Reproduced under Creative Commons CC BY licence

4.5   Surface X-Ray Diffraction (SXRD)

One application of X-ray diffraction that can really only be pursued successfully using synchrotron radiation is in the determination of the structure of the outermost few (as little as two or three) atomic layers of a solid surface. The basic problem is not only that this represents a very small number of scatterers, and thus a very weak scattering signal, but also that these few layers have below them a far larger number (typically by a factor of at least 10^6) of atomic layers of the underlying solid, so this weak signal is superimposed on a very much stronger scattering signal from the substrate. There are two rather different ways of overcoming this problem. One is to conduct the experiments using a grazing incidence angle below the critical angle for total reflection (a condition described more fully in Chapter 3 in the context of mirrors of optics). Under these conditions the scattering signal from the substrate is greatly suppressed because the penetration depth of the incident radiation is limited to that of the evanescent wave at the surface.

The second way to minimise the effect of strong substrate scattering is to investigate the scattering intensity in regions of k-space corresponding to scattering vectors that arise from the surface, but not from the bulk. There are two variants of this approach. If the surface layers have a different (longer) periodicity, parallel to the surface, than the underlying bulk, as is commonly the case for ordered atomic and molecular overlayers or reconstructed surfaces, then the larger real-space periodicity gives rise to a smaller periodicity in reciprocal space, and thus to diffracted beams occurring at extra values of $\Delta\mathbf{k}_\parallel$ (the component of the scattering vector parallel to the surface) in addition to those from the substrate. These beams are often referred to as 'fractional order' beams, because if the substrate beams are labelled with integer order labels, these 'extra' beams appearing between them must be labelled with fractional orders. It is possible to obtain important structural information on the surface by recording these beams alone. However, a significant limitation of using only the intensity of these beams is that the scattering leading to these beams arises *only* from the atomic layer or layers that have this larger real-space periodicity than the substrate. Scattering from the unreconstructed substrate does not contribute to the intensity of these beams, so the resulting structure determination can establish the relative location of the atoms within these layers, but not the registry of these surface layers to the underlying bulk. To gain this information one can also measure the intensity of beams with the same values of $\Delta\mathbf{k}_\parallel$ as the scattering vectors from the substrate, but to avoid the signal being dominated by the scattering from the many layers of the substrate, these measurements must be made at values of $\Delta\mathbf{k}_\perp$ (the perpendicular component of the scattering vector) that do not correspond to bulk diffracted beams. To understand this it is important to recognise that a surface layer (or layers) is only two-dimensionally periodic, meaning that the third Laue condition is fully relaxed. As a result, the reciprocal lattice points in the Ewald sphere construction (Fig. 4.15(a)) must be replaced by reciprocal net 'rods' (Fig. 4.15(b)). This relaxation of the third Laue condition clearly leads to far more diffracted beams being allowed at any specific value of $\mathbf{k}$ (i.e. a specific wavelength and incidence direction), but as these surface

(a)   (b)   (c)

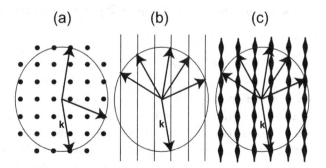

Figure 4.15 (a) and (b) show a comparison of the Ewald sphere construction for three- and two-dimensionally period structures, respectively. ((a) is a repeat of Fig. 4.2(a)). (c) provides a qualitative illustration of the diffracted beam intensity to be expected as one samples different points on crystal truncation rods.

layers are superimposed on a bulk crystalline substrate, we may expect strongly enhanced scattered intensity when the value of $\Delta \mathbf{k}_\perp$ corresponds to the condition for a bulk diffracted beam. Of course, formally the Ewald sphere construction is simply a representation of the requirements of energy and momentum conservation, and tells us nothing about the intensity of scattered beams. However, Fig. 4.15(c) illustrates schematically how the intensity of scattering at fixed values of $\Delta \mathbf{k}_\parallel$ may be expected to vary with $\Delta \mathbf{k}_\perp$.

It is rather straightforward to provide a slightly more quantitative description of these intensity variations along these diffraction 'rods' by considering diffraction from a crystalline solid that is terminated at a surface. In the discussion of crystal diffraction so far it has been effectively assumed that the crystal is infinite in size, although it has been pointed out that a relaxation of the three Laue conditions resulting from a finite-sized crystal can be accounted for in a simplistic way by regarding the reciprocal lattice points in the Ewald sphere construction have a finite size that is inversely proportional to the crystal size. To quantify this a little further it is helpful to recast Equation (4.1), writing the scattering momentum transfer in terms of components in the directions defined by $\mathbf{a}$, $\mathbf{b}$ and $\mathbf{c}$ as $\Delta \mathbf{k} = \mathbf{q}_1 + \mathbf{q}_2 + \mathbf{q}_3$, to give

$$A_{\Delta k} = \sum_{n_1=1}^{N_1} \sum_{n_2=1}^{N_2} \sum_{n_3=1}^{N_3} \exp\left[i(q_1 n_1 a + q_2 n_2 b + q_3 n_3 c)\right] F_{\Delta k}, \tag{4.16}$$

where

$$F_{\Delta k} = \sum_j f_j \exp\left(i\, \Delta \mathbf{k} \cdot \mathbf{r}_j\right). \tag{4.17}$$

The modulus squared of this amplitude gives the scattered intensity

$$I_{\Delta k} = F_{\Delta k}^{\,2} \frac{\sin^2\left(\frac{1}{2} N_1 q_1 a\right)}{\sin^2\left(\frac{1}{2} q_1 a\right)} \frac{\sin^2\left(\frac{1}{2} N_2 q_2 b\right)}{\sin^2\left(\frac{1}{2} q_2 b\right)} \frac{\sin^2\left(\frac{1}{2} N_3 q_3 c\right)}{\sin^2\left(\frac{1}{2} q_3 c\right)} \tag{4.18}$$

and at the three-dimensional Laue conditions $q_1 a = 2\pi h$, $q_2 b = 2\pi k$ and $q_3 c = 2\pi l$ this becomes

$$I_{hkl} = F_{hkl}^2 N_1^2 N_2^2 N_3^2. \tag{4.19}$$

Now consider the scattering from the surface of a large crystal. If the lateral dimensions are large the first two of these Laue conditions may be satisfied, but perpendicular to the surface the crystal is not truly periodic because of the presence of the surface. This statement is not only formally mathematically correct, but in practice the presence of the surface means that the intensity of the incident X-rays must decrease as they penetrate the solid. This may occur because of absorption (inelastic scattering), but must also be a consequence of elastic scattering, because if X-rays are scattered out of the surface by successive layers of the crystal this also attenuates the incident beam. Suppose that the amplitude of the incident X-ray wavefield is attenuated with a characteristic decay length δc (i.e. the distance δ is measured in units of the appropriate unit cell dimension perpendicular to the surface). The scattered amplitude can then be written as

$$A_{\Delta k} = \sum_{n_1}^{N_1} \sum_{n_2}^{N_2} \exp\left[i(q_1 n_1 a + q_2 n_2 b)\right] \sum_{n_3}^{N_3} \exp\left(-n_3/\delta\right) \exp\left[i(q_3 n_3 c)\right] \qquad (4.20)$$

and proceeding with the summation to N_1 and N_2 parallel to the surface, but to infinity (i.e. to convergence of the damped series) perpendicular to the surface, yields an intensity expression when the first two Laue conditions are satisfied of

$$I_{hk} = F_{hk}{}^2 N_1{}^2 N_2{}^2 \frac{1}{\left\{\left[1 - \exp\left(-1/\delta\right)\right]^2 + 4\exp\left(-1/\delta\right)\sin^2\left(\frac{1}{2}q_3 c\right)\right\}}. \qquad (4.21)$$

The final term in this expression still peaks at the third Laue condition ($\sin\left(\frac{1}{2}q_3 c\right) = 0$), but describes the finite intensity along the 'crystal truncation rods' (as shown schematically in Fig. 4.15(c)) away from this three-dimensional diffraction condition. This peak intensity has a value

$$I_{hkl} = F_{hkl}^2 N_1^2 N_2^2 \delta^2. \qquad (4.22)$$

The value of N_3, the number of layers in the finite-sized three-dimensionally period crystal (Equation (4.19)) being replaced by δ, the effective number of contributing layers. An important result that may be derived from Equation (4.21) is the intensity midway between the conditions that satisfy the third Laue condition, $\sin\left(\frac{1}{2}q_3 c\right) = 1$. Here the intensity from a truly three-dimensionally periodic crystal is zero, but in the presence of the surface this becomes

$$I_{hkl} = F_{hk}^2 N_1^2 N_2^2 / 4. \qquad (4.23)$$

The significance of this is that it is essentially the same as the intensity of scattering from a single atomic layer (strictly, the intensity from one half of a layer). This is important, because if the crystal is not ideally bulk terminated, but rather has a modified surface layer (possibly a reconstructed layer but also, perhaps, an atomic or molecular overlayer), then the amplitude of the scattering from the truncated substrate and the surface layer is similar, so interference between them, influenced by their relative atomic positions, will lead to clearly observable changes in the measured intensity. Measurements of the scattered intensity along the crystal trunca-tion rods, well away from the bulk three-dimensional diffraction conditions, thus provide the data required to determine the surface/substrate registry and layer spacing.

Notice that one implication of the relaxation of the third Laue condition is that diffracted beam intensities must be measured over continuous ranges of $l = \Delta\mathbf{k}_\perp$. An example of such a crystal truncation 'rod scan' is shown in Fig. 4.16, corresponding to the (01) beam from an InSb(001) surface that has undergone a 'c(8×2)' reconstruc-tion (i.e. a reconstruction leading to a centred rectangular surface mesh of dimensions 8×2 relative to the lateral periodicity of the underlying substrate). The dotted line

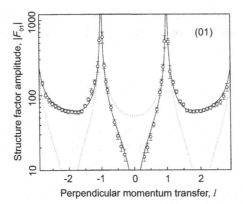

Figure 4.16 SXRD rod scan derived from the intensity along a (01) diffracted beam from a reconstructed InSb(001) surface. The dotted line is the result of calculations for an ideal bulk termination. The full lines are the results of calculations for the preferred model of the surface reconstruction. After Jones et al. (1998) copyright (1998) with permission from Elsevier

superimposed on Fig. 4.16 shows the relative measured structure factor modulus (i.e. the square root of the measured intensity) to be expected from an ideal bulk termination of the surface with no reconstruction or relaxation. Clearly the measured values differ greatly away from the conditions for bulk (three-dimensional) diffraction conditions at odd integral values of l. The full line that matches the data corresponds to calculated intensities for a particular structural model of the reconstruction. Notice that the ordinate of Fig. 4.16 has a logarithmic scale; the intensity of scattering at the conditions for diffraction from the three-dimensionally periodic substrate is many orders of magnitude larger than that from the one or a few layers in the surface. It is the need to measure this very weak surface scattering that is the most important reason for the need for synchrotron radiation to perform SXRD experiments. Armed with these measurements, structure determination is commonly achieved using trial and error simulations of different structural models compared with the experimental data, although the use of a two-dimensional Patterson function, producing a self-convolution of the surface structure projected onto a plane parallel to the surface, can also be used to guide this process. More sophisticated direct methods have also been developed and used with some success (Marks et al., 2001).

Of course, there are alternative methods available to solve surface structures that do not require the use of synchrotron radiation, of which the most directly comparable to SXRD is low energy electron diffraction (LEED). The scattering cross-section of atoms for low energy electrons (typically $\sim30-300$ eV) is many orders of magnitude higher than for X-rays, while strong inelastic scattering ensures that diffracted electrons only emerge from the outermost few layers, so the technique is intrinsically surface specific. However, the large scattering cross-sections mean multiple scattering is far more important than for X-rays, leading to simulations ('dynamical' theory, rather than 'kinematical' theory) being computationally far more demanding. SXRD therefore has a significant advantage when investigating more complex, larger surface mesh, structures. Even more significant, however, is the possibility of using the SXRD technique to determine the structure of buried interfaces, including solid/electrolyte interfaces, that are intrinsically inaccessible to truly surface specific methods.

4.6   X-Ray Standing Waves (XSW)

4.6.1   XSW Principles

A rather different crystal structure technique that is also based on X-ray diffraction exploits the X-ray standing wave (previously illustrated in Fig. 3.14) that is created by the interference of incident X-rays with a Bragg diffracted beam (Zegenhagen, 1993; Woodruff, 2005). Fig. 4.17 illustrates the consequences of this effect schematically. The standing wave has a periodicity in intensity that is equal to the periodicity of the atomic scattering planes associated with the Bragg reflection. Of course, for a non-absorbing crystal, at the condition corresponding to a single beam that is diffracted back out of the crystal, the fact that the amplitude of this beam relative to the incident beam is unity means that the X-ray penetration is limited; the presence of the surface means that the conservative process of elastic scattering nevertheless leads to an attenuation of the incident beam with depth into the crystal. This finite penetration leads to a finite range of the diffraction condition, in angle or in X-ray wavelength (and thus in photon energy), over which the reflectivity is unity. If one scans through the Bragg condition by varying the incidence angle then this flat-topped (unity reflectivity) peak in the 'rocking curve', known as the Darwin reflectivity curve, has a finite width in angle; a similar scan in photon energy (wavelength) at fixed angle leads to a finite width of the reflectivity peak in energy. Evidently, if unity reflectivity exists over this range, so does the standing wave, but the phase of this standing wave relative to the atomic scattering planes shifts through this range in an entirely predictable fashion. Consider, now, the impact of this on the atomic absorption at an atom in (or on) the crystal. If the nodal plane of the standing wave lies on this atom, there will be no absorption, but if an antinodal plane lies on the atom, the absorption will reach a peak. This is a classic 'two-beam' interference problem (like Young's slits in optics) with an intensity variation between two equal amplitudes out of phase $(1 - 1)^2 = 0$ and two equal amplitude exactly in phase $(1 + 1)^2 = 4$. Outside this total reflectivity range the absorption intensity is determined only by the incident beam (intensity 1 in the same notation). The consequence of this effect is that if one measures the X-ray absorption at a particular atomic species as one scans through the Bragg condition, the variation of this absorption leads to a profile that is characteristic of the location of the absorber relative to the atomic scattering Bragg planes of the crystal.

Fig. 4.17(c) shows a schematic representation of this effect in a non-absorbing crystal while Fig. 4.17(d) shows calculations for a specific example corresponding to a photon energy scan through the Bragg condition close to normal incidence to the (111) scattering planes of Cu(111). Because this crystal is not non-absorbing, the

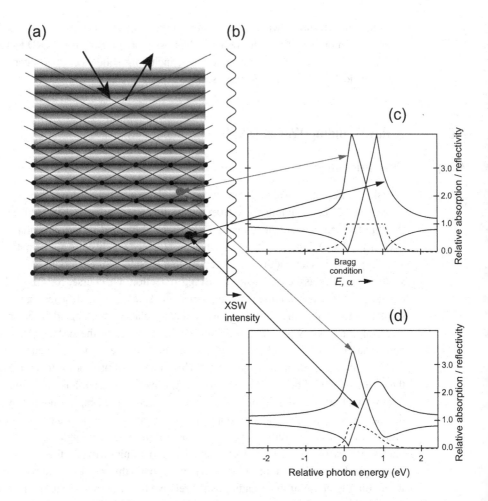

Figure 4.17 (a) Schematic diagram showing the formation of an X-ray standing wavefield generated by the interference of an incident and Bragg-reflected X-ray beam, represented by planar wavefronts, scattered from the periodic array of atoms (small black discs) in a crystal. (b) sketches the intensity of the standing-wave field as a function of the depth into the crystal. (c) reflectivity (dashed line) and X-ray absorption (full lines) from a non-absorbing crystal, at absorbing atoms lying on, and midway between, the crystal atomic scattering planes, as represented by that large black and grey discs in (a), respectively. (d) shows the results of calculations for the specific case of Cu(111) in which bulk absorption does occur (see Woodruff et al. (1988))

reflectivity is not actually unity across the full range but is significantly lower on the high energy side; this is because at the high energy side of the range the standing wave has its antinodal planes on the Cu atomic planes of the crystal, leading to enhanced absorption and thus reduced reflectivity. The absorption curve for an atom lying in these scattering planes clearly shows this effect; at the low energy end of the range, where the nodal planes of the standing wave lie on the atoms, the absorption is zero. By contrast, for an absorbing atom lying midway between the atomic scattering planes, the absorption profile is inverted, the strongest absorption in this case occurring at the low energy end of the standing wave range. Notice that this effect occurs not only for an absorber atom within the crystal (be it a constituent atom of the periodic lattice or an impurity or solute atom), but also for absorbing atoms located above the surface of the crystal, because the overlap of incident and scattered waves, and thus the standing wave, extends far above the surface. In fact it is in studies of surface structure that the XSW technique has come to be most widely exploited.

To understand the benefits and limitations of the technique more fully, it is helpful to outline the underlying theory more quantitatively. The X-ray standing wave intensity can be written as

$$I = |1 + (E_H/E_0) \exp(-2\pi i z/d_H)|^2, \tag{4.24}$$

where E_H and E_0 are the amplitudes of the electromagnetic field of the scattered and incident beams respectively, z is the perpendicular distance from the scatterer planes and d_H is the separation of these scatterer planes. The scattered amplitude is given in terms of the geometrical structure factors F_H and $F_{\bar{H}}$ for the H and $\bar{H}$ reflections (where H represents hkl)

$$E_H/E_0 = -(F_H/F_{\bar{H}})^{\frac{1}{2}}\left[\eta \pm (\eta^2 - 1)^{\frac{1}{2}}\right]. \tag{4.25}$$

The parameter η is a measure of the exact displacement in incidence angle, $\Delta\theta$, or photon energy, ΔE, from the nominal Bragg condition. For a scan in grazing incidence angle, θ, from the nominal Bragg angle, θ_B, the displacement parameter, η, is given by

$$\eta = [\Delta\theta \sin(2\theta_B) + \Gamma F_0]/|P|\Gamma(F_H F_{\bar{H}})^{\frac{1}{2}}. \tag{4.26}$$

P is a polarisation factor which is unity for σ polarisation and $\cos(2\theta_B)$ for π polarisation, F_0 is the geometrical structure factor for forward scattering (the (000) reflection), and

$$\Gamma = (e^2/4\pi\varepsilon_0 mc^2)\lambda^2/\pi V \tag{4.27}$$

where V is the volume of the unit cell, e and m the charge and the mass of the electron, respectively, ε_0 the permittivity of free space and c the speed of light (the bracketed term being the classical radius of an electron). Equation (4.25) shows that if the crystal is non-absorbing (i.e. if the structure factors F are all real), then there is a range of the displacement parameter η between the values -1 and $+1$ over which the reflectivity is unity, as described above. Equation (4.26) shows how this range depends on a physical parameter, the displacement in incidence angle, $\Delta\theta$, from the Bragg condition. Notice that Equation (4.26) shows that the value of this angle at the midpoint of this total reflectivity range, when $\eta = 0$, is not zero (when $\theta = \theta_B$), but is offset from the nominal Bragg angle by an amount $\Delta\theta = \Gamma F_0/\sin(2\theta_B)$; this is due to the effects of multiple ('dynamical') scattering, the importance of which is determined by the magnitude of F_0. The range of $\Delta\theta$ over which the total reflectivity condition persists can also be obtained from Equation (4.26)

$$\text{range}(\Delta\theta) = \pm\left[|P|\Gamma(F_H F_{\bar{H}})^{\frac{1}{2}}/\sin(2\theta_B)\right]. \tag{4.28}$$

For example, evaluating the magnitude of this range for a Si(111) reflection at a photon energy of 10.5 keV (wavelength, 1.18 Å; Bragg angle, 10.8°) leads to a value of approximately 20 μrad or 4 arc seconds. Indeed, because $F_0 > (F_H F_{\bar{H}})^{\frac{1}{2}}$, the nominal Bragg condition actually lies outside the range of total reflectivity.

Evidently, this extremely narrow rocking curve means that the measurement and control of the incidence angle must be extremely precise, setting important constraints on the instrumentation. In addition, however, it means that the substrate crystal used in the experiment must be extremely perfect; it must have a crystal mosaicity (i.e. a range of scattering plane angles due to small angle grain boundaries) which is smaller than, or at worst comparable to, this rocking curve width; otherwise the effect of the standing wave would be smeared out. Because of this constraint the great majority of early XSW experiments were performed on one of the very small number of semiconductor materials (especially silicon) that can be prepared to a high degree of perfection. Typical metal single crystals, by contrast, have a mosaicity of the order of tenths of a degree.

The implication that one must be able to rotate the crystal with extreme precision can actually be overcome rather readily using synchrotron radiation, by keeping the incidence angle fixed and scanning through the Bragg peak by varying the photon energy, and thus the X-ray wavelength. Equation (4.26) can then be recast in terms of the change in the photon energy, ΔE, as

$$\eta = \left[-2(\Delta E/E)\sin^2\theta_B + \Gamma F_0\right]/|P|\Gamma(F_H F_{\bar{H}})^{\frac{1}{2}}. \tag{4.29}$$

The range of photon energy over which the standing wave exists is then

$$\text{range}(\Delta E) = \pm\left[E|P|\Gamma(F_H F_{\bar{H}})^{\frac{1}{2}}/(2\sin^2 2\theta_B)\right]. \tag{4.30}$$

To overcome the requirement of extreme levels of perfection in the sample crystals, one can choose to work specifically very close to normal incidence to the scatterer planes. Substituting this condition ($\theta_B = 90°$) into Equation (4.28) leads to a predicted rocking curve width that is infinite! This unphysical result is a consequence of the breakdown of some approximations used in deriving this equation, but the implication that the rocking curve width becomes very large near normal incidence to the scatterer planes is correct. The origin of this effect is simple; the standard Bragg condition, $n\lambda = 2d\sin\theta$, passes through a turning point at $\theta = 90°$ (when the local gradient with respect to θ is zero) and the scattering condition is therefore only weakly dependent on θ (i.e. it is independent to first order). Close to $\theta = 90°$ the rocking curve width can be as much as 1° or so, ensuring that near normal incidence the experiment is tolerant of a substantial degree of crystal mosaicity (Woodruff et al., 1988).

This variant of the SXW technique, usually referred to as Normal Incidence XSW (NIXSW), is now the most commonly used form of the method (Woodruff, 2005). Of course, one limitation of working at near-normal incidence to the Bragg planes is that the interlayer spacing of these planes fixes the photon energy range, which in turn influences the ability to detect absorption at atoms of interest, as the photon energy must exceed the binding energy of core electron states and correspond to reasonably large photoionisation cross-sections at these atoms. These lower photon energies also impact on the method of detecting the X-ray absorption in an atom-specific fashion; specifically, one can either detect the photoelectrons emitted as a direct result of the photoionisation, or detect the fluorescent X-ray or Auger electrons resulting from the refilling of the core hole created by the photoionisation event (see Fig. 5.16 and the associated text for a fuller explanation of these processes). Because the inelastic scattering mean-free-paths of electrons with energies of a few keV or less are quite short – typically no more than a few nm (see Chapter 6, Fig. 6.2), the detection of the photoemission or Auger electron emission leads to detection of absorption only in the near-surface region. By contrast, fluorescent X-rays with energies of a few keV have much longer absorption path lengths and so detect absorption to much greater depths. A number of factors therefore favour the application of the NIXSW technique to the investigation of near-surface structure. The relatively low associated X-ray energies favour low photoelectron energies, which in turn lead to shorter inelastic mean-free-paths and thus a higher degree of surface specificity in the measurement. Moreover, photoelectron binding energies are sensitive not only to the elemental species of the emitter, but also to the local chemical bonding state of these atoms, so the resulting structural analysis is both element- and chemical-state-specific. All of these aspects of the photoemission technique are described more fully in Section 6.1. Although a complication arises in photoelectron detection arises due to the intrinsic (atomic) angular distribution of the photoelectrons, that can lead to the variation in detected photoemission not being directly proportional to the photoabsorption (Fisher et al., 1998; Vartanyants and Zegenhagen, 1999; Lee et al., 2001), this can be accounted for and has led to this form of NIXSW experiment becoming an important technique for the determination of a range of surface molecular adsorption and interface structural problems.

While the alternative method of detecting the atomic adsorption in XSW studies using X-ray fluorescence does allow sampling to much greater depths, exploiting the technique to determine bulk crystalline structural effects also has some associated problems. One of these is highlighted by the first experiment to explore the XSW effect by Batterman (1964), who rocked a Ge crystal through the Ge(220) Bragg condition (using an X-ray wavelength of 0.71 Å and grazing incidence to the scatterer planes) and measured both the reflectivity and the X-ray fluorescence from the Ge atoms. Although the reflectivity showed the expected Darwin reflectivity curve, the fluorescence signal was effectively the inverse of the reflectivity curve, with the fluorescence yield showing a decrease across the full range of the reflectivity curve. This effect is a consequence of the fact that the sampling depth of the fluorescence is significantly larger than the extinction depth of the X-rays under the reflectivity condition, together with the fact that the absorption cross-section is small. As a result, the fluorescence obtained with no X-ray reflectivity was from a much greater depth of the crystal than when the high reflectivity led to the fluorescence being generated only in the thin near-surface layer within the extinction length. Batterman did show, however, that if the fluorescence sampling depth was reduced by detecting this signal only at grazing emission angles, the expected XSW absorption profile was obtained. Indeed, for 'bulk' measurements, a further complication arises that depends on the relative size of the fluorescence sampling depth and the extinction length. Specifically, in Figs. 4.17(c)–(d), the two X-ray absorption profiles (corresponding to different sites of the absorbing atoms) are calculated (as in Equation (4.24)) as being due to the interference of the incident X-rays and the X-ray scattered *out of* the crystal. This is therefore the wavefield just above the surface. A key difference between Figs. 4.17(c)–(d), however, is that in part (d) account is taken of the absorption of some of the X-rays in the constituent atoms of the crystal (Cu in this case). One manifestation of this is that the reflectivity is reduced on the high energy side of the Darwin reflectivity curve, where the standing wave antinodes lie on the Cu atom sites, leading to enhanced absorption. This in turn reduces the amplitude of the standing wave at the surface at this condition, but must further decrease the standing wave amplitude below the surface at this condition. The standing wave amplitude across the rocking curve is therefore depth dependent, and if fluorescence is detected from these deeper-lying layers, this effect must be taken into account. Of course, 'bulk' structural effects commonly persist to within a few atomic layers of a surface, so suitable choices of sampling depth, in some cases through the investigation of thin epitaxial films, can still be interpreted in a relatively simple fashion.

4.6.2   XSW Structure Determination, Triangulation and Imaging

As described so far, the XSW technique and its NIXSW variant provide a means to determine the layer spacing of an X-ray-absorbing atom relative to the scattering planes associated with a Bragg reflection. This one coordinate is clearly not a complete structure determination, although in some cases, such as if it is the height of an adsorbed large planar (organic) molecules above a surface, this one parameter alone can provide valuable information on the nature of the chemical bonding at the interface. More generally, identifying the location of an atom completely requires three non-parallel and non-coplanar distances, although in some situations the crystal symmetry may reduce this requirement to only two such distances. Exploiting this requires a more complete understanding of exactly what structural information is contained in a single measurement. For this purpose it is helpful to recast Equation (4.24) by noting that the reflectivity, $R = |E_H/E_0|^2$, so one can write $E_H/E_0 = \sqrt{R}\exp(i\Phi)$, whence

$$I = \left| 1 + \sqrt{R}\exp(i\Phi - (2\pi i z/d_H)) \right|^2 \tag{4.31}$$

or

$$I = 1 + R + 2\sqrt{R}\cos\left(\Phi - (2\pi z/d_H)\right). \tag{4.32}$$

This expresses the wavefield intensity at a single well-defined height, z, relative to the scatterer planes. In practice, the absorbing atoms will always have a distribution of heights, if only due to thermal vibrations, but also possibly due to static disorder or the co-occupation of more than one single site. This distribution of heights can be represented by $f(z)$ so the fraction of atoms within an interval dz is $f(z)dz$, the distribution being normalised such that $\int_0^{d_H} f(z)dz = 1$. The XSW intensity is then

$$I = 1 + R + 2\sqrt{R} \int_0^{d_H} f(z) \cos\left(\Phi - (2\pi z/d_H)\right)dz \tag{4.33}$$

or

$$I = 1 + R + 2f_{co}\sqrt{R} \cos\left(\Phi - 2\pi p\right) \tag{4.34}$$

which defines two parameters that relate the measurement to the structure, namely the *coherent position*, p and the *coherent fraction*, f_{co}. Notice that p can be related to a distance $D = pd_H$. The reason for the name of this second parameter is perhaps clearer if Equation (4.34) is expanded as

$$I = f_{co}\left(1 + R + 2\sqrt{R} \cos\left(\Phi - 2\pi p\right)\right) + (1 - f_{co})(1 + R), \tag{4.35}$$

so the first bracketed term is the ideal coherent term if all absorbers were at a height D, multiplied by the coherent fraction, while the second term is the sum of the incoherent intensity $(1 + R)$, multiplied by $(1 - fco)$. Notice that implicit in this derivation is that

$$f_{co} \exp\left(2\pi i p\right) = \int_0^{d_H} f(z) \exp\left(2\pi i z/d_H\right)dz, \tag{4.36}$$

so the coherent fraction and coherent phase correspond to the amplitude and phase of one Fourier component of the one-dimensional atomic distribution along the direction perpendicular to the scattering planes H.

This contrasts in a significant way with the quantity that is measured in a conventional X-ray diffraction experiment. In X-ray diffraction the diffracted beam *amplitudes* correspond to a Fourier transform of the real-space structure (in three dimensions), but the measurement is of the diffracted beam *intensities*, leading to the well-known *phase problem*. In an XSW experiment one measures both the amplitude and phase, so there is no equivalent phase problem, although a single measurement determines only a single Fourier component. However, by measuring the XSW absorption profiles for several different Bragg reflections, from differently-oriented scattering planes, it is possible to create a real-space image of the distribution of the structure of the atomic absorbers using a simple Fourier transform approach. The first demonstration exploiting this idea by Cheng et al. (2003), in a study of the bulk mineral muscovite, used XSW measurements with X-ray fluorescence detection from a series of higher order $(00\ n)$ reflections to obtain the first eight Fourier components of the distribution of Ba, Ti, Fe, Mn, Al, Si and K atoms along the direction perpendicular to the planar layers of this material. Subsequently three-dimensional images of lower resolution have been obtained using this technique from a small number of non-colinear scattering vectors from surface adsorption and bulk solution structures (Okasinski et al., 2004; Zhang et al., 2004; Lee et al., 2010).

Fig. 4.18 shows the results of the investigation of Zn^{2+}, Sr^{2+} and Y^{3+} ions adsorbed at a rutile $TiO_2(110)$ interface with an aqueous electrolyte by Zhang et al. (2004). The X-ray absorption at these atoms (and the Ti atoms of the substrate) was monitored by detecting their K_α X-ray fluorescence while scanning through the (110), (111), (200), (101) and (211) Bragg scattering conditions. The measured coherent fractions and positions were then used to define the amplitude and phase of the set of Fourier series components to construct a 3-dimensional map of the location of these atomic species. Two-dimensional cuts parallel to the surface through these maps at heights (relative to the TiO planes of the crystal structure) corresponding to the main peaks are shown in Figs. 4.18(a)–(d) for the four constituent atomic species. Fig. 4.18(a) shows the resulting map of the Ti atoms within this plane (height 0 Å) providing a lateral site reference for the mapping of the adsorbed species. Fig. 4.18(e) shows a plan view of the $TiO_2(110)$ surface distinguishing the bridging oxygen atoms (BO) and the terminal oxygen atoms (TO) that lie direct above Ti atoms. Figs. 4.18(f)–(g) show perspective views of the inferred adsorption sites of the three electrolyte solute ions. While the effective spatial

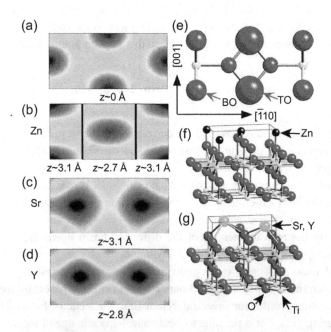

Figure 4.18 Results of XSW 'imaging' of adsorbates at a $TiO_2(111)$/electrolyte interface. Two-dimensional cuts through the derived three-dimensional image at heights above the Ti-O plane of the substrate corresponding to the different constituent species are shown in (a)–(d), while (e)–(g) show the derived structural model. After Zhang et al. (2004), copyright (2004) with permission from Elsevier

resolution of these images is only ~1 Å, this direct imaging can provide the initial structure for refinement by comparing measured and simulated absorption profiles.

More commonly, however, XSW, and particularly NIXSW, has been used to investigate the structure of atomic and molecular adsorbates on surfaces using a single X-ray reflection from the scattering planes parallel to the surface, in some cases combined with a second or more reflections from planes not parallel to the surface. The first of these measurements provides a value of the height of the absorber atom above the surface, or more precisely, relative to the extended bulk scattering planes (which differs if the surface layers are relaxed perpendicular to the surface); this quantity alone can provide important information on the bonding at the interface. However, with the addition of further measurements using differently oriented scattering planes, the lateral registry of the absorber can be determined by triangulation of the several coherent positions. A particularly simple and fruitful example has been a number of investigations of adsorbates on fcc (111) surfaces for which (111) NIXSW measurements determine the height of the absorber above the surface, while complementary $(\bar{1}11)$ NIXSW (in the same photon energy range) provides the data required to triangulate the lateral registry of the absorber. Fig. 4.19 shows how the three high symmetry adsorption sites, atop (directly above a surface layer atoms), hcp hollow (directly above a second layer atom) and fcc hollow (directly above a third layer atoms) can be distinguished by this measurement. For an atom at the same height above the surface (the same value of $D_{(111)}$, the heights above the $(\bar{1}11)$ planes are quite different.

In interpreting such data, however, it is important to recognise that the coherent fraction should not simply be regarded as an order parameter. For example, if an absorbing atom occupies a single well-defined low-symmetry site on such a surface (i.e. a site that lacks the three-fold rotational symmetry of the surface, such as a

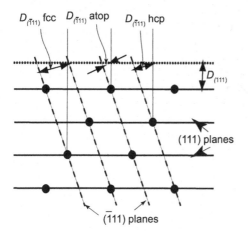

Figure 4.19. Schematic diagram showing how the three high symmetry atomic adsorption sites on an fcc(111) surface can be triangulated from NIXSW measurements using the (111) and $(\bar{1}11)$ reflections.

two-fold coordinated bridging site) then the coherent positions, relative to the different symmetrically-equivalent $\{11\bar{1}\}$ scattering planes, differ. This leads to a distribution of different z values in the $(11\bar{1})$ NIXSW measurement and thus a lower *fco* value, despite there being only a single occupied site. Notice that Equation (4.36), which shows how a summation over the different contributing z values can be accounted for, can be conveniently represented on an Argand diagram (Woodruff, Cowie and Ettema, 1994). Specifically, the left hand side of this equation can be represented as a vector of length f_{co} having a direction determined by the phase angle $2\pi p$ relative to the positive real axis (see Fig. 4.20(a), while the right hand side of the equation is then a summation (integral) over component vectors of length $f(z)$ and phase angle $2\pi z/d_H$. This representation is especially useful when $f(z)$ comprises a series of essentially discrete values. A particularly simple example of this situation is the case of two equally-occupied values of z, say z_1 and z_2, when $p = (z_1 + z_2)/2d_H$ and $f_{co} = |\cos(\pi(z_1 - z_2)/d_H)|$(Fig. 4.20(b)). One can also visualize the effect of a narrow distributions of z values about a mean, such as occurs as a consequence of thermal vibrations. The resultant vector has the direction of the mean of these z values, but has a reduced length (f_{co}) defined by a Debye-Waller factor (Fig. 4.20(c)). A particularly clear illustration of the fact that the coherent fraction is not a simple order parameter is given by the fact that if $(z_1 - z_2) = d_H/2$ then the two components partially cancel one another, leading to a much shorter resultant vector (reduced f_{co}) – see Fig. 4.20(d). Indeed, if these two z values are equally occupied then $f_{co} = 0$; namely, two equally occupied sites differing in height by one half of the interlayer spacing leads to a coherent fraction of zero, despite the system being perfectly ordered.

An example of the application of the use of photoemission monitoring of NIXSW as a means of adsorbate structure determination is provided by Fig. 4.21. The system

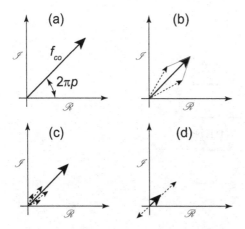

Figure 4.20 Argand diagram representation the way the coherent fraction and coherent position of co-occupied sites can be used to illustrate the measured values can be determined. Details of the four components (a)–(d) are given in the main text.

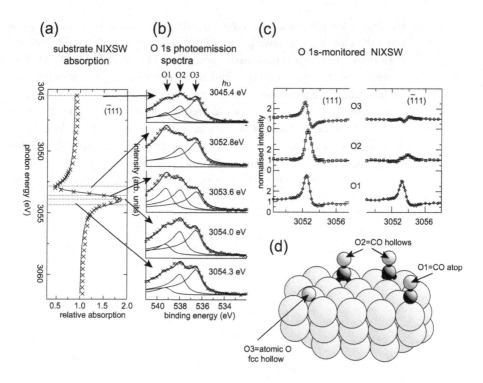

Figure 4.21 Results of a NIXSW investigation of the local adsorption structure of co-adsorbed atomic O and CO on Ni(111). (a) shows the NIXSW absorption profile in the Ni(111) substrate while (b) shows a selection of O 1s photoemission spectra recorded at different energies within this NIXSW range, each showing three chemically-distinct species labelled O1, O2 and O3. (c) shows the absorption profiles extracted from these three components in both the (111) and ($\bar{1}$11) NIXSW geometries. The extracted values of the fitting parameters of the coherent position and the coherent fraction (corresponding to the smooth lines through the experimental data points) lead to the structural assignment illustrated schematically in (d).

studied by Lee et al. (reported in detail by Lee, 2002) is the coadsorption of atomic oxygen and carbon monoxide on the Ni(111) surface. The O 1s X-ray photoemission spectra (XPS) from this surface recorded at the X-ray energy corresponding to the (111) 180° backscattering condition ~3,050 eV clearly shows three distinct components (see Fig. 4.21(b)) labelled O1, O2 and O3. These small 'chemical' shifts in the O 1s binding energy, as measured in photoemission, are a key feature of XPS and provide spectral fingerprint of different coexisting bonding environments of atoms of the same elemental species, as discussed more fully in Chapter 6. In NISXW, however, these chemical shifts also allow structural information to be obtained in a chemical-state-specific fashion. Recording this spectrum as the incident photon energy is stepped through the NIXSW substrate absorption profile (Fig. 4.21(a)) clearly shows changes in the relative intensity of these different components that must be due to the different location on the surface of the O species corresponding to these different components. By recording a complete set of these XP spectra through the standing wave photon energy range the intensity variations of the three components show the characteristic absorption profiles (Fig. 4.21(c)) of the three atomic species. Performing this experiment using both the (111) and $(\bar{1}11)$ NIXSW conditions provides the data to triangulate the lateral registry of their adsorption sites, as well as their atomic heights above the surface. The resulting structure is shown schematically in Fig. 4.21(d). The O3 atoms correspond to atomic oxygen located in the 'fcc' three-fold coordinated hollow sites on the surface, directly above third layer Ni atoms. The O2 oxygen atoms are CO molecules occupying with equal probability the two inequivalent three-fold coordinated hollow sites (fcc and hcp – directly above second layer Ni atoms), while the O1 emission arises from CO molecules adsorbed directly adopt surface layer Ni atoms.

5 Local Structural Techniques

5.1 Introduction

The structural techniques described in Chapter 4 exploit the special way in which crystalline materials – materials in which the constituent atoms are arranged on a three-dimensionally periodic lattice – scatter ('diffract') X-rays that have a wavelength comparable to typical interatomic distances. Not all materials have this property, and some that do cannot be grown as crystals that are sufficiently large to enable a complete structure determination, even by powder diffraction, a technique designed for the study of small crystals. Nevertheless, non-crystalline materials may possess *local* atomic-scale ordering; for example a particular elemental species in the material may arrange to have a specific number of atoms of another element at a preferred separation. Not only are there X-ray scattering techniques that cast light on this ordering, but there are also *electron* scattering techniques that provide this information but rely on the supply of X-rays of tuneable energy to deliver these electrons; specifically, they exploit the photoelectrons emitted within the material by these incident X-rays. X-ray scattering may also be used to determine structural and morphological aspects of materials on a significantly longer distance scale. This chapter outlines the principles of some of these methods that have been found to benefit from, or are totally dependent on, synchrotron radiation.

5.2 X-Ray Scattering from Non-Crystalline Samples: General Considerations

In Chapter 4 the constraints imposed on the scattering vector Δk by X-ray scattering from a three-dimensionally periodic (crystalline) solid were discussed in terms of the reciprocal lattice. Fig. 5.1 shows the basic definition of the scattering vector with no reference to whether or not the scatterers are in a crystalline material. Notice that by convention the angle between k_{in} and k_{out} is defined as 2θ. This provides a direct connection to Bragg's law in diffraction from a crystalline material, in which θ is the grazing incidence angle relative to the crystal scattering planes, the specular reflection from these planes leading to a scattering angle of 2θ. The short-dashed line in Fig. 5.1 shows where this specular reflection plane would lie, with the scattering vector perpendicular to this line. The modulus of the scattering vector is commonly assigned the letter q or Q, and, as shown by simple geometry in Fig. 5.1,

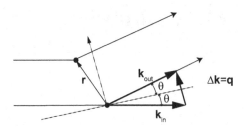

Figure 5.1 Generic diagram of X-ray scattering from two scattering centres related by a vector **r** showing the scattering vector Δ**k** or **q**.

$q = \Delta k = 2k\sin\theta = 4\pi\sin\theta/\lambda$. Of course, even in the absence of a crystalline lattice, and hence an absence of a reciprocal lattice, q is still a reciprocal space quantity. Large values of q allow one to probe short distances in real space, while to probe large distances in real space one must focus on scattering data obtained at small values of q. The latter conclusion is manifest in crystal diffraction from structures with large unit cells (such a macromolecular crystals), because the reciprocal lattice spacing is very small so diffracted beams are very close together. On the other hand, if one wants to determine closely spaced atomic positions within this large unit cell one needs scattering data to high q. In solving such structures it is common to refer to the solution being 'to x Å', where x defines the effective resolution of the structure determination. Ultimately (if other factors such a crystal quality do not limit the resolution), x is inversely related to the highest value of q sampled.

In Fig. 5.1 scattering through the same angle is shown from two different scattering centres separated by the vector **r**. In all discussion so far these centres have been individual atoms, and if they are near-neighbours in a condensed material or a molecule (i.e. their separation is of atomic dimensions) then strong variations in the scattering path lengths relative to the wavelength (leading to changes in the interference conditions) will only occur at high values of q and thus at large scattering angles. However, these two scattering centres could correspond to different regions of a larger object, with dimensions of hundreds of nm or even μm. If the question of interest is the size and shape of these objects, then to detect meaningful variations in the scattering interference from different parts of the object one must work at low q and thus small scattering angles. At large scattering angles there will be large variations of scattering pathlengths from different parts of the object, and averaging over these will lead to no net constructive or destructive interference. Investigating the size and shape of objects on these size scales is achieved through the use of Small Angle X-Ray Scattering (SAXS).

5.3   Determining Atomic Pair Distribution Functions from X-Ray Scattering

In discussing the complete structural solution of crystalline solids using X-ray diffraction (Chapter 4) it was stressed that because one measures only diffracted

beam *intensities* from a single crystal, rather than *amplitudes*, crucial phase information is lost. Specifically, this precludes the use of a simple Fourier transform of
the diffraction measurements to achieve a direct imaging of the complete structure.
Nevertheless, a Fourier transform of the intensities, the Patterson function, does
give an image of all interatomic *vectors*, a useful starting point in building possible
models of the structure. In the same way, a Fourier transform of diffracted beam
intensities obtained from a material that is not in the form of a single crystal (be it a
powder of crystalline material, or amorphous material, or even a liquid) can give
information on interatomic *distances*. The orientation information of the Patterson
function from a single crystal is lost due to the many different orientations of the
material in this form, but the interatomic distance information is maintained (i.e.
one obtains values of the scalar values of r but not the vectors $\mathbf{r}$). Moreover, this
information stems from the coherent interference of the scattering from pairs of
atoms separated by a radial distance r, so there is no requirement on the scattering
material to be crystalline. These values of the quantity r define a reduced pair
distribution function (PDF), $G(r)$,

$$G(r) = 4\pi r[\rho(r) - \rho_0] = 4\pi r\rho_0[g(r) - 1], \tag{5.1}$$

where in this case $\rho(r)$ is the atomic pair density function and ρ_0 is the average
atomic number density. The PDF is a measure of the probability of finding pairs
of atoms separated by the distance r. $g(r)$ is known as the atomic partial
distribution function. Notice that notation in this field is quite variable; for
example in some work $G(r)$ is written as $D(r)$, while $g(r)$ is written as $G(r)$,
and even some of the terminology varies. Here the notation follows that of
Egami & Billinge (2012).

Recall that in discussing X-ray diffraction from a single crystal in Chapter 4 the
scattered amplitude was shown (Equations (4.6) and (4.7)) to be the geometrical
structure factor or form factor of a single unit cell multiplied by the number of unit
cells in the material

$$A_{hkl} = NF_{hkl} = N\sum_j f_j \exp\left(2\pi i(\mathbf{G}_{hkl}.\mathbf{r})\right) \tag{5.2}$$

at the specific values of $\mathbf{q} = \mathbf{G}_{hkl}$, when the scattering contributions from the N unit
cells are all in phase. The Patterson function is derived from a Fourier transform of
$|F_{hkl}|^2$.

More generally, removing this constraint on the value of $\mathbf{q}$ that is specific to a
single crystal, the scattered amplitude over a volume of material, V, may be written as

$$A_q = \sum_V f_j \exp\left(2\pi i\mathbf{q}.\mathbf{r}_i\right), \tag{5.3}$$

so the scattering intensity is

$$I(q) = \sum_{V,i}\sum_{V,j} f_i^* f_j \exp\left(2\pi i\mathbf{q}.\mathbf{r}_{ij}\right), \tag{5.4}$$

where $\mathbf{r}_{ij} = \mathbf{r}_i - \mathbf{r}_j$. Averaging over all orientations (as in a crystalline powder, but also intrinsically in amorphous or liquid material) leads to

$$I(q) = \sum_{V,i} \sum_{V,j} f_i^* f_j \exp\left(2\pi i q r_{ij}\right). \tag{5.5}$$

To extract the pair distribution function one defines the total-scattering function (sometimes referred to as the structure factor, although this function is quite different from the geometrical structure factor in crystal diffraction)

$$S(q) = I(q) / N_a \langle f(q) \rangle^2, \tag{5.6}$$

where N_a is the total number of atoms. $S(q)$ converges to unity at high values of q. It is then the reduced structure function

$$F(q) = q[S(q) - 1] \tag{5.7}$$

that is Fourier transformed to give the pair distribution function

$$G(r) = \frac{2}{\pi} \int_{q_{\min}}^{q_{\max}} F(q) \sin\left(qr\right) dq. \tag{5.8}$$

The standard technique of powder diffraction, applied to crystalline powders and described in Section 4.4, involves the measurement of the scattered intensity as a function of q. For a crystalline powder this intensity function is dominated by intense sharp peaks due to the crystal diffraction. In this case it is easy to distinguish this intensity, that arises from coherent interference of scattering in the sample, from other sources of background intensity, such as scattering from the sample container, Compton scattering and X-ray fluorescence. The same measurement from material that lacks crystalline order shows much weaker variations due to the coherent interference of scattering from different centres, but this still provides structural information. However, in this case it is important to apply a range of corrections to the raw experimental data due to these other sources of measured intensity in order to isolate the coherent scattering signal that contains the structural information. Applying these corrections leads to a determination of $S(q)$ and thus to the calculation of the pair distribution function.

Of course, as in other X-ray scattering or diffraction methods for atomic-scale structure determination the PDF technique can be exploited with standard laboratory X-ray sources and detectors. However, synchrotron radiation offers the possibility of going to much higher photon energies (e.g. 100 keV) than are routinely available from laboratory sources, and higher photon energies allow one to access much higher values of q, leading to higher resolution in the data. As in all Fourier transform methods, a wider range of q also reduces the problems caused by truncation of the q range to be transformed.

Figs 5.2 and 5.3 show an example of the application of this approach from work by Masadeh et al. (2007) and Masadeh (2016). In Fig. 5.2(a) is shown a powder

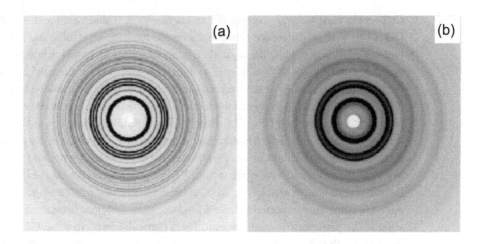

Figure 5.2 (a) Powder diffraction pattern from CdSe crystals and (b) comparable data from CdSe QD3 nanoparticles of average diameter ~36 Å. After Masadeh (2016), reprinted by permission of the publisher (Taylor & Francis Ltd, www.tandfonline.com)

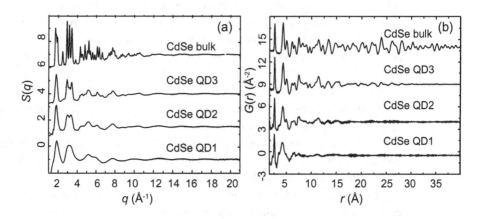

Figure 5.3 (a) $S(q)$ and (b) $G(r)$ from crystalline CdSe and three different sizes of CdSe nanoparticles as described in the text. After Masadeh (2016), reprinted by permission of the publisher (Taylor & Francis Ltd, www.tandfonline.com)

diffraction pattern of crystalline CdSe recorded using a photon energy of 87 keV (wavelength ~0.14 Å), while Fig. 5.2(b) shows similar data taken from CdSe nanoparticles (quantum dots) with an average diameter of approximately 36 Å (referred to here as CdSe QD3). Bulk crystalline CdSe, an important II-VI semiconductor, has the Wurtzite structure, but this compound is also of interest in the form of quantum dots, so the structure of very small particles of CdSe is of significant continuing interest (e.g. Blanc, 2018). Fig. 5.3 shows the $F(q)$ and $G(r)$ functions extracted from the raw data of Fig. 5.2, together with additional results obtained from two sets of smaller

CdSe nano particles with average diameters of ~29 Å (CdSe QD2) and ~21 Å (CdSe QD1). An obvious difference between the results from the crystalline and nanocrystalline materials is that while the crystalline material shows sharp diffraction peaks in the powder pattern and the resulting $F(q)$ function, the rings in the raw 2D data and the resulting peaks in $F(q)$ are smeared out in the data from the nanocrystals, with the effect becoming increasingly pronounced as the particle size is reduced. This effect is particularly clear in the PDFs. $G(r)$ from the crystalline material shows sharp peaks to the largest values of r, reflecting the long-range order of the crystals. By contrast, the features at higher r values in $G(r)$ from the nanocrystals are suppressed, although at low r all the materials have quite similar PDFs. This reflects the short-range order of the nanocrystalline material. Of course, even elemental nanoparticles (such as those of Au – see below) are expected to have a fundamentally different structure from the bulk solid if the particles are small enough, but a particular issue in these CdSe nanoparticles is whether the small size leads to a core structure that adopts the alternative zinc blende structure of other II-VI materials. Distinguishing these in small particles is challenging, because these structures differ only in their atomic layer stacking sequence; the nearest neighbour distances are the same in the two structures although they differ in more distant neighbours. In fact simulations of the PDFs based on these two structures gave a strong preference for the wurtzite structure for the crystalline powder material, but comparably good fits for the two alternative models to the nanocrystalline data, implying that in these particles neither model is completely correct. The best fit was found to be for a wurtzite structure with a significant density of stacking faults.

A further example of the application of PDFs is in a study of a different type of nanoparticle, namely thiol-terminated 'magic' gold clusters. Au clusters with a very specific 'magic' number of Au atoms, terminated by 'staples' comprising an Au-SR-Au species (with R being the organic part of a deprotonated thiol – RSH), have been found to be particularly stable. An important question to answer is: what are the details of their structural ordering? In effect, these small clusters have a very high proportion of surface atoms, so the structure is expected to differ from that of the bulk (face-centred cubic – fcc) crystalline material, but 'how small is small?' and at what size do they adopt the fcc structure? One of the largest of these clusters is Au_{144} and the structure of these clusters has been investigated using the PDF technique by Jensen et al. (2016). Fig. 5.4 shows the structural model that has been proposed for this species by Bahena et al. (2013); this consists of a 54-atom icosahedral core (Fig. 5.4 (a)) covered by a further 60 gold atoms (Fig. 5.4(b)) terminated by the Au-SR-Au 'staples', shown in Fig. 5.4(c) as the larger dark Au spheres and the small S spheres (the organic component of the thiol attached to these S atoms is omitted to allow clear vision of the Au atoms of the cluster).

The results of the PDF study of the structure of this cluster in which the SR species was hexanethiolate ($S(CH2)_5CH_3$, abbreviated to SC6) are shown in Fig. 5.5. In particular, Fig. 5.5(c) shows the experimentally-derived pair distribution function compared with a simulation based on the proposed structural model of Fig. 5.5. Clearly, this model gives an excellent description of the data. However, these results

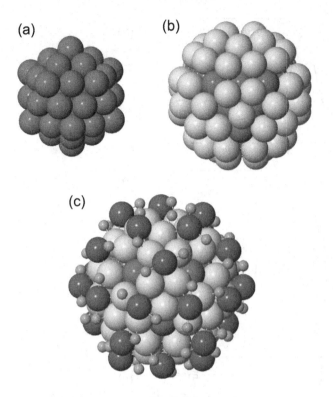

Figure 5.4 Proposed structural model of the $Au_{144}(SR)_{60}$ cluster. (a) shows the 54-atom icosohehral Au core, (b) the location of the surrounding 60 atom Au shell, (c) the complete cluster including the Au-SR-Au 'staples' (but with the organic R groups omitted). The small spheres are the S headgroup atoms of the thiols. Adapted from Jensen et al. (2016) under Creative Commons Licence (http://creativecommons.org/licenses/by/4.0/)

are only part of the wider study of Jensen et al., which included measurements from other $Au_{144}(SR)_{60}$ clusters with different thiols. Not all of these were similarly consistent with the Bahema et al. model. In particular, clusters terminated by *para*-mercaptobenzoic acid (*p*-MBA) were found to adopt a dodecahedral structure, while for those terminated by other shorter and longer chain thiols (SC4 – butanethiol and SC12 – dodecanethiol) both structures were found to coexist, indicating surprising polymophism in these clusters.

Of course, the PDF technique can also be applied to the local structure of liquids and amorphous solids. A longstanding issue is the structure of liquid water, but also of different amorphous forms of water ice produced at elevated pressures. Because H atoms are very weak scatterers of X-rays the total scattering of X-rays from these materials is dominated by the contribution from the oxygen atoms, although by contrast in the complementary technique of neutron scattering the scattering cross-section from hydrogen atoms is very large. Information gained from neutron scattering can therefore be used to remove from X-ray scattering PDF data from water phases the roles of H-H and O-H contributions, allowing the atomic partial distribution function

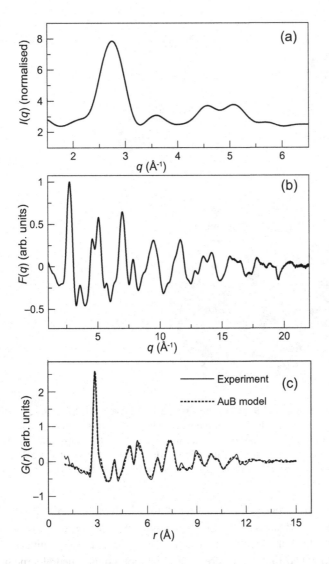

Figure 5.5 Results of the PDF study of $Au_{144}(SC6)_{60}$ clusters. (a) shows the scattering intensity as a function of q in the low q range. (b) shows the extracted $F(q)$ over a wider range, while (c) shows the resulting experimentally-derived pair distribution function as a thin continuous line, compared with a simulation of the expected PDF for the Bahema et al. (2013) model. Adapted from Jensen et al. (2016) under Creative Commons Licence (http://creativecommons.org/licenses/by/4.0/)

for O-O, $g_{O-O}(r)$, to be extracted. A recent example of this approach is the work of Mariedahl et al. (2018) investigating the structure of several amorphous phases of ice of different densities prepared under high pressure and low temperature conditions, labelled LDA (low-density amorphous), HDA (high-density amorphous) and VHDA (very high-density amorphous). The resulting values of $g_{O-O}(r)$ over two different ranges of r, are shown in Fig. 5.6.

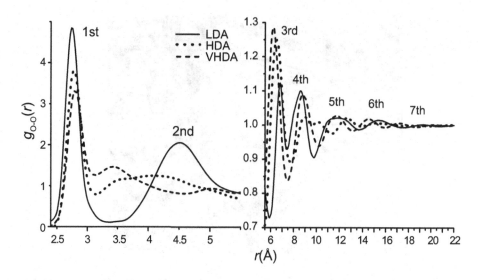

Figure 5.6 O-O X-ray atomic partial distribution functions obtained from different amorphous phases of water ice at three different densities, low, high and very high. After Mariedahl et al. (2018) under Creative Commons Licence (http://creativecommons.org/licenses/by/4.0/). Data courtesy of Katrin Amann-Winkel

Somewhat counter-intuitively the results show the first (nearest neighbour) shell distance increases from 2.750 Å to 2.803 Å as the density increases, an effect known as the density-difference paradox. The authors identify some interesting comparisons with the properties of liquid water under certain conditions in the nearer shell distances although at higher distances this is no longer the case. Notice that, unsurprisingly, the modulations in $g_{O-O}(r)$ at the highest distances become weaker and are plotted on an expanded ordinate scale; these show significant differences in the different phases.

5.4   SAXS (Small Angle X-Ray Scattering)

A rather different technique, which is nevertheless based on, qualitatively, the same type of experimental measurement, and indeed on similar underlying theory, is SAXS. As outlined in Section 5.2, if the X-ray scattering pattern is recorded at small scattering angles, and thus for small values of q, the structural information is on a much larger length scale than the atom-atom scale of atomic PDFs or crystal diffraction. How small is small? For example, if the photon energy is ~8 keV, as in traditional laboratory-based experiments (Cu K_α radiation), then for scattering objects separated by 1,000 Å the 'diffraction' angle is ~0.05°, corresponding to a lateral displacement from the unscattered beam of ~8 mm at a sample-to-detector distance of 10 m. Evidently, with sample-to-detector distances of 10–20 m (see, e.g. Fig. 5.7) the

Figure 5.7 The I22 SAXS beamline at Diamond Light Source, showing the 300 mm diameter evacuated tube to minimise gas scattering of the small-angle scattered X-rays that must travel a long distance (9.9 m) to the detector. The SAXS detector is in the distance, while in the foreground is a second detector for WAXS, with the sample off the right of the photograph. Courtesy Nick Terrill, Diamond Light Source

intrinsically highly collimated nature of synchrotron radiation is a considerable advantage relative to conventional laboratory-based X-ray sources.

The basic underlying theory of SAXS, of coherent interference between scattering from different locations in the sample, is the same as in crystal diffraction and in scattering experiments from non-crystalline material to extract atomic PDFs. However, the fact that no high-q data are collected means that the atomic-scale information is lost and the scattering factors are determined by the average electron density over significantly longer distances. In effect, many objects become (atomically) homogeneous, and the structural information is determined by 'contrast', defined as the difference between the average electron density of the object and that of the surrounding fluid (which may be simply air) in which it is contained. In this context, large biochemical molecules are essentially homogeneous objects, often freely rotating in a liquid; the information to be gained from SAXS is their size and shape.

In considering this coherent interference of scattering over larger distances the issue of the coherence of the incident radiation can play an important role in the design of the experiment. As discussed in Section 2.7, the ability to detect coherent interference between different scattering objects using synchrotron radiation depends on the size and divergence of the source of radiation in the experiment. This defines the 'transverse coherence dimension' $\xi_T = \lambda D/s$, where s is the lateral dimension of the source (typically defined in a SAXS experiment by a slit or pinhole), and D is the distance between this source and the scattering object. Evidently, if one is attempting to determine the separation of scattering objects that are widely spaced, it is important that this transverse coherence is larger than this separation of the scatterers. If this is not the case, coherent scattering will only be detected over smaller, potentially homogeneous regions. The size of the source aperture and its separation from the sample thus defines the length scale over which coherent scattering within the sample can be detected.

The scattering amplitude from a homogeneous object with an electron density $\rho_e(r)$ and a size determined by its volume V can be written as

$$A_{object}(q) = \int_V \rho_e(r)e^{-iqr}dr. \tag{5.9}$$

A particularly simple example is a homogenous sphere of radius R within which $\rho_e(r)$ is constant

$$A_{sphere}(q) = 4\pi \int_0^R \rho_e(r)\left(\frac{\sin qr}{qr}\right)r^2dr = \frac{4}{3}\pi R^3 \rho_e\left(\frac{3(\sin qR - qR\cos qR)}{(qR)^3}\right). \tag{5.10}$$

For a sphere surrounded by a fluid of finite average electron density, ρ_e is replaced by the difference in electron density or 'contrast' $\Delta\rho_e$. The numerator of Equation (5.10) is oscillatory, and differentiating with respect to q and equating to zero leads to the conditions for maxima and minima as $\tan(qR) = qR$. Of course, the intensity (given by the square of the amplitude) also oscillates. This effect can be seen strikingly in SAXS from monodispersed spherical particles in a fluid, as illustrated by the example of Fig. 5.8. These results come from experiments to characterise new ways of synthesising polymer spheres, based on poly

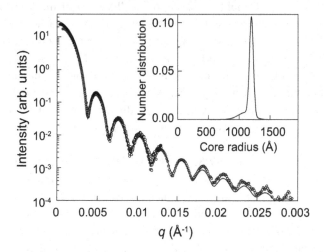

Figure 5.8 SAXS recorded from a sample of near-monodispersed PEG-grafted fluorinated colloidal spheres in an aqueous solution. The experimental data are shown as circles while the line is a simulation based on a size distribution shown in the inset. Adapted from Jonsson et al. (2017) under Creative Commons Licence (http://creativecommons.org/licenses/by/4.0/)

(heptafluorobutyl methacrylate) (pHFBMA) with a poly(ethylene glycol) graft, by Jonsson et al. (2017). In fact this work also made use of SAXS measurements based on a laboratory source, but the data shown here, recorded at the ESRF, are clearly superior and were able to exploit a very large sample-detector distances of 10 m to obtain data at q values greater than ~0.013 Å^{-1}, while a larger distance of 20 m was used to record the data at lower q values; the X-ray wavelength was 0.995 Å. The first minimum defined by the condition $\tan(qR) = qR$ occurs for a value of qR of 4.493 and Fig. 5.6 shows that this occurs at $q \approx 0.0035 \text{ Å}^{-1}$ leading to a value for the particle size of $R \approx 1{,}280$ Å, consistent with the peak of the size distribution used to model the whole scattering curve. Notice that Equation (5.10) relates to a single spherical particle, but the experimental data of Fig. 5.6 corresponds to scattering from many near-identical particles. Equation (5.10) provides a valid mathematical description of the experimental situation because the scattering from separate particles is effectively incoherent if the particles have no long-range order, so the intensity resulting from scattering from many particles simply corresponds to the sum of the intensities from each one.

One important advantage of using synchrotron radiation for SAXS studies is that the intrinsically highly collimated and high intensity beam makes it possible to follow time-dependent processes in real time. One such example is the crystallisation of metal-organic frameworks (MOFs) in solution. MOFs are nanoporous solids comprising metal ions or inorganic centres joined together by organic linkers to form three-dimensionally periodic networks containing pores of dimensions ~3−30 Å. They have applications in molecular selectivity of absorption that can be exploited for gas purification, gas storage, heterogeneous catalysis and drug delivery. Under suitable conditions mixing solutions containing the inorganic centres and the organic linkers can lead to rapid formation (crystallisation) of the MOF network. Fig. 5.9(a) shows an example of this effect being followed in a kinetic SAXS experiment. In this case the MOF being formed is known as ZIF-8 (ZIF stands for zeolitic imidazolate framework) from solutions of zinc acetate, and of 2-methylimidazole, both in methanol. The metal salt solution was circulated using a peristaltic pump through a capillary in the X-ray beam path, and the organic linker solution was injected remotely, synchronised with the start of data acquisition. SAXS spectra were recorded every 2 seconds but for clarity the figure shows a subset of these at intervals of 20 seconds.

The emergence of the oscillatory SAXS profile clearly indicates the formation of near-spherical particles whose size can readily be determined. Of course, the MOF also has periodicity on a nanometre scale, which can lead to Bragg scattering at larger q transfers. These data can be obtained in essentially the same experimental arrangement as SAXS by using a second detector at a much smaller sample-detector distance. This technique is generally known as WAXS – wide angle X-ray scattering – although the basic measurements is equivalent to that in powder diffraction and indeed, in experiments to obtain atomic PDFs from non-crystalline material. At the I22 beamline of Diamond Light Source (Fig. 5.7) both SAXS and WAXS can be measured simultaneously, the narrow cone of small-angle scattered X-rays passing through a

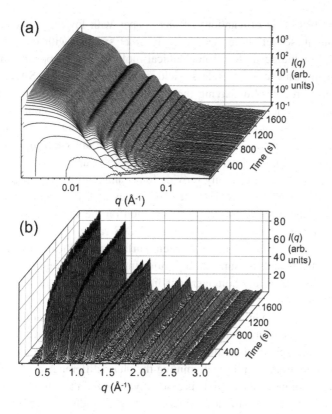

Figure 5.9 Kinetic (a) SAXS data and (b) WAXS data following the crystallisation of the MOF ZIF-8 as a function of the time after the mixing of the methanol solutions containing the Zn ions and the organic linkers. Data courtesy Andy Smith, Diamond Light Source

small aperture in the WAXS detector. The matching kinetic WAXS data (Fig. 5.9(b)) show that the emergence of the small particles in the SAXS data is accompanied by the growth of Bragg diffraction peaks, demonstrating the crystalline character of these particles. Further analysis of the widths of the Bragg peaks in the WAXS data allows the size of the crystalline regions to be determined (the broadening effect of crystal size and crystal strain can be separated using a procedure first demonstrated by Williamson & Hall (1953)). Application of this analysis revealed that the crystal sizes correspond to the size of the particles identified in SAXS, showing that the particles are, in effect, small crystals.

While the SAXS scattering curves in these two examples are richly structured, this is a consequence of the very narrow size distribution and the near-spherical shape of the scattering particles. Indeed, other scattering curves obtained in the experimental study of Jonsson et al. (2017) for less monodispersed polymer spheres (i.e. with a wider size distribution) showed significant damping of these modulations. One can derive more complex expressions than Equation (5.10) to describe the scattering from particles of other shapes, but notice that these derivations must average over all possible orientations of any shapes with lower symmetry than a sphere. In general, scattering curves for many other systems with less symmetric particle shapes, including those of biochemical molecules, can appear almost featureless. Such spectra can, nevertheless, yield valuable information. SAXS scattering curves can be regarded as having three different regions that carry somewhat different information, low q, medium q and high q, although of course these are relative terms within a range that is entirely low q in the more general context of X-ray scattering experiments. These regions, through the inverse of q, define three different spatial regions over which the scattering provides information. At the lowest values of q, the data can only provide long-range information regarding the size of the scattering objects. At intermediate values of q the spatial information becomes slightly more local, providing information on the shape of these objects, whereas at the highest values of q within the SAXS range the increasingly local information becomes most sensitive to their surface properties – the interfaces between the particles and the surrounding fluid. In all of these regions the objects can be regarded as essentially homogeneous in electron density.

The lowest q region can be treated by Guinier theory, first presented by André Guinier in 1939 (see also Guinier and Fournét, 1955), and the form of the scattering curve in this region is determined by the particle size: but which dimension defines this size for a non-spherical particle? For a non-spherical particle in a specific orientation the important quantity determining the phase difference between scattering from the extremities of the particle is evidently the size in the direction of $\mathbf{q}$. For a randomly oriented set of particles, which is being considered here, the relevant quantity is some average over different orientations that proves to be the *radius of gyration*, R_g. This is a quantity that is commonly used to define the moment of inertia of a body about its centre of mass, the moment of inertia corresponding to the product of the total mass and the radius of gyration; it is defined as the root mean square of its various parts from the centre of mass. In the context of SAXS the mass density is replaced by electron density. The Guinier theory then gives the scattered intensity as

$$I(q) = I(0) \exp\left(-q^2 R_g^2/3\right), \tag{5.11}$$

so a plot of $\ln I(q)$ against q^2 (a Guinier plot) should lead to a straight line, the value of R_g being extracted from the gradient. The low-q range over which Guinier theory can be used extends to values of qR_g in the range 0.9–1.3 depending on the shape of the particles, which must be monodispersed and sufficiently separated and non-interacting to ensure there is no long-range order. Deviations from linearity of a Guinier plot are taken to imply that these constraints are not satisfied.

Information about both the size and shape of the particles can be obtained from a Fourier transform of the scattering curve. In Section 5.3 it was shown that a similar Fourier transform of scattering data taken to high values of q leads to the pair distribution function, describing the distribution of local *interatomic* distances. An essentially equivalent procedure applied to the data of a SAXS experiment (which is intrinsically restricted to lower values of q) provides a pair distance distribution function, $p(r)$

$$p(r) = \frac{1}{2\pi^2} \int_0^\infty q^2 I(q) \frac{\sin qr}{qr} dq. \tag{5.12}$$

This is a measure of how often each distance appears within the individual particles (of nominally homogeneous electron density), a parameter that clearly depends on the shape of the particles. In particular, $p(r)$ very clearly distinguishes between particles in which three orthogonal dimensions are very similar and those that have much lower shape symmetry (interestingly, $p(r)$ for spheres and cubes are very similar, perhaps unsurprisingly for randomly oriented or rotating particles).

Fig. 5.10 shows an example of SAXS data and their analysis from a simple biological system, namely an enzyme called glucose isomerise. This is a commonly used protein standard (with a known structure) for crystallography and protein SAXS, and in this case is at a concentration of 2 mg/ml in an aqueous salt solution (PBS or phosphate-buffered saline). A representation of this protein (with a molecular weight of ~44 kDa) is shown highlighted in black in Fig. 5.10(c), but in practice it forms a tetramer (shown in dark grey in Fig. 5.10(c)) with a molecular weight of ~170 kDa. The gradient of the Guinier plot (Fig. 5.10(b)) obtained from the very low-q part of the whole SAXS spectrum of Fig. 5.9(a) leads to a radius of gyration of 33.3 Å, while the fact that the pair distance distribution function (Fig. 5.10(d)) is approximately a normal distribution is consistent with the near-spherical shape of the tetramer. Superimposed on the raw SAXS data are computed simulations, using the FoXS suite of programs (https://modbase.compbio.ucsf.edu/foxs/about.html) based on monomers (dashed lines) and tetramers (continuous line) in which the structures are taken from protein crystallography. These data clearly show much better agreement for the tetramer. Of course, one might ask why one would use the relatively low resolution SAXS technique to investigate the structure of such a protein that can be crystallised, allowing a higher resolution analysis by X-ray crystallography (as in Chapter 4). The problem with biological systems, in particular, is that these molecules do not function naturally in a crystalline form. Specifically, protein-protein interactions in a living organism are crucial to their functionality, so knowledge of the preferred multimer structure in a dilute aqueous environment is highly relevant to understanding their biological function.

The third piece of information that can be extracted directly (without modelling) from the scattering data is in the higher q part of the (intrinsically low-q) SAXS, often referred to as the Porod region. Porod (1951) first proposed a theory for this region, in which the information content is primarily concerned with the roughness of the interface between the particles and their surrounding fluid, under favourable circumstances. Specifically, if the surfaces are smooth then

$$I(q) \propto q^{-4}, \tag{5.13}$$

where the constant of proportionality provides information on the surface area. It is therefore common to perform a Porod plot of $\ln\left(I(q)\right)$ versus $\ln\left(q\right)$ for this portion of

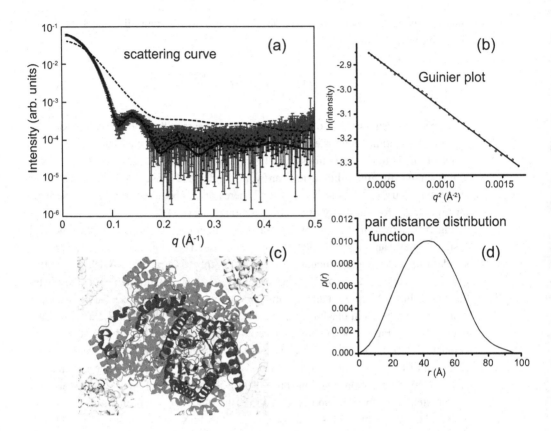

Figure 5.10 Results of a SAXS experimental study of glucose isomerise in an aqueous saline solution as described more fully in the text. (a) shows the experimental scattering curve together with simulations assuming the molecules are monomers (dashed line) and tetramers (continuous line). (b) and (d) show the Guinier plot and pair distance distribution functions derived from these data. (c) shows a model of the structure of the monomer (black) and tetramer (dark grey) superimposed on a background of other distributed molecules in the solution. Courtesy Nathan Cowieson, Diamond Light Source

the data (in principle as $q \rightarrow \infty$!) for which the slope should be -4 if the interface is smooth. Deviations from this exponent value indicate rougher surfaces, and in the case of fractal interfaces an exponent of $-(6 - D)$ is expected for a fractal dimension of D, but lower values of the magnitude of the exponent are found for some specific particle shapes, such as rigid rods.

While these simple analysis methods can prove valuable for the case of dilute monodispersed particle systems, in practice SAXS can prove to be effective in studying a much wider range of systems for which these constraints do not apply. Indeed, as is clear from the foregoing discussions, the basic measurement of the scattered X-ray intensity as a function of q, which is performed in SAXS, is the same as that used in studies of atomic PDFs and in powder diffraction from small crystalline particles.

In discussing X-ray diffraction from single crystals it was shown that the scattered amplitude could be written as (with Δk of Equation (4.1) re-written here as q)

$$A_q = \sum_{n_1=1}^{N_1} \sum_{n_2=1}^{N_2} \sum_{n_3=1}^{N_3} \exp\left(i\, q{\cdot}\mathbf{r}_c\right) \sum_j f_j \exp\left(i\, q{\cdot}\mathbf{r}_j\right). \tag{5.14}$$

The first three summations take account of coherent interference between the scattering from the many equivalent unit cells of the crystal (which leads to the conclusion that these are all in phase when $\mathbf{q}$ is a reciprocal lattice vector). The fourth summation on the right is over the scattering interference between the atoms within each unit cell. The scattering from more general samples can be expressed in essentially the same fashion as

$$A_q = S(q)F(q), \tag{5.15}$$

where $S(q)$ and $F(Q)$ are referred to as the structure factor and the form factor, respectively. The structure factor thus describes the effect of scattering interference between different objects and the form factor describes the effect of scattering interferences within these individual objects. Notice that this nomenclature is potentially slightly confusing because this structure factor is quite different from the 'geometrical structure factor' in crystal diffraction, which is actually the fourth summation in Equation (5.14) (i.e. it is the 'crystal form factor').

In the SAXS experiments considered so far, namely those conducted on distributions of particles that are at sufficiently large separations to ensure that there is no significant interaction between them, and thus no long-range ordering of them, $S(q) = 1$. The scattering intensities from the separate particles then simply add, and the measurement is of the form factor directly. If the particles interact significantly, however, there may be at least partial long-range ordering, and $S(q)$ can lead to crystal-like Bragg scattering, albeit at much lower q values than in crystal diffraction, because the length scale of the ordering is much longer. Notice, too, that if the angular range of a SAXS-type experiment is increased to WAXS, the data include high q transfers that can then lead to atomic-scale structural information (such as that of the atomic PDF technique) being included. In these studies of non-dilute systems the simple direct extraction of quantitative structural information from methods like the Guinier plot is no longer possible, and explicit modelling of the structure of the sample and the expected resultant scattering is necessary to get the most out of the technique.

5.5 Photoelectron Scattering Techniques: Basic Principles

Two very different synchrotron radiation techniques that provide quantitative local structural information rely not on the scattering interferences of X-rays, but the scattering interferences of electrons. However, high-intensity soft or hard X-rays of variable energy, such as can only be effectively provided by synchrotron radiation, are essential for these experiments, in which the source of electrons – photoelectrons excited from atomic core levels by the incident X-rays – is located within the material of interest, rather than being provided by an electron source outside the sample. The fact that this source of electrons is an atom within the material is what allows these techniques to provide structural information that is local to the environment of these photoemitting atoms.

Fig. 5.11 shows schematically the photoelectron scattering interferences involved in the two different techniques, namely XAFS (X-ray Absorption Fine Structure) and photoelectron diffraction. In both techniques an incoming X-ray photon core-ionises an atom in the material leading to the ejection of a photoelectron which emerges from this atom as an outgoing spherical wave (described by some combination of spherical harmonics that depend on the nature of the core level that is excited).

Conceptually, the second of these techniques, in which the resulting photoelectron emission is detected outside the sample, is the simpler one to understand. In this case the detected signal corresponds to the coherent sum of the directly-emitted wavefield and components of the same wavefield elastically scattered by surrounding atoms into the detected direction. Evidently all of the scattering path-lengths depend on the location of the emitter (absorber) atom relative to the surrounding atoms, and are thus specific to this local structure. If one now changes the photon energy, and thus the photoelectron energy and its corresponding photoelectron wavelength, the relative phases of the different scattering paths change. Scanning the photon energy thus causes these scattering paths to switch in and out of phase, producing modulations in the detected photoemission signal. Appropriate analysis of these measured modulations, described in more detail below, allows the structural environment of the local emitter atom to be determined. However, an important limitation is placed on this technique by the requirement to detect directly the angle-resolved photoemission. In the photoelectron energy range of interest up to ~300−400 eV, for which the associated electron wavelength is ~1 Å, the mean-free-path for inelastic scattering in most solids is only ~3−10 Å, so one can only detect emission from the near-surface region.

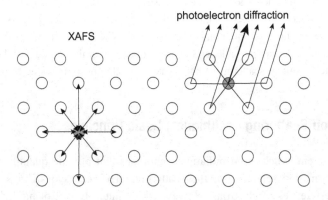

Figure 5.11 Schematic diagram showing some of the photoelectron scattering paths in the techniques of XAFS and photoelectron diffraction.

As such, photoelectron diffraction is essentially only a surface technique, used to determine the local structure of the outermost few atomic layers of a crystalline solid.

In XAFS, on the other hand, the photoelectron scattering detector is not outside the solid, but is the emitter atom itself. This possibility arises because the cross-section, σ, for ionisation of the core level, is determined by a matrix element involving the initial state, $|i\rangle$ – the wavefunction of the occupied core level, and the final state, $\langle f|$ – the outgoing photoelectron wavefield at the atom.

$$\sigma = \langle f|\mathbf{A} \cdot \mathbf{p}|i\rangle, \tag{5.16}$$

where $\mathbf{A}$ is the vector potential of the incident X-rays and $\mathbf{p}$ is the electron momentum operator. The key aspect of this equation is that the total photoabsorption cross-section scales with the amplitude of the photoelectron wavefield at the atom. Now consider the photoelectron scattering paths shown for XAFS in the schematic diagram of Fig. 5.11. These scattering paths go from the emitter to a near-neighbour scattering atom and back to the emitter. Their total pathlength is thus simply $2r_j$, where r_j is the distance between the emitter atom and the jth atomic neighbour. In this case, too, scanning the photon energy leads to changes in the photoelectron wavelength that causes these scattering paths to switch in and out of phase, increasing or decreasing the photoelectron wavefield amplitude at the emitter atom, thereby modulating the total absorption cross-section. Insofar as XAFS can be measured by a conventional absorption technique (the attenuation of X-ray passing through a sample) this is very much a bulk, as opposed to a surface technique, and is applicable to a wide range of physical and biological materials.

Notice that because both techniques exploit the photoemission from atomic core levels, which have binding energies that are characteristic of the atomic species, both techniques provide structural information that is element-specific, namely, they provide local structural information on the local environment of atoms of a particular elemental species. In the following sections more details of both techniques are provided.

5.6   X-Ray Absorption Fine Structure (XAFS)

As remarked above, absorption in general is typically measured in a transmission experiment, and a particularly simple case of an XAFS measurement based on transmission through a metal foil, as illustrated in Fig. 5.12(c). Figs. 5.12(a) and (b) show the results of such measurements for Ni and Fe foils. The absorption coefficient, expressed in terms of material thickness, t, as in Fig. 5.12(c), is given by

$$\mu = -t^{-1} \ln (I/I_0). \tag{5.17}$$

The absorption spectra of Fig. 5.12 show sharp rises in adsorption ('absorption edges') at photon energies of ~7,100 eV for Fe, and ~8,330 eV for Ni, corresponding to the thresholds for producing photoemission from the 1s states of these elements into the empty states above the Fermi level. At higher photon energies there are clearly

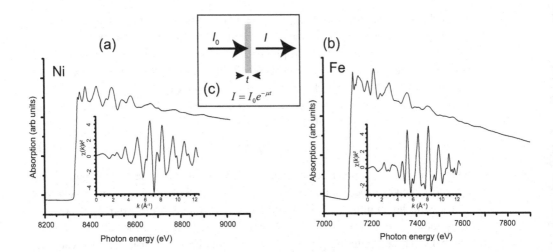

Figure 5.12 Absorption spectra of Ni (a) and Fe (b) foils measured as a function of incident photon energy around the Ni and Fe K-edges. Courtesy of Andy Dent, Diamond Light Source. (c) shows the measurement schematically. The insets in (a) and (b) show the EXAFS modulation functions extracted from the adsorption spectra.

modulations in the absorption coefficient that decrease in amplitude as the photon energy is increased. This 'fine structure' is generally separated into two regions; for photon energies greater than the absorption edge by more than ~20 – 30 eV, and for photon energies closer to the absorption edge. The higher energy range is known as the extended X-ray fine structure – the EXAFS – while closer to the edge this part of the spectrum is commonly referred to as XANES (X-ray Absorption Near-Edge structure) or, particularly in the case of molecular materials, as NEXAFS (Near-Edge X-ray Absorption Fine Structure). In the early days of observation of these effects the near-edge region was referred to as Kossel structure while the higher-energy range was referred to as Kronig oscillations after the individuals who offered the initial theoretical descriptions of the effects (Kossel, 1920; Kronig, 1931). Kronig's theory contains many of the same ideas included in current theories of EXAFS. The name EXAFS seems to have been first used in a paper by Sayers, Stern & Lytle (1971) who presented the first modern theory, although Lytle himself, in a personal historical review of the technique (Lytle, 1999), credits Prins (1934) with the invention of the name.

5.6.1   EXAFS

The theory of EXAFS is based on the scattering interference argument presented in the introductory Section 5.5. The fundamental physical processes governing the near-edge region can be described in the same way but are commonly thought of as reflecting the electronic structure of the absorbing atom's environment. In the case of molecules, for example, the structure here can reflect the energies of the unoccupied molecular orbitals, while in solids the energies of other excited states influence the near-edge spectrum. Of course, it is possible to describe the electronic structure of materials in terms of scattering (as in methods to calculate the band structure of solids), so both phenomena can be modelled computationally in a qualitatively similar way, but in practice the spectral signature of the near-edge region is often taken as a fingerprint of the local electronic structure. In molecular systems, however, the NEXAFS is also exploited to gain specific structural information.

To analyse experimental EXAFS data one must first extract the fine structure function from the measurement of μ

$$\chi(k) = \frac{\mu - \mu_0}{\mu_0}. \tag{5.18}$$

Of course, μ is measured experimentally as a function of the photon energy, but it is straightforward to convert this into the photoelectron wave vector amplitude k, determined by the energy above the adsorption edge, including an offset (an 'inner potential') to define the photoelectron kinetic energy inside the material. In Equation (5.18) μ_0 is nominally the atomic absorption in the absence of the EXAFS backscattering; in practice it is the function produced by drawing a stiff spline through the experimental measurement of μ in the material being investigated. Examples of extracted $\chi(k)$ functions, here multiplied by k^2 (see below), are shown as insets in Figs. 5.12(a) and (b). For the case of measurements corresponding to photoionisation of an initial s-state (most commonly a 1s state corresponding to a K-edge), and assuming only the single 180° back-scattering events depicted in Fig. 5.11 contribute to the EXAFS oscillations, the theoretical description then yields (assuming the sample is spherically averaged as in a liquid, an amorphous solid or a crystalline powder) an expression for the fine structure function

$$\chi(k) = -\sum_j \frac{N_j}{kr_j^2} \left| f_j(\pi, k) \right| \sin \left[2kr_j + \psi_j(k) \right] e^{-2\sigma_j^2 k^2} e^{-2r_j/\lambda_j(k)}, \tag{5.19}$$

in which the summation is over 'shells' containing N_j atoms of the same atomic species at a distance r_j from the X-ray-absorbing atom. $e^{-2\sigma_j^2 k^2}$ is a Debye-Waller factor accounting for the de-phasing of the backscattering paths due to atomic vibrations of mean-square amplitude σ_j^2, while $e^{-2r_j/\lambda_j(k)}$ is a damping term to account for the effects of the inelastic scattering of the photoelectrons having an associated mean-free-path of λ_j. $f_j(\pi, k)$ are the atomic scattering factors of the jth atom at an electron wavevector amplitude k through a scattering angle π (i.e. 180° backscattering). The term $2kr_j$ is the phase change associated with an electron path from the absorber atom to the jth backscatterer and back to the absorber, while $\psi_j(k)$ is the phase shift associated with the scattering events. Specifically,

$$\psi_j(k) = \phi_j(k) + 2\delta(k), \tag{5.20}$$

where $\phi_j(k)$ is the phase shift experienced by the photoelectron in its 180° backscattering from the jth atom and $\delta(k)$ is the phase shift experienced by this electron as it escapes from the absorbing atom and again as it penetrates back to the centre of this atom.

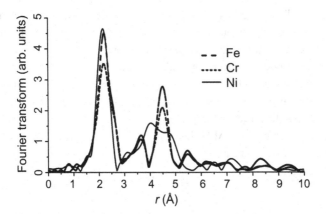

Figure 5.13 Fourier transforms of EXAFS obtained from the Fe and Ni foils as shown in Fig. 5.12, together with similar results from a measurement from a Cr foil. Courtesy Andy Dent, Diamond Light Source

An important feature of this equation for the fine structure function, recognised by Sayers, Stern & Lytle (1971) is that providing that $\psi(k)$ is approximately linear in k (i.e. $\psi_j(k) \approx \alpha k + \delta_0$), a Fourier transform of this measured function leads to peaks at distances $(2r_j + \alpha)$. In practice, this Fourier transform is applied to the measured value of $\chi(k)$ multiplied by some power of k, to offset the damping of the modulations that occurs at higher values of k due to the Debye-Waller factor and to the decrease in $f_j(k, \pi)$ with increasing energy that is characteristic for most atoms.

Fourier transforms of this type are shown in Fig. 5.13 obtained from EXAFS measurements on Fe, Cr ad Ni foils. These are not corrected for scattering phase shifts, so the actual values of the neighbour distances are offset by systematic errors. Nevertheless, Fig. 5.13 shows clearly the effect of the difference in the atomic shell structure beyond the nearest neighbours of the associated bulk structures bulk structures (Fe and Cr are bcc, Ni is fcc). Of course, to extract the true interatomic distances, r_j, from this Fourier transform it is necessary to correct for the scattering phase shifts. One way to do this, exploited in many early experiments, was to compare the results from a sample of an unknown structure with those from a reference sample of a known structure having the same atomic ingredients (emitter and scatterer species). In this case the unknown and reference samples are influenced by the same scattering phase shifts. Indeed, a range of refinements were developed to extract structural information from more distant shells using Fourier filtering and back-transform methods (e.g. Lee et al., 1981). However, these atomic scattering phase shifts can also be calculated rather easily and this approach is now more widely used. Clearly, the ability to extract local structural information *directly* via Fourier methods is attractive. However, more reliable and complete quantitative information is generally extracted nowadays using modelling computer programs that take account of a number of significant refinements not included in the early theory and which also take proper account of the scattering phase shifts.

Two particular refinements incorporated into this modelling approach are to account for the role of 'curved wave' effects and multiple scattering. In the original formulation of the scattering model described above, it was assumed that the atoms are point scatterers and that the photoelectron wavefield scattering off an atom can be assumed to be represented by a plane wave. In reality, of course, the scattering potential of the atoms are not localised at a point, but can be better represented in a solid by a 'muffin tin' potential, while the close proximity of the photoelectron source atom and the scattering atoms means that the curvature of the incoming spherical wavefront must be accounted for. The second original assumption, implicit in the diagram of Fig. 5.11, was that one need only consider single 180° backscattering events, and need not include other electron paths back to the emitter involving scattering by several atoms ('multiple scattering'). In this regard it is well-established that a single scattering description fails badly in the surface structural technique of LEED – Low Energy Electron Diffraction – a technique conceptually similar to X-ray diffraction, but using backscattering of electrons of energy $\sim 50-400$ eV originating from an electron gun outside the solid. Multiple scattering is known to be extremely important in LEED (atomic elastic scattering cross-sections for low energy electrons are many orders of magnitude larger than those for X-rays of a similar wavelength). As EXAFS involves electron backscattering in exactly the same energy range, one might expect multiple scattering to be comparably important. In fact this proves not to be the case; one reason can be attributed to the fact that in EXAFS the electron source is local, decaying in amplitude as r^{-1}. In addition, however, the most important multiple scattering effects (in LEED and EXAFS) are those involving only one single backscattering, but in addition one or more small angle scattering events, for which the scattering cross-sections are large. In EXAFS this means that multiple scattering is only very important in sampling backscattering events from atoms in more distant atomic shells that are aligned with atoms in closer shells.

A particular example of this situation is in crystalline Cu, for which the imaginary component of the Fourier transform of the EXAFS shows a complete inversion of the phase associated with the fourth shell contribution. The explanation of this effect was addressed in the theoretical treatment of Lee & Pendry (1975). It is characteristic of the fcc structure and can also be seen in the imaginary part of the Fourier transform of EXAFS from the Ni foil (Fig. 5.12(a)), presented in Fig. 5.14. Atoms in the fourth shell are exactly aligned with atoms in the first shell, so fourth-shell atoms are illuminated not only by the directly-emitted photoelectrons but also by 0° forward scattering from the first shell atoms. The same effect occurs for the electrons back-scattered to the emitter. The additional scattering phase shifts in the multiple scattering components leads to the inversion in the Fourier transform. Implicit in this early theoretical paper, and explicit in the later theoretical papers of Rehr et al. (1986) and Rehr & Albers (1990), was that the basic EXAFS Equation (5.19) can still take account of curved wave and multiple scattering effects if the atomic scattering factors f are replaced by effective scattering factors f_{eff}. A review of these developments may be found in the paper by Rehr & Albers (2000). The outcome of these theoretical developments and further refinements was a series of different EXAFS modelling

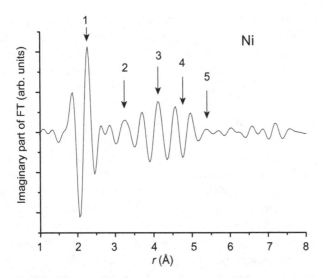

Figure 5.14 Imaginary part of the Fourier Transform of EXAFS from the Ni foil shown in Fig. 5.12(a). The arrows show the expected location of the 1st, 2nd, 3rd, 4th and 5th near-neighbour shells. Data courtesy of Andy Dent, Diamond Light Source

codes that are now exploited to provide far more complete and reliable interpretation of experimental data in terms of atomic distances, shell occupancy and vibrational amplitudes. The most widely used computer codes have been EXCURVE (https://ccpforge.cse.rl.ac.uk/gf/project/excurv/), deriving originally from the paper by Lee and Pendry with subsequent computational development by Gurman, Binstead & Ross (1984, 1986), and FEFF (http://feff.phys.washington.edu/), deriving from the work of Rehr and co-workers, although more user-friendly implementations of FEFF have become most favoured in recent years (notably ARTEMIS https://bruceravel.github.io/demeter/artug/index.html).

The experimental arrangement for standard EXAFS measurements, based on direct measurements of the absorption, are generally based on the use of ion chamber detectors. Different gases and gas pressures can be used to optimise the performance at different photon energies, but the basic set-up has remained largely unchanged for several decades. Because ion chambers have a high transmission probability one can use essentially identical detectors to measure both the incident and the transmitted radiation intensities. Measurements can be taken by stepping the photon energy through the range of interest, from below the adsorption edge to several hundred eV above it. Notice that, particularly to extract reliable results on more distant scattering shells, it is important to measure as wide an energy range as possible. The measurable limit is ultimately determined by the underlying physics, because the amplitude of the EXAFS modulations decays with increasing energy due both to the photoelectron wavelength dependence of the Debye-Waller factor and the fact that backscattering cross-sections become smaller at higher electron energies. This is particularly true for scattering atoms of low atomic number, which have small scattering cross-sections even at quite low energies. EXAFS spectra recorded from materials in which all the local backscattering atoms are of low atomic number (e.g. C, N, O) show very much weaker modulations that those of Fig. 5.12. Obtaining data by this standard step-scan procedure can be quite slow, with total data collections times of as much as tens of minutes. However, in many modern EXAFS studies the objective is to follow structural changes that may occur on much shorter timescales as in, for example, *in operando* investigations of heterogeneous catalysts under reaction conditions. In this situation significant rapid structural changes may occur as a result of small changes in the reaction conditions of temperature and pressure. This has led to the development of a Quick EXAFS or QEXAFS strategy in which the photon energy scan (which requires mechanical movements by stepper motors of monochromator crystals and, in some cases, also of undulator gaps) is effected by continuous, or quasi-continuous, movement. In this mode of operation data collection times can be reduced to a few seconds.

For even faster data collection, a different strategy of Energy Dispersive EXAFS has been developed. Fig. 5.15 is a schematic diagram showing how this is achieved. In particular, the usual double-crystal monochromator used to step the photon energy delivered to the sample is replaced by a 'polychromator' consisting of a bent crystal, illuminated by an angular spread of 'white' (or 'pink') synchrotron radiation. Single Bragg reflections from this crystal occur at different wavelengths depending on the incidence angle, which varies across the crystal. The curvature also leads to these different wavelengths being brought to a focus at a point at which the sample is located. Beyond the sample these different wavelengths are dispersed to produce a spread of wavelengths on a position-sensitive detector. The EXAFS spectrum thus appears directly on this detector with all wavelengths measured simultaneously. The absorption spectrum for the Ni foil shown in Fig. 5.12(a) was recorded in this way. Using this approach EXAFS spectra have been recorded in time-scales as short as a few tens of ns, ideal for following quite rapid changes in *in operando* studies, and for

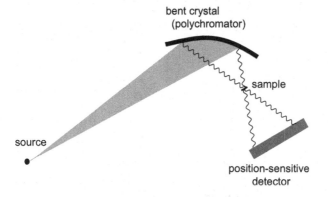

Figure 5.15 Schematic diagram showing the method of implementing energy-dispersive EXAFS measurements. The different wavelengths of X-rays arriving at the detector is shown in an exaggerated fashion at the two ends of the detector.

pump-probe experiments to study photochemical processes. An added advantage of this approach is that for small or inhomogeneous samples there is no problem associated with small physical movements of the incident radiation during an energy scan, such as those that can arise using conventional double-crystal monochromators. A potential disadvantage is that it is not possible to measure the photon energy-dependent incident and absorbed flux simultaneously. However, time-dependent incident flux variations such as those associated with operation of the storage ring source in the 'top-up' mode, do not influence EXAFS spectra collected in the energy dispersive mode. By contrast, in a conventional energy-scanned transmission EXAFS experiment, it is essential to have careful continuous measurement of these changes in incident flux intensity to remove the resulting changes in the transmitted intensity.

While the 'A' in EXAFS stands for 'Absorption', not all EXAFS measurements are made by measuring absorption directly. In particular, if the objective is to determine the structural environment of an atomic species of low concentration in the material, either because it is in a very dilute solution (liquid or solid) or because it is localised only in the outermost few atomic layers of the surface, then a conventional absorption experiment would be unlikely to be able to measure the very weak absorption arising from this small fractional content (superimposed on the much stronger absorption of the majority species in the material). The solution to this problem is to measure not the process of the photon absorption itself, but the energy emission resulting from the refilling of the atomic core hole created in the absorption event. There are two competing processes involved in this core hole decay, namely X-ray fluorescence and Auger electron emission. These two processes are illustrated schematically in Fig. 5.16, which also includes in the same figure the initial photoabsorption/photo-emission process that creates the core hole. A quantum-mechanical branching ratio determines the relative probabilities of these two alternative decay processes, but for

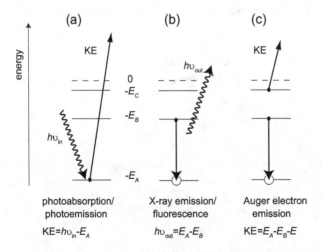

Figure 5.16 Schematic energy level diagram showing the processes of photoabsorption/photoemission, X-ray fluorescence and Auger electron emission.

deep core holes (particularly having binding energies, E_A in excess of ~10 keV), X-ray emission is the dominant decay process, whereas Auger electron emission dominates at low core-hole binding energies.

In practice, though, which detection method is used is commonly dominated by a further consideration. In particular, in the core level energy range in which Auger electron emission is the main decay mechanism the resulting Auger electrons will have kinetic energies of no more than a few keV. The inelastic scattering of these relatively low energy electrons means that the electrons that are detected outside the material at the characteristic kinetic energy of the atomic species will escape only from emission in the near-surface region without energy loss. Auger electron emission detection is thus a probe of the near-surface region alone. This is the basis of Surface EXAFS, or SEXAFS, used to obtain structural information from the near-surface region of a material. Notice that both the photoelectrons and Auger electrons that are emitted above an absorption edge undergo multiple inelastic collisions in the solid, ultimately leading to the emission of many low energy secondary electrons from the surfaces. The escape length of these electrons is much longer than that of the unscattered photoelectrons and Auger electrons, so detecting the intensity of this low energy electron emission actually provides an alternative 'near-bulk' method of determining the EXAFS. To investigate the structural surroundings of a dilute species in the bulk of a material X-ray fluorescence is used, the emitted X-rays being able to escape from much deeper in the material. The energy of the fluorescent X-ray is, of course, also characteristic of the atomic species whose structural surroundings are being investigated, so energy-specific detection is required to obtain element-specific data, typically achieved using a solid-state detector.

Fig. 5.17 shows the results of an EXAFS investigation using fluorescence detection of transition metal dopants in the topological insulator material Bi_2Se_3, by Figueroa et al. (2015). The data in the figure are Cr K-edge EXAFS from a sample containing 4.6% Cr. These experimental data were fitted to optimised models using tools from the IFFEFIT software package (Ravel and Newville, 2005). Based on this analysis the Cr atoms were identified as occupying an octahedral environment of Se atoms, although for this particular sample, having the lowest Cr concentration, two slightly different Cr-Se first shell distances were found. A more complete understanding of these materials was obtained from the combination of this EXAFS result and measurements of the near-edge structure described in the following section.

As remarked above, EXAFS can also be detected by measurements of the variation with photon energy of the intensity of the Auger electrons emitted by the alternative core hole decay mechanism. The strong inelastic scattering in solids of electrons with energies up to a few keV, with associated mean-free-paths mainly in the 5–10 Å range (see Section 6.1), ensures that the resulting data relates only to the properties of the outermost few atomic layers of a solid. This mode of detection is therefore optimal for studies of sub-monolayer adsorption of atoms and molecules at surfaces. A significant challenge in such SEXAFS studies is achieving acceptable levels of signal-to-noise ratios. Several effects conspire to make this a problem. Firstly, the amount of material

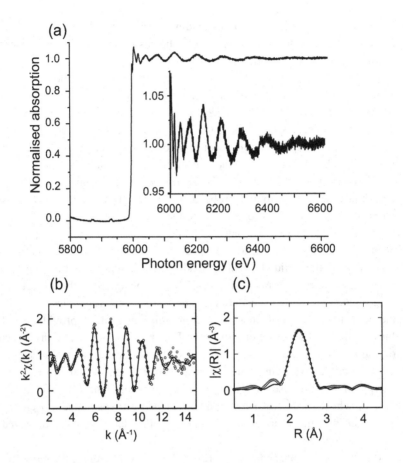

Figure 5.17 Cr K-edge EXAFS from Bi_2Se_3 doped with 4.6% Cr. (a) shows the raw transmission data with the inset covering the range above the edge on a magnified scale. (b) shows the experimental data points of the extracted modulation function (multiplied by k^2) as a function of electron wavevector amplitude k, while (c) shows the resulting modulus of the Fourier transform. Smooth lines superimposed over the experimental data points are the results of structure model fitting. Adapted with permission from Figueroa et al. (2015) (copyright American Chemical Society) from data kindly supplied by the authors

in such a surface layer is very small, so the emitted signal is necessarily weak. In addition, submonolayer coverages of atoms on a surface means that the number of nearest neighbour atoms in the underlying surface is significantly smaller than if these same atoms were embedded in the bulk; this means the EXAFS modulation amplitudes are small. Finally, Auger electron emission is generally superimposed on a much more intense background of inelastically scattered and secondary electrons resulting from the original emission of photoelectrons in the photoabsorption event, so the signal to background ratio is generally poor. Somewhat perversely, it sometimes proves to be more effective to extract the SEXAFS data from measurements of the intensity of this background electron emission; the signal to background ratio of the absorption edge-jump is necessarily worse than that measured by detecting the Auger emission in an energy-selective fashion, but the total signal is much larger, so the resulting signal-to-noise ratio of the EXAFS can prove to be better.

A consequence of these problems is that in SEXAFS it is often only possible to extract reliable nearest-neighbour distances, making identification of the local geometry – the 'adsorption site' – difficult. However, as most of these surface science experiments are conducted on single crystal surfaces, this site information can often be obtained from the dependence of the amplitude of the EXAFS modulations on the direction of the polarisation **A**-vector of the incident (linearly polarised) radiation. In particular, in K-edge EXAFS dipole selection rules ensure that the electrons photoemitted from the 1s state emerge in the form of a p-wave oriented along the direction of the incident radiation **A**-vector. By changing the incident geometry to vary the direction of **A**, this outgoing p-wave can be swept through the surrounding atoms, causing geometry-dependent amplitude variations of the back-scattered amplitude and thus also of the EXAFS modulations. This effect is often referred to as the 'searchlight effect', although a p-wave is a very broad searchlight! Nevertheless, the effect can be used to distinguish different possible near-neighbour geometries. The EXAFS Equation (5.19) is based on the assumption that the local environment of the emitter has no specific orientation in the experiment, as in a liquid, an amorphous sample, a crystalline powder or a polycrystalline solid with no crystallographic 'texture'. For a specific oriented local structure, such as on a single crystal surface, N_j, in Equation (5.19) is replaced by an effective number of neighbours in each shell, N_j^*

$$N_j^* = \sum_{i=1}^{N_i} 3\cos^2\theta_{r,i} = 3N_j\left(\cos^2\theta_p \cos^2\beta_i + 0.5\sin^2\theta_p \sin^2\beta_i\right), \qquad (5.21)$$

where β_j is the angle between the surface normal and the vector linking the emitter to one of the scatterers in the jth shell and θ_p is the angle between the surface normal and the polarisation vector **A** of the incident radiation. Varying the incidence direction thus changes θ_p, resulting in a change in N_j^* and a resultant change in the amplitude of the EXAFS modulations. Measuring this amplitude change for two or more different incidence angles allows one to determine β_j; this angle, combined with the determined nearest neighbour distance, distinguishes alternative adsorption sites.

5.6.2 Near-Edge Absorption Fine Structure

While there is no change in the fundamental physics between the near-edge and extended energy ranges of an X-ray absorption spectrum insofar as both can, in principle, be described in terms of the scattering by surrounding atoms of electrons emitted from atomic core levels, alternative descriptions can prove to be more helpful in understanding the information content of such spectra. In particular, at very low photoelectron kinetic energies it is often more helpful to think of the final state of the excited photoelectron as occupying unoccupied local electronic states of the atom, molecule or solid in which the photo-absorbing atom is situated. The fact that one approach to the theoretical computation of these excited states can be based on multiple scattering provides a formal link to EXAFS.

In general near-edge investigations of molecular species, especially when adsorbed on surfaces, are referred to as NEXAFS. While detailed multiple scattering simulations of such data can be performed, NEXAFS measurements can often be used to gain some quantitative structural information, particular regarding molecular orientations, without the need for such detailed quantitative modelling. In these systems the final state into which the core electron is excited is a discrete unoccupied bound or quasi-bound molecular orbital, also describable as a multiple scattering resonance. As in EXAFS, electric dipole transitions dominate, so if the initial state is of s symmetry the final state must be of p symmetry. For simple diatomic molecules, but also for more complex molecules that are essentially planar, these final states have either π-symmetry, with the p-orbitals perpendicular to either the diatomic molecular axis or the molecular plane, or σ-symmetry, with the p-orbitals lying within this axis or plane. In order to excite these transitions the polarisation vector, **A**, of the incident radiation must have a component parallel to these final p-orbital directions. Thus, for example, to observe a transition into a state of π-symmetry (a so-called π-resonance) of a planar molecule, **A** must have a component perpendicular to this plane; if **A** lies parallel to the plane, no π-resonance will be observed. To observe a σ-resonance **A** must have a component lying parallel to the molecular plane; no such resonance is observed if **A** is perpendicular to the molecular plane. These selection rules therefore provide a relatively simple means of using NEXAFS to determine the orientation of oriented molecules.

Fig. 5.18 shows two examples of experiments exploiting this potential, both involving O K-edge NEXAFS from molecular adsorbates on single crystal metal surfaces. In Fig. 5.18(a) data are shown from the results of an experiment studying the interaction of formic acid, HCOOH, on Cu(110). The molecule loses its acidic H atom on reaction with the surface to produce an adsorbed formate species, HCOO-, which is found to adsorb with its molecular plane perpendicular to the surface and aligned in the $[\bar{1}10]$ azimuthal (close-packed) direction on the surface. This is deduced from the NEXAFS spectra shown, with the sample surface in a vertical plane and incident (horizontally) linearly polarised incident synchrotron radiation at different incidence angles. At normal incidence the **A**-vector of the radiation lies in the surface plane and only excites the π-resonance strongly when it lies in the [001] direction. At normal incidence with **A** in the $[\bar{1}10]$ direction, and at grazing incidence such that **A** is near-perpendicular to the surface, the π-resonance is absent or extremely weak.

In Fig. 5.18(b) is an example of the molecule furan, C_4H_4O, which adsorbs intact on a Pd(111) surface with its molecular plane essentially parallel to the surface. In this case the π-resonance is excited at grazing incidence but not at normal incidence. Notice that in the normal incidence spectrum there is a sharp feature very close to the absorption edge at a slightly different photon energy from that of the furan π-resonance seen at grazing incidence. This feature is assigned to the π-resonance of small amounts of coadsorbed CO resulting from partial

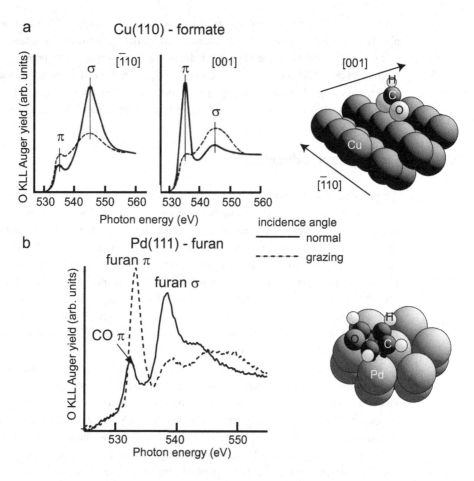

Figure 5.18 Oxygen K-edge NEXAFS from (a) formate, HCOO-, adsorbed on Cu(110) and (b) furan, C$_4$H$_4$O, adsorbed on Pd(111), measured in several different photon incidence directions. Schematic diagrams show the associated adsorption geometries. Spectra in (a) are reprinted from Puschmann et al. (1985), copyright 1985 by the American Physical Society and in (b) from Knight et al. (2008), copyright 2008, with permission from Elsevier

fragmentation of the furan. CO adsorbs on this surface with its molecular axis perpendicular to the surface, its π orbitals thus lying parallel to the surface so that the associated π-resonance is most strongly excited at normal incidence. Similar conclusions may be drawn from the polarisation angle dependence of the intensity of the σ-resonance, but such studies generally rely most heavily on the behaviour of the π-resonance, as these peaks are generally much sharper; it is also easier to extract reliable intensity measurements because π-resonances are very close to the absorption edge. There is some evidence that the exact energy of σ-resonances relative to the absorption edge can provide some measure of intramolecular bondlengths in a molecule through an EXAFS-like mechanism (Stöhr, Sette, and Johnson 1984), but at these low energies an accurate determination of this excitation energy relative to the zero-kinetic energy reference (that lies below the absorption edge) seems to have discouraged further pursuit of this idea.

In inorganic solids, on the other hand, the near edge absorption measurement is usually referred to as XANES (X-ray absorption near-edge spectra) and is regarded as a spectral fingerprint of the electronic bonding environment of the absorbing atom. As such, this is described in Chapter 6 with other techniques to investigate electronic structure.

5.7 Photoelectron Diffraction

As introduced in Section 5.5, an alternative to EXAFS as a way to extract structural information from the elastic scattering interferences of photoelectrons is to detect the photoelectrons directly as they emerge from the solid. However, as remarked in Section 5.6.1 in discussing the SEXAFS technique (and described more fully in Section 6.1), electrons with energies of less than a few keV have a very high probability of inelastic scattering in solids, with associated mean free paths of only ~5–20 Å. This means that photoelectron diffraction is exclusively a technique for investigating the structure of the outermost few atomic layers of a solid, because only those electrons that have not lost energy can contribute to the coherent interference effects that underpin the technique.

As shown in Fig. 5.11, both the EXAFS and photoelectron diffraction techniques exploit the coherent interference between the directly emitted component of the photo-electron wavefield from an emitter atom and components of the same wavefield elastically scattered by the surrounding atoms. In EXAFS the detector of this scattering interference is the emitter atom itself, and changes in the phase relationship between different scattering paths only arise from changes in the photoelectron energy (and thus the photoelectron wavelength). In photoelectron diffraction, however, these scattering interference conditions also depend on the direction in which the photoemission is detected, so structural information can be obtained either from measurements of the photoemission in a particular direction as a function of photon (and thus photoelectron) energy, as in EXAFS, or from measurements of the photoemission as a function of detection direction at fixed photon energy. These two alternatives are commonly referred to as energy-scan photoelectron diffraction (given the acronym PhD) and angle-scan photoelectron diffraction. In the energy-scan mode the recorded data are in the form of photoemission peak intensities as a function of photoelectron energy, which can be reduced to modulation spectra, $\chi(E)$, as in Equation (5.18) for EXAFS. This mode of data collection has been used primarily for structural problems of atoms and molecules adsorbed *on* (i.e. above) a surface, in which all the main scattered compon-ents of the outgoing photoelectron wavefield involve backscattering (i.e. scattering angles > ~90°) from the underlying solid. Because atomic backscattering cross-sections decrease with increasing kinetic energy, these experiments are generally performed at photoelectron kinetic energies of no more than ~300–400 eV. Notice that it is this energy range that shows the strongest modulations in EXAFS for which the dominant scattering events are through an angle of ~180°. In both techniques the modulation amplitudes at higher energies are strongly suppressed by the Debye-Waller effect due to thermal vibrations, as in EXAFS. The angle-scan mode is commonly referred to as XPD (X-ray photoelectron diffraction), although this notation originates from experiments conducted using a standard laboratory source of X-rays, typically leading to measurements being conducted at higher photoelectron kinetic energies (~1 keV). Under these conditions only forward scattering (i.e. scattering angles close to 0°) is strong, such that the angular distribution can be dominated by 'zero order diffraction', namely, scattering involving zero path length difference between the directly emitted and elastically scattered electron wavefields.

Fig. 5.19(a) shows how the angular dependence of the electron scattering cross-section by a Cu atom varies at different photoelectron energies; there is a forward scattering peak at all energies, but at higher energies this peak dominates particularly strongly. Fig. 5.20(b) shows schematically the angular distribution resulting from interference of the directly emitted wavefield at high energy with the component scattered from a single nearby atom. The presence of the zero order scattering peak is energy independent (there is no scattering path difference), so at energies at which this peak dominates little is to be gained from using the variable energy of synchrotron radiation. However, an increasing number of experiments are now performed using

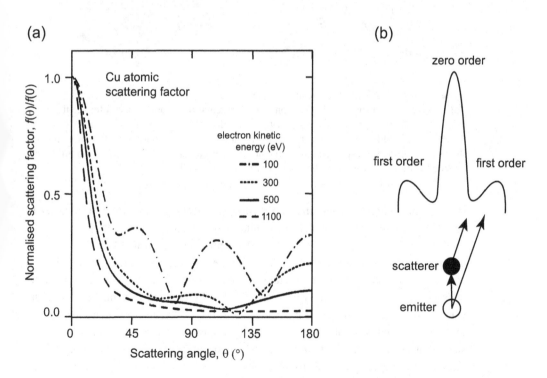

Figure 5.19 (a) shows the angular dependence of the Cu atomic scattering factor for electrons at several different energies, normalised to the value at zero scattering angle. (b) shows schematically the photoelectron diffraction angular distribution resulting from small-angle scattering from a single nearby atom.

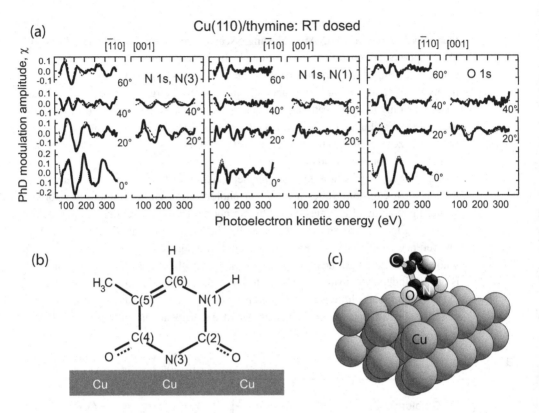

Figure 5.20 Results of an energy-scan photoelectron diffraction study of thymine adsorbed on a Cu(110) surface. (a) shows a comparison of experimental modulation spectra for the three different photoemission peaks studied, with the results of simulations for the best-fit structure. (b) shows a simplified schematic of the adsorption geometry labelling, the different atoms. (c) shows a model of the resulting model structure (with H atoms omitted). After Allegretti, Polcik & Woodruff (2007) copyright 2007, with permission from Elsevier

synchrotron radiation at intermediate photoelectron energies at which it is easier to detect the first order (and higher order) interference peaks.

While the mode of data collection does provide one way of classifying two different forms of the photoelectron diffraction technique, a more meaningful classification is based on the type of structural problem to be addressed. Specifically, is the location of the emitter atom, whose local geometry is to be determined, above the surface or below the surface? If it is above the surface one must exploit backscattering interferences and work at low ($< \sim 400$ eV) photoelectron energies. If it is below the surface it is more likely to be helpful to exploit near-forward scattering which dominates at higher photoelectron energies.

Backscattering photoelectron diffraction, in the energy-scan (PhD) mode generates modulation spectra that are qualitatively similar to those of (S)EXAFS. There are, however, important differences. In particular, PhD modulation spectra differ when the direction of photoelectron detection relative to the crystalline substrate is changed, providing a real-space directional probe of the atomic environment around the emitter atoms. By contrast, EXAFS provides information on the orientationally averaged scattering environment. In effect, SEXAFS is a kind of orientationally averaged PhD. The consequences of this can be appreciated by comparing the EXAFS Equation (5.19), incorporating the scattering phase shift expansion (Equation (5.20)) and polarisation angle dependence of the scattering amplitude (Equation (5.21)) (Woodruff, 2007)

$$\chi(k) = -\sum_{j} \frac{3\cos^2\theta_r f_j(\pi, k)}{k r_j^2} \sin\left[2kr_j + 2\phi_e(k) + \delta_j(\pi, k)\right] W_j(\pi, k) e^{-2r_j/\lambda_j(k)}$$

(5.22)

in which the Debye-Waller factor is now written simply as $W_j(\pi, k)$, with the equivalent expression for PhD assuming (as in the EXAFS equation) only single scattering

$$\chi(k) = -\sum_{j} \frac{f_j(\theta_j, k)}{r_j} \cos\left[kr_j(1 - \cos\theta_j) + \delta_j(\theta_j, k)\right] W_j(\theta_j, k) e^{-L_j/\lambda_j(k)} \quad (5.23)$$

Here θ_j is the scattering angle from the jth scatterer; the path-related phase difference of EXAFS, $2kr_j$ (due to travel to a scatterer and back to the emitter) is replaced by $kr_j(1 - \cos\theta_j)$, while the inelastic scattering damping length of $2r_j$ for EXAFS is replaced by the scattering path length L_j for PhD. A further difference, however, is in the initial amplitude term. The EXAFS equation has an additional factor in the denominator of kr_j, so for a nearest neighbour contribution with r_j of ~2.5 Å and a mid-range k value of ~6 Å^{-1} the EXAFS modulation amplitude is about an order of magnitude less than that of the PhD. This can be attributed to the effect of the angular averaging in EXAFS. Thus, typical PhD modulation amplitudes of ~ $\pm 10-40\%$ compare with those in SEXAFS of ~$\pm 1-3\%$. This, combined with the real-space directionality of the PhD measurement, does make the technique more incisive in the specific case of adsorbate structure determination on oriented single crystal surfaces. However, the method of data analysis of in PhD is more complex than in SEXAFS. In particular, as is clear from Equation (5.23), simple Fourier Transforms, which can provide first-order information in SEXAFS, cannot provide comparatively helpful information in PhD because the relative phase differences introduced in different scattering paths are influenced not only by the scattering pathlengths, but also by scattering phase shifts that depend on the scattering angle as well as the photoelectron energy and the atomic species. A further complication is that multiple scattering (not included in Equation (5.23)) is much more important in PhD than in SEXAFS, further scrambling the scattering pathlength information that might have emerged from a Fourier Transform approach.

Nevertheless, model calculations taking account of these complications can be undertaken to simulate both angle-scan and energy-scan experimental data; by simulating the photoelectron diffraction expected from different structural models and comparing them with the experimental data it is possible to identify well-defined best-fit models. This 'trial-and-error' iterative approach is typical of almost all methods of surface structure determination. Fig. 5.20 shows the results of one such study using the scanned-energy (PhD) mode by Allegretti, Polcik & Woodruff (2007), applied to the interaction of the nucleobase molecule thymine with the Cu(110) surface. Fig. 5.20(a) shows a comparison of the experimental PhD modulation spectra obtained from three different core level photoemission peaks with computationally simulated spectra for the best-fit structure shown in (c), while (b) shows a schematic of the adsorbed molecule with the different constituent atoms labelled. Notice that in panel (c) no H atoms are included in the diagram; H atoms are very weak scatterers and have no core levels, so the technique is 'blind' to the location of these atoms. As shown in Fig. 5.20(c) the molecule comprises a 6-membered ring of four C atoms and two N atoms, with two carboxylic O atoms and one methyl group attachment. In the free molecule both N atoms have a H atom attached, but interaction with the Cu(110) surface leads to deprotonation of the N(3) atom. The identification of both this deprotonation and the way in which the molecule bonds to the metal surface, as shown in Figs. 5.20(b)–(c), is achieved through the results of the PhD experiment, but is aided by exploiting a further feature of the photoelectron energy spectra, described more fully in Chapter 6 but summarised here.

The technique of X-ray photoelectron spectroscopy (XPS), which involves measurement of the photoelectron kinetic energy spectrum emitted from a surface illuminated by monochromatic X-radiation, is a standard surface analysis method performed in home laboratories using standard X-ray sources based on electron bombardment of a target anode (see Section 4.1). Most commonly it is performed with Al K_α or Mg K_α radiation (photon energies 1,486.6 eV and 1,253.6 eV, respectively) but of course synchrotron radiation removes the energy constraint of available atomic emission lines and both much softer and harder X-rays are used for particular reasons, as discussed in Chapter 6. Fuller details of the synchrotron XPS technique are given in Section 6.1.1, but an important feature of the technique, relevant to the present discussion and exploited in this PhD structural study, is the existence of 'chemical shifts' in the measured photoelectron binding energies of core level emission from individual atomic species. The binding energies of core level states in an atom are, of course, characteristic of the elemental species and the XPS technique exploits this to determine surface composition. However, atoms of the same elemental species in different bonding environments do show small differences (in some cases as much as a few eV) in their binding energy as measured by photoemission. It is this chemical shift, and the consequential appearance of two different spectral components in the N 1s photoelectron spectrum from adsorbed thymine on Cu(110), that provide the initial indication that one, and only one, of the two N atoms in the thymine molecule is deprotonated and bonded to the Cu. A further important consequence of this chemical shift is that it makes it possible to collect separate PhD modulation spectra from the two distinct N atoms, as shown in Fig. 5.20(a); this figure also includes spectra recorded from the O 1s emission showing only a single peak, implying that the two O atoms in the adsorbed molecule are symmetrically equivalent. Notice that the PhD modulation spectra from the N(3) and O atoms are very similar, a consequence of the fact that these atoms have very similar local sites relative to the underlying Cu atoms. By contrast, the modulations seen from the N(1) atom are much weaker due to this atom being much further from the strongly backscattering Cu atoms.

Of course, EXAFS also provides a means to determine local structure surrounding an atom in an element-specific fashion, but the chemical-state specificity of photo-electron diffraction is not possible in EXAFS; chemical shifts do occur in the edge-jump energy (as exploited in XANES), but the EXAFS from the coexisting distinct chemically-inequivalent emitters overlap at higher energies and thus cannot be separated. Chemical state specific atomic site determinations can also be achieved using the NIXSW technique described in Chapter 4; an example is given in Fig. 4.21.

As remarked above, somewhat different experimental conditions are favoured when using photoelectron diffraction to determine the structural environment of emitter atoms that lie below the outermost surface atoms. In this case photoelectrons can be forward-scattered by the higher-lying atoms en route to the detector, so by using higher energies one can exploit the intense peak in the atomic scattering cross-section that occurs for scattering angles around $0°$ (Fig. 5.19(a)). At the higher photoelectron energies of ~1 keV, commonly used in XPD experiments based on laboratory X-ray sources, many studies have focused on the zero-order forward scattering peaks (Fig. 5.19(b)) that are seen in measured angular distributions. These identify the near-neighbour interatomic *directions*, but provide no direct information on interatomic distances. However, the angles at which first (and higher) order true diffraction features appear do allow this additional information on interatomic distances to be extracted. Notice that the schematic diagram in Fig. 5.19(b) shows the angular distribution within a single azimuthal plane passing through the emitter-scatterer direction. The integral order diffraction peaks form cylindrically-symmetric arcs about this direction (although the intensity around the arc is influenced by the intrinsic angular distribution of the atomic photoemission). This valuable additional information may be more readily observed at intermediate photoelectron energies at which the forward scattering peak in the atomic scattering factor is broader. While detailed structural deductions can be best obtained at all energies through the use of multiple scattering simulations of the experimental data based on different structural models, there has been longstanding interest in possible computational schemes to invert the experimental data to obtain an 'image' of the real space structure.

In this context, one aspect that has attracted a lot of interest is the recognition that angle-scan photoelectron diffraction data can be regarded as a *photoelectron hologram* (Szöke, 1986). There is a direct analogy with optical holograms, which comprise an interference pattern between a reference wave and a scattered wave from an object; in the photoelectron hologram the 'object' is the atomic surroundings of the emitter atom and the reference wave is the directly-emitted photoelectron wave. This analogy and terminology might lead one to expect that it should be possible to 'reconstruct' an 'image' of the scattering object in a simple fashion using Fourier transforms. Indeed, the first attempts to develop this idea by Barton (Barton 1988, 1991) were based on such a procedure, specifically using Helmholtz-Kirchoff formulae, and applying the method to simulated data for a backscattering mode experiment. The underlying problems of using Fourier transforms to analyse such data have already been mentioned above, and are twofold. Firstly, scattering path phase differences that determine the interferences arise in electron scattering not only from pathlength differences, but also from the atomic scattering events themselves, which are dependent on the atomic species, the scattering angle and the electron energy. Secondly, because at these energies elastic scattering is strong, multiple scattering events also have a significant influence on the detected angular distribution. Neither of these problems arises in optical holography, in which interference phases are exclusively determined by the pathlength differences. While a number of approaches have been proposed to circumvent these problems there do not appear to have been any demonstrations of reliable structure determinations being achieved from data recorded in backscattering geometries by these direct inversion methods. However, an alternative approach to obtaining real-space structural images from such data, recorded in forward-scattering geometries, has been shown to be much more successful by Daimon, Matsushita, Matsui and co-workers (e.g. Daimon, 2018; Matsui et al., 2016; Matsushita et al., 2018). The underlying image extraction technique involves the use of algorithms using maximum entropy and L1-regularised regression together with combinations of two-atom scattering calculations, rather than Fourier transform inversion.

Some of the results from an example of the application of this approach by Tsutui et al. (2017), to identify the local structure around multiple As dopant sites in Si, are shown in Fig. 5.21. The total As concentration in the (001) near-surface region investigated was 0.3 atom %. Wide angle photoelectron angular distributions (photoelectron holograms) of the Si 2p and As 3d photoemission were recorded at kinetic energies of ~600 eV and are presented in the form of grey-scale stereographic projections. Fig. 5.21(a) shows this hologram for the Si 2p emission, while planar sections of the real-space image obtained from these data, shown at different heights relative to the emitting atom, are shown in Figs. 5.21(c–e). The location of the emitter atom is at the centre of the $z = 0$ Å section, Fig. 5.21(c), marked by the superimposed square. Notice that the bulk structure of Si comprises a fcc lattice with an atomic basis of two atoms at relative fractional coordinates of (0, 0, 0) and (0.25, 0.25, 0.25) (shown differently shaded in the structural model of Fig. 5.21(b)). The local structural environment of these two sites (referred to here as the A and B sites) differ for the nearest neighbours and other odd-numbered distant neighbours, but are the same for even-numbered distant neighbours. Because the experimental data consists of an incoherent sum of these two different emitters, this results in odd-numbered distant neighbours appearing as weaker features than those of the even-numbered distant neighbours in the reconstructed images. These images do therefore provide a meaningful representation of the location of Si atoms in the known real-space structure of Si. The As 3d photoemission spectrum displayed in Fig. 5.21(f) clearly shows three distinct chemically-shifted 3d doublet contributions labelled by the authors as BEH, BEM and BEL. These must correspond to As local sites that differ in some way. The chemical-state specific As 3d photoelectron holograms are shown in Figs. 5.21(g–i). For the BEH species the hologram is very similar to that of the Si 2p emitters, while many of the main features of the BEM hologram are also similar, but weaker. The

BEL hologram is almost featureless. The real-space reconstructed images for the BEH and BEM species (shown in the original publication) are qualitatively similar to that of the Si 2p, clearly indicating As occupation of substitutional Si sites, but there are some significant differences. In particular, in the reconstructed images obtained from the BEH data, the features associated with the nearest (1st) neighbour atoms were much weaker, suggesting a distribution of positions for these atoms. This was reconciled by

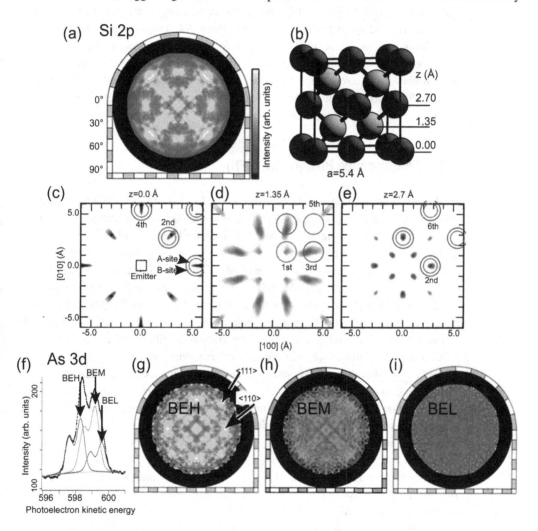

Figure 5.21 Results of a photoelectron holography investigation of the local sites adopted by As dopant atoms in Si. (a) shows the Si 2p emission hologram obtained for the (001) surface, (b) shows the bulk unit cell on Si with the two inequivalent local atomic sites in different shading. (c)–(e) show different sections of the real-space structure obtained from (a). (f) shows the As 3d XP spectrum and the fit to three distinct spin-orbit split components. (g)–(i) show the photoelectron holograms extracted from these three As 3d components. Adapted from Tsutui et al. (2017), https://pubs.acs.org/doi/10.1021/acs.nanolett.7b03467 with permission. Further permission to reproduce this material must be directed to the American Chemical Society

the authors as being due to enhanced local vibrations of these nearest neighbour atoms, a conclusions supported by molecular dynamics calculations. For the BEM images, on the other hand, all the neighbouring atom features were radially elongated, suggesting that in this case it is the As atoms that suffer enhanced fluctuations in their position. This was interpreted as indicating that the BEM As atoms had nearest-neighbour Si vacancies, an effect that could also account for the photoelectron binding energy chemical shift. For the BEL As atoms no meaningful real space images could be obtained, leading to the conclusion that these As atoms occur in disordered or amorphous regions, or occupy a multiplicity of different sites. These conclusions could be reconciled with complementary information regarding As-doped Si from other studies by different methods. Notice, though, that the chemical-state specificity of the information obtained contrasts with the more averaged structural information provided by the EXAFS technique.

5.8   X-Ray Fluorescence Holography (XFH)

As described above, photoelectron diffraction is intrinsically a surface technique. Its strength for determining surface structure is that the data that emerge are confined to the outermost few atomic layers, but this necessarily means that it is no use for determining bulk, or even significantly sub-surface, structure. Structural data for bulk and thin film materials can, however, be obtained from the diffraction and holography of X-ray fluorescence, methods that rely on essentially the same local elastic scattering interferences as photoelectron diffraction, but for emitted X-ray fluorescence photons instead of photoelectrons. When a core hole is created in an atom in a solid, the hole can be refilled by the emission of an X-ray photon, as illustrated in Fig. 5.16. This process, resulting from core hole creation by energetic incident electrons, is how a conventional laboratory X-ray source gives rise to monoenergetic 'characteristic' X-rays, as described in Section 4.1. The same decay process occurs if the incident core ionisation is achieved by incident X-rays, in which case the X-ray emission is referred to as fluorescence. These emitted photons can be elastically scattered by the atoms surrounding the emitter atom, and the coherent interference of the directly emitted and elastically scattered components of the photon wavefield gives rise to variations in the angular distribution of the emitted X-ray photons (see Fig. 5.22(a)). As first also pointed out by Szöke (1986), this leads to an X-ray fluorescence hologram, akin to the photoelectron hologram. One important difference is that X-rays can penetrate (and escape from) solids over much longer distances than electrons of a similar energy, so this measurement leads to an essentially bulk technique. Moreover, because X-ray scattering does not involve significant angle-dependent scattering phase shifts, and is sufficiently weak that multiple scattering is not significant, Fourier transform methods have the potential to be effective in inverting the hologram to reconstruct a real space image. However, the fact that the scattering is weak means that the intensity modulations displayed in the hologram are very weak – typically ~0.1%. As a result, the first experimental demonstration of the effect by Tegze & Faigel (1996) imaging the Sr atoms in $SrTiO_3$ using monochromated Mo K_α radiation from a standard laboratory X-ray, required almost 2 months of acquisition time. Image reconstruction was achieved using the Fourier transform-based method of Barton originally developed for photoelectron holography (Barton, 1991).

One feature of this inversion approach, pointed out by Barton, is that for data recorded at a single energy the resulting images contain spurious twin features, whereas if holograms recorded at multiple energies can be used, this problem is removed and the image resolution improves. However, when using the scattering of the X-ray fluorescence of a single atomic species, the emitted X-ray energy is fixed by the core level binding energy. To overcome this problem, Gog et al. (1996) proposed, and demonstrated, an alternative inverse XFH technique based on the idea of time reversal of X-ray fluorescence holography, to image the local structure of Fe atoms in haematite (Fe_2O_3), detecting the ('total') Fe K_α fluorescence. Specifically, instead of measuring the angular dependence of emitted fluorescent X-rays, one can measure the (nominally angle integrated) fluorescence as a function of the incident direction of the illuminating X-rays (Fig. 5.22(b)). This approach can readily be achieved at multiple incident energies using synchrotron radiation.

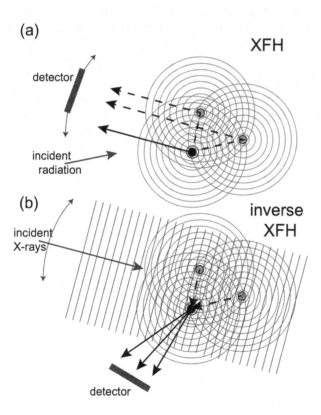

Figure 5.22 Schematic diagram of (a) XFH and (b) inverse XFH. The X-ray emitter/absorber atom is shown as a dark-shaded disc, two scatterer atoms are shown as lighter shaded discs.

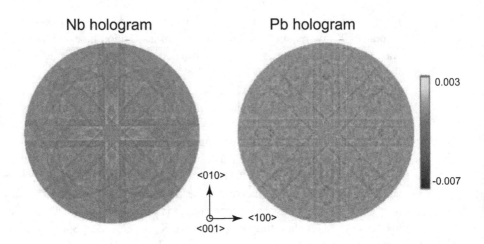

Figure 5.23 X-ray fluorescence holograms, obtained in the inverse mode, from the Nb and Pb sites in the relaxor ferroelectric PMN. Reprinted with permission from Hu et al. (2014). Copyright (2014) by the American Physical Society

An example of X-ray fluorescence holograms, obtained in the inverse mode, from the Nb and Pb sites in the prototypical relaxor ferroelectric $Pb(Mg_{1/3}Nb_{2/3})O_3$ (PMN) measured by Hu et al. (2014), was used to identify the atomic surroundings of the Pb^{2+} and Nb^{5+} ions in this material. Data were collected using incident X-ray energies of 19.0–23.5 keV in 0.25 keV steps, but Fig. 5.23 shows the single holograms collected at an energy of 20.0 keV. Notice that the grey-scale range of the amplitude modulations is from -0.7% to $+0.3\%$.

6   Probes of Electronic Structure

6.1   Photoemission

Photoemission provides a particularly direct probe of the occupied electronic states of atoms, molecules and condensed matter. The discovery of this effect by Albert Einstein (1905) led to the award of the Nobel Prize in Physics in 1921 (a less controversial discovery than his work on relativity). The key conclusion of this work is that incident radiation can be described as discrete energy packages (photons) with an energy $h\nu$, and these can give up all of their energy to an electron in a bound state with a binding energy E_B; this leads to the ejection of the electron into the vacuum with a kinetic energy $KE = h\nu - E_B$. Fig. 6.1 shows a simple energy-level diagram representation of this basic process. Bound-state electrons, be they in the occupied valence band (shown in the figure as a shaded region, with the Fermi level E_F below the vacuum level E_{vac}), or core levels, E_{B1} etc., are displaced up in energy by an amount $h\nu$. This gives rise to a photoelectron energy spectrum in which the energies of the peaks allow identification of the binding energies of the occupied bound states. It is this photoelectron kinetic energy spectrum that is measured in a photoemission experiment.

In traditional photoemission experiments performed using laboratory sources of photons the photoemission technique is typically divided into two distinct energy ranges. Using lower energy ultraviolet photons, most commonly He discharge line sources with photon energies of 21.2 eV (He I) and 40.8 eV (He II) the technique is known as UPS (Ultraviolet Photoelectron Spectroscopy). At higher energies, particularly using Al K_α ($h\nu = 1,486.6$ eV) and Mg K_α ($h\nu = 1,256.6$ eV) radiation, the technique is known as XPS (X-ray photoelectron spectroscopy). Evidently, UPS can only access relatively shallowly bound valence states, while XPS can produce photoemission from more strongly bound core levels, so the traditional emphasis of the two laboratory-based techniques is on these two distinct binding energy ranges. However, the availability of a continuous range of photon energies from the ultra-violet through soft X-rays to hard X-rays has not only blurred these boundaries, but also opened up a range of important new capabilities. Nevertheless, the objectives of such studies remain focused on either core levels or valence levels, so this division remains an appropriate way to discuss two essentially distinct techniques.

In photoemission studies of condensed matter it is important to recognize that peaks in the emitted electron energy spectrum arise only from the near-surface region of the material, because the mean-free-path for inelastic scattering is very short. Indeed, the

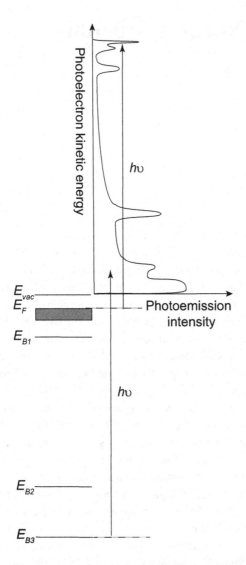

Figure 6.1 Schematic energy level diagram showing how the photoemission experiment simply shifts all the bound state energies of a system up by the photon energy $h\nu$, to produce a photoelectron kinetic energy spectrum. Notice that the peaks in this spectrum lie on a background of inelastically-scattered and secondary electrons.

depth sampled in a photoemission experiment is also influenced by strong elastic scattering of the photoelectrons in the material, because elastic scattering by atoms as a photoelectron travels to the surface from the sub-surface lengthens the escape path and thus increases the probability of inelastic scattering. Fig. 6.2 shows the results of experiments undertaken by Seah & Dench (1979) to determine the resulting attenuation length for a wide range of materials. The significant scatter of the data is a consequence of this broad range of materials, but clearly shows a general trend with

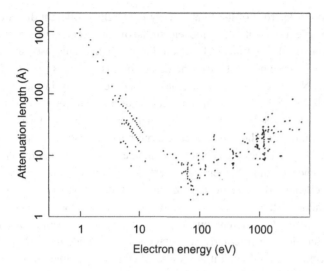

Figure 6.2 Measured values of the attenuation length due to inelastic scattering as a function of electron energy measured by thin film deposition of a wide range of materials onto substrates with different characteristic electron emission energies. After Seah & Dench (1979). Copyright (1979) Heyden & Son Ltd.

typical attenuation lengths of ~5–10 Å for photoelectrons in the energy range ~50–1,000 eV. The main mechanisms of the inelastic scattering are electron-hole pair creation and electron-plasmon scattering; the steep rise in the attenuation length at much lower energies is a consequence of the threshold energy for plasmon creation being ~20 eV. The key conclusion to be drawn from the data of Fig. 6.2 is that most detected photoemission from solids and liquids arises from the outermost few (~2–5) atom layers at the surface. This is clearly an advantage if one is primarily interested in the specific properties of the near-surface region, but does have implications for the extent to which measurements may be characteristic of the underlying bulk.

One other general issue to address, before discussing the details of the different ways in which synchrotron radiation photoemission is used, is that of detectors. In earlier chapters some of the key developments in X-ray detectors have been briefly described, with an increasing emphasis on detectors that can detect signals in an intrinsically digital fashion with a capability to detect multiple 'channels' in parallel, specifically in many different scattering directions through the use of two-dimensional 'image' detectors. There have also been somewhat similar developments in parallel detection for photoelectrons. As implied by Fig. 6.1, the key measurement in a photoemission experiment is a spectrum displaying the number of electrons, emitted in a particular direction or range of directions, as a function of their kinetic energy. Much the most common way to achieve this is using an electrostatic dispersive analyser, in which the electrons are passed between two electrodes that deflect the electrons by different amounts depending on their kinetic energy. A number of different electrode geometries can be used that all have some focusing properties, namely, that allow electrons of the same energy but slightly different directions to be focused at a particular position, while electrons of other energies are focused at different positions in the exit plane. In much the most commonly used design, these electrodes are concentric hemispheres. The design of this device is shown schematically in Fig. 6.3. Notice that the focusing of the analyser is only in the plane of the section shown in Fig. 6.3(a). Electrons emitted out of this plane arrive at the exit also out of this plane. However, by placing a two-dimensional detector at the exit (typically a channel plate multiplier followed by a 2D charge-pulse detector conceptually similar to many 2D X-ray detectors) one obtains an 'image' that represents the energy and angular dependence of the photoemitted electrons within a certain range of each parameter. Specifically, displacements of the arriving electrons within the dispersive plane correspond to different electron energies, while displacements perpendicular to this plane correspond to different out-of-plane angles of emission. This mode of parallel data collection is particularly advantageous for valence band photoemission, as shown in Section 6.3, although it is also beneficial, for example, in the photoelectron diffraction technique described in Chapter 5.

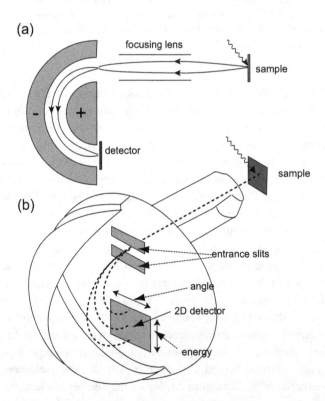

Figure 6.3 Schematic diagram of a concentric hemispherical electron energy analyser as used in photoemission experiments. (a) shows the basic device in cross-section, including one set of single-energy electron trajectories passing through the focusing lens and the dispersive analyser. (b) shows a cut-away view of the same device including a two-dimensional detector to record simultaneously electrons emitted in a range of emitted energies and angles.

6.1.1　Core Level Photoemission

Because the binding energy of electrons in atomic core states is characteristic of the elemental species, a widely exploited aspect of XPS is the determination of the elemental composition of the near-surface region of materials. In this regard, the ability to vary the incident photon energy over a wide range, offered by synchrotron radiation, allows one to vary the degree of surface specificity of the analysis. By choosing a photon energy that leads to photoelectron energies of ~50 eV, the surface specificity can be enhanced. By contrast, if much higher photon energies of 10 keV or more are used, the sampling depth is increased sufficiently that it is possible to obtain compositional information on shallowly buried interfaces.

Despite this utility of core level XPS to analyse near-surface composition, the earliest use of the technique was not to study surfaces, but to obtain 'chemical' or electronic state information from what was effectively assumed to be 'bulk' material. Specifically, the technique, using laboratory X-ray sources, was known as ESCA – Electron Spectroscopy for Chemical Analysis, introduced by Kai Siegbahn (for which he was awarded the Nobel Prize in Physics in 1981). Here, 'Chemical' means 'chemical state' rather than elemental identification. In particular, while core level binding energies are characteristic of the elemental species, small variations in the binding energy measured in photoemission arise from atoms of the same element in different charge states. This effect is also seen in the XANES technique described in Chapter 5. However, to understand this, and its implications, in more detail, it is important to recognise that photoemission, like all spectroscopic techniques, does not measure a ground state energy, but rather the energy difference between an initial state (which is this case is the ground state) and an excited state – in this case a core hole and an electron above the vacuum level. The 'chemical' shift in the measured photoelectron binding energy, defined as the difference between the incident photon energy and the detected electron kinetic energy, is influenced by electronic or 'chemical' effects in both the initial and final state. What is *not* measured is the one- electron ground state binding energy of the state, a quantity often referred to in XPS parlance as the Koopman's energy.

This 'chemical shift' is often assumed to be entirely due to the influence of different electronic environments on the initial state binding energy and on this basis is commonly attributed to the charge state of the photoemitting atom in condensed matter and, indeed, in free molecules. For example, in stoichiometric titania (TiO_2) the titanium atoms are generally regarded as being in a Ti^{4+} charge state, but in other Ti-O and Ti-OH bonding situations lower charge states are believed to occur, while in titanium metal one has neutral Ti^0 atoms. XPS of materials in which a range of different Ti-O bonding occurs can be used to identify the fraction of Ti in each charge state. Identification of the chemical shifts of the different charge states is based on reference spectra recorded from different materials in which one of these charge states is widely accepted as being dominant. In practice some variation of these photoelectron binding energies are found in different studies, but if multiple chemical shifts are present their relative values are rather reliable. As an example, Fig. 6.4 shows the

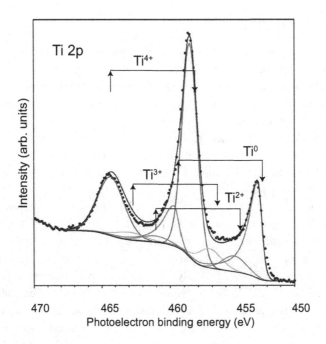

Figure 6.4 Ti 2p XP spectra recorded from a sample of heat-treated Ti-apatite. The black curve is a fit to the experimental data points based on a sum of 4 sets of doublet ($2p_{1/2}$ and $2p_{3/2}$) components, shown as grey lines, with associated Ti charge states as shown by the pairs of arrows. After Biesinger et al. (2010), copyright (2010) with permission from Elsevier

results of an application of this approach to a sample of heat-treated Ti-apatite (a modified calcium phosphate mineral containing Ti in different oxide and hydroxide states). XP spectra from Ti 2p core levels comprise spin-orbit split doublets corresponding to emission from the $2p_{3/2}$ and $2p_{1/2}$ states with an intensity ratio corresponding to the electron occupation ratio of 2:1. This energetic splitting is constant for different chemically-shifted states, so the spectrum from this material comprises a set of such doublets, identified by the pairs of arrows. Analysis of this spectrum allows the relative concentrations of the different charge states to be determined.

Early investigations of these 'chemical shifts' in XPS from a range of different simple molecules in the gas phase showed that in these systems there is essentially perfect agreement between the observed *relative* shifts in the photoelectron binding energy and simple calculated one-electron binding energies. This is clearly shown in Fig. 6.5 which presents a comparison of measured and calculated C 1s binding energies ranging from a simple pure hydrocarbon with no double bonds (C_2H_6) to the fully fluorinated CF_4 in which very significant charge transfer occurs to the highly electronegative F atoms. However, while this strong correlation between theory and experiment is illustrated by the straight line fit with a gradient of unity, there is a consistent shift of 15 eV between computed and measured energies. The origin of this shift is intra-atomic relaxation – a final state effect. While the initial state is a ground state atom or ion (depending on the nature of the interatomic bonding in the material

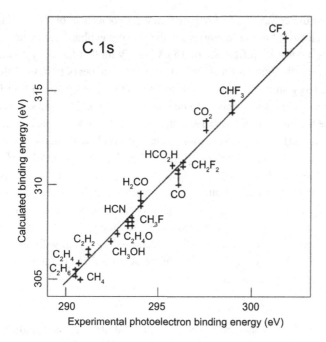

Figure 6.5 Comparison of experimental XPS C 1s binding energies with calculated ground state energies for C in a range of molecules adapted from data presented by Shirley (1973). The systematic comparison is excellent as indicated by the straight line of unity gradient but experimental and theoretical values differ consistently by 15 eV.

as described above), in the final state a core hole appears while the electron from this state is emitted into the continuum above the vacuum level. While the core hole is eventually filled by an electron from a more weakly bound state leading to the release of a fluorescent X-ray or an Auger electron, at the time of the photoelectron emission the more weakly bound electrons of the atom now effectively 'see' a screened nucleus with one extra positive electron charge, and therefore 'relax' to more strongly bound states. In effect, the removal of the core electron means that the effective nuclear charge experienced by the weakly bound states, occupied by electrons more distant from the nucleus, is increased from Z to $Z + 1$, in which the binding energy of these states is larger. This reduction in energy of the core-ionised atom thus makes extra energy available to the escaping electron, leading to a reduction in the measured photoelectron binding energy.

In the absence of this final state relaxation the photoelectron kinetic energy would be

$$E_{kin} = h\upsilon - E_B, \tag{6.1}$$

where E_B is the ground state one-electron binding energy (the Koopman's energy) but as a result of this final state relaxation the actual measured photoelectron kinetic energy is

$$E_{kin} = h\upsilon - E_B + E_a. \tag{6.2}$$

Notice that in the comparison of experimental photoelectron binding energies and calculated one-electron binding energies for the set of gas-phase molecules shown in Fig. 6.4 there is a *constant* difference of 15 eV, implying that this may be only intra-atomic relaxation – the bonding to H, N. O or F atoms appears to have little, if any, effect on this energy shift. For this reason the relaxation energy shift E_a in Equation (6.2) has been given the subscript 'a' to denote an intra-atomic shift. Were this to be universally true the origin of the observed chemical shifts in XPS would, indeed, be a purely initial state effect. However, more generally, and particularly for studies of metallic systems, including adsorbed atoms and molecules on metal surfaces, and interfaces with metals, there is also an *interatomic* relaxation effect and associated shift, E_r. This arises because, despite the intra-atomic relaxation, the creation of the core hole by the photoemission process still leaves the initial atom or ion with an excess of positive charge, and electrons from adjacent bonding atoms, and particularly from the large density of mobile conduction electrons in a metal, will move to screen the core hole and thereby increase the magnitude of their binding energy. The measured photoelectron kinetic energy is then

$$E_{kin} = h\upsilon - E_B + E_a + E_r. \qquad (6.3)$$

The experimentally measured shift in the photoelectron binding energy $(E_{kin} - h\upsilon)$ is thus a sum of the change in the one-electron binding energy of the initial state, E_B, due to a change in the chemical or electronic environment, and in the interatomic relaxation energy, E_r. There is no entirely reliable way of separating these initial and final state shift contributions experimentally, although estimates based on certain assumptions can be obtained from comparisons of the measured values of the photoelectron binding energy and the kinetic energy of Auger electrons associated with refilling of the core hole (see, e.g., Gaarenstroom and Winograd, 1977; Moretti, 1998). For most purposes the measured chemical shift is used simply as a spectral fingerprint, although it is now possible using some modern density functional theory (DFT) computer codes to calculate the predicted values of these shifts for different bonding environments as a means of extracting the full infor-mation content of these measurements.

Notice that this discussion effectively assumes that the core ionised atom has sufficient time to fully relax before the photoemitted electron becomes distant from the atom; this is known as the adiabatic approximation. In practice, the photoelectron may leave too quickly to accommodate all the excess energy provided by this relaxation, and may leave the ion in an excited state. This leads to satellite features in XP spectra with reduced kinetic energy, generally referred to as 'shake-up' (or, in the case of double ionisation, 'shake off'). The energies and intensities of these features also provide information on the electronic environment of the emitting atom.

While the data of Figs. 6.4 and 6.5 were obtained using laboratory sources of X-rays, there are major advantages in the use of synchrotron radiation that are now widely exploited. In particular, the ability to vary widely and continuously the photon energy allows one to identify the optimal conditions for photoionisation cross-section, spectral resolution and (for studies of surfaces) surface specificity. As shown in

Fig. 6.1 the inelastic scattering mean free path for electrons in solids is shortest for electrons in the energy range around ~50 eV, and tunability of the photon energy allows this photoelectron kinetic energy to be selected to minimise the depth from which the recorded XP spectra originate. More generally significant is the potential to achieve higher spectral resolution, particularly important in the study of subtle chemical shifts. In part, this arises from the beamline monochromators that can typically achieve much better resolution than the intrinsic line-width of characteristic K_α radiation from laboratory sources, while it is far easier and more efficient to monochromate the very narrow core of synchrotron radiation than the wide angular distribution of radiation from a conventional electron-impact X-ray source. As monochromators typically have an approximately constant resolving power $(E/\Delta E)$ one can also select a low photon energy (close to photoionisation threshold) to achieve the best resolution (ΔE). In addition, however, choosing a low photoelectron kinetic energy allows one to optimise the energy resolution of the detected photoelectrons, a dispersive electron energy analyser also having a constant resolving power.

One example of the benefits of this improved resolution in studies of surfaces is shown in Fig. 6.6, taken from an investigation by Smedh et al. (2001) of CO adsorption on a Rh(111) surface. Fig. 6.6 shows XP spectra recorded in the photoelectron energy range of the C 1s emission from two different coverages of the molecule. At low coverages all the CO molecules occupy adsorption sites directly atop surface Rh atoms, while at higher coverages co-occupation of three-fold

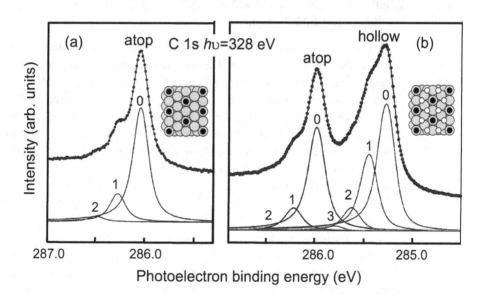

Figure 6.6 C 1s XP spectra recorded from CO adsorbed on Rh(111) at (a) low and (b) high coverage. At low coverage all CO molecules occupy atop sites (shown in black in the inset model of (a), while at high coverage CO molecules in hollow sites (open circles in the inset of (b)) are also occupied. After Smedh et al. (2001), copyright (2001) with permission from Elsevier

coordinated hollow sites occurs. The inset diagrams show schematic plan views of the surface with atop site CO molecules shown in black and hollow site CO molecules shown as open circles. The XP spectra clearly show a C 1s chemical shift of approximately 0.75 eV between these two states. In addition, however, it is clear that the spectral features from each state contain more than a single peak. Specifically, as shown in the figure, the atop peak can be fitted by 3 components while the hollow peak has 4 components. These satellite features are due to vibrational excitations to one or more excited states above the ground state, the number of excitations being shown by the number labels in the figure; each is associated with excitation of the C-O stretching vibrational mode. As is well known from more conventional vibrational spectroscopies of adsorbed molecules (reflection-absorption infrared spectroscopy – RAIRS and high-resolution electron energy loss spectroscopy – HREELS) the energy associated with this mode differs for singly and multiply coordinated adsorption sites; this energy difference is also found in fitting these XP spectra. However, because these are vibrational excitations of a CO molecule in which the C atoms have a core hole, leading to reduced screening of the nucleus due to this missing core electron, the vibrational energies correspond closely to those of adsorbed NO on this surface, rather than CO; this is also a manifestation of a final state effect, in which the C atom of nuclear charge Z^+ $(Z = 6)$ is effectively replaced by the adjacent atom in the Periodic Table with a nuclear charge of $(Z + 1)^+$.

6.1.2 Valence State Photoemission and ARPES (Angle-Resolved Photoelectron Spectroscopy)

Evidently photoemission from shallowly-bound valence states offers a more direct probe of the electronic structure that underlies a range of electronic properties of materials and their exploitation, not only in semiconductor, metal, insulator and superconducting-based devices, but also in the chemistry of surfaces and interfaces. As the states of interest mostly have binding energies of no more than ~10 eV, these can be readily accessed by photoemission using photon energies of no more than a few tens of eV. As remarked in the introduction to this section, typical laboratory-based UPS experiments use gas-discharge line sources of radiation in the ~20–40 eV range. However, the wider range of accessible energies at high resolution, and the controlled polarisation offered by synchrotron radiation prove to offer important advantages for these measurements.

A key consideration of particular relevance to studies of crystalline solids is that in photoemission one must not only conserve energy (as given by the Einstein relationship described above), but also momentum. This is an important issue because the momentum of an electron at a given energy is vastly larger than the momentum of a photon of the same energy. In most cases it is sufficient to regard the photon momentum as being zero when establishing momentum conservation in photoemission. It is for this reason that a free electron cannot absorb a photon; electron-photon energy exchange with a free electron can only occur through Compton scattering (see Section 6.3.3). Photoemission from an electron bound to a free atom or molecule,

(a)   (b)

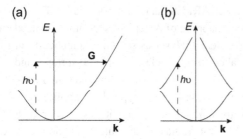

Figure 6.7 Energy-momentum diagram showing the conditions for energy and momentum conservation for photoemission from a nearly free electron solid in the (a) extended and (b) reduced zone scheme. **G** is a reciprocal lattice vector, **k**, is the electron momentum.

however, can occur because the photoelectron momentum can be balanced by recoil of the much more massive atom with minimal implications for the energies. This recoil momentum can take any value required to ensure conservation of momentum. In a crystalline solid, however, the periodic structure means that the momentum recoil can only have discrete values, equal to the reciprocal lattice vectors of the material. This is illustrated in Fig. 6.7 for the simplest case of a nearly free electron solid. Notice that in the reduced (Brillouin) zone scheme, photoemission is seen as a 'vertical' or 'direct' (**k**-conserving) transition, like all optical transitions. This important conservation law can be exploited to allow experimental 'band mapping' by measuring both the energy and the emission direction of the photoelectrons using angle-resolved photoelectron spectroscopy (ARPES).

Strictly, this argument is only true for a three-dimensionally periodic material. In practice, the short attenuation length (Fig. 6.2) experienced by the emitted photoelectrons means that only a thin near-surface layer of the solid is sampled in an experiment, and formally this 'surface slab' is only two-dimensionally periodic, parallel to the surface. Atomic layers at different depths do not contribute equivalently, so the system is not truly periodic perpendicular to the surface, even if the interlayer atomic spacings remain identical to those in the underlying bulk. In this case the formal selection rule is conservation only of the component of the electron momentum parallel to the surface, $k_{\parallel}$.

This does, of course, provide all the relevant momentum information for systems that are intrinsically two-dimensional (2D) in character, such as surface states (electronic states localised at the surface) and 2D materials like graphene. Providing that the photoemission final state can be assumed to be free-electron like, a measurement of the kinetic energy E_{kin} (in eV) and emission angle of photoelectrons relative the surface normal, θ, allows the final state value of $k_{\parallel}$ (in Å^{-1}) to be calculated

$$k_{\parallel} = 0.5123 E_{kin}^{1/2} \sin \theta. \tag{6.4}$$

If this value falls outside the first surface Brillouin zone knowledge of the surface reciprocal net vectors allows one to determine the appropriate value of the initial state within this zone (as shown in Fig. 6.7).

Fig. 6.8 shows two examples of ARPES data from essentially 2D systems. In both cases what is shown is the direct mapping of occupied electronic bands taken from the 2D detector of a concentric spherical analyser as shown schematically in Fig. 6.3(b). Fig. 6.8(a) displays results taken from a clean Au(111) surface and shows the dispersion of the intrinsic surface state on this surface; this is a state in which electrons can move freely parallel to the surface with typical parabolic free-electron-like dispersion, but cannot couple to bulk states because it lies in a projected band gap of the bulk band structure. As a metal Au has no absolute band gap at the Fermi level, but there is a band gap in the direction perpendicular to the (111) surface. The clear evidence of splitting into two distinct states is a consequence of spin-orbit coupling, an effect made possible by the absence of inversion symmetry at the surface (LaShell, McDougall and Jensen, 1996). The results presented in Fig. 6.8(b) are from bilayer graphene grown on SiC(0001) and show the classic 'Dirac cone' of a linearly dispersing band around the K point of the Brillouin zone; the exact energy, E_{Dirac}, of the tip of this cone depends on the extent of interactions of the graphene with the substrate or with dopants.

Both of these examples correspond to 2D localised states, for which $k_{\parallel}$ is the only component of the electron momentum influencing the binding energy. The localisation perpendicular to the surface means that there can be no energy dispersion with

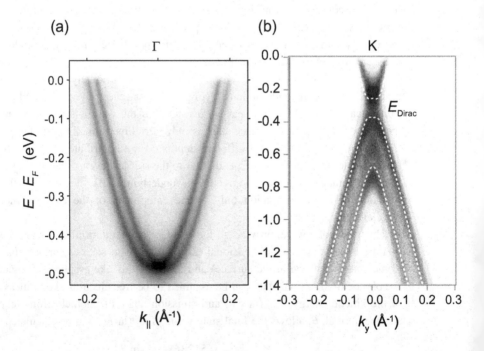

Figure 6.8 Results of ARPES band mapping from (a) the surface state on clean Au(111) (courtesy Phil King, University of St Andrews) and (b) bilayer graphene grown on SiC(0001). After Razado-Colombo et al. (2018) under a CC BY 4.0 licence (https://creativecommons.org/licenses/by/4.0)

variations in $k_\perp$. For bulk electronic states, of course, this is not true, and while formally only $k_\parallel$ is conserved in these photoemission experiments, the sampling depth of the experiments is sufficient for a strong vestige of periodicity perpendicular to the surface to be evident, leading to significant evidence of $k_\perp$ conservation. To access different values of $k_\perp$ in the emitted photoelectrons at any particular emission angle it is necessary to change the photoelectron energy, and thus also the photon energy. If a photoemission peak appears at a constant binding energy as the photon energy is changed the initial state must be localised perpendicular to the surface. If this is not the case, the binding energy change with $k_\perp$ allows one to map the dispersion of such a bulk state. To exploit this potential, the wide-range tunability of monochromated synchrotron radiation is required.

Fig. 6.9 shows the results of an experiment on the transition metal dichalcogenide WSe_2 undertaken by Riley et al. (2014) that exploits this potential and shows coexistence of both 2D and 3D electronic states. Fig. 6.9(a) shows the real-space structure of this layered semiconductor material, while Fig. 6.9(b) shows the reciprocal space three-dimensional Brillouin zone with the different symmetry points labelled. The symmetry points labelled with a bar over the symbol (e.g. $\bar{\Gamma}$) correspond to the two-dimensional zone in which energies can be mapped as a function of $k_\parallel$ alone. Fig. 6.9 (c) shows an 'image' recorded from the 2D detector of the electron energy analyser equivalent to those shown in Fig. 6.8, at an incident photon energy of 120 eV. Notice that while the data from the 2D states shown in Fig. 6.8 have intensity maxima confined to distinct dispersing bands, in Fig. 6.9(c) there are also regions of more distributed intensity, particularly around $\bar{\Gamma}$. This arises because some of the 3D electronic states of WSe_2 disperse in $k_\perp$. For a fixed photon energy, varying the detected emission angle θ changes not only $k_\parallel$, but also $k_\perp$. Thus, Equation (6.4) gives the value of $k_\parallel$, while the value of the final state $k_\perp$ *inside* the crystal is given by

$$k_\perp = 0.5123 \sqrt{(E_{kin} \cos^2\theta + V_0)}, \tag{6.5}$$

V_0 being the *inner potential* (i.e. the difference in kinetic energy of the electrons inside and outside the crystal), which is usually taken to correspond to the sum of the Fermi energy (the energy of the Fermi level above the bottom of the valence band) and the work function (the energy of the vacuum level above the Fermi level). Strictly, therefore, the detector 'images' of Figs. 6.8 and 6.9 correspond to simultaneous variation of both $k_\parallel$ and $k_\perp$ according to Equations (6.4) and (6.5), but in the case of the 2D systems of the Au surface state and the graphene layers there is no dispersion in $k_\perp$ so there is no resulting energy shift. For a 3D state this is not true, and the results do show a vestige of $k_\perp$ conservation, but because ARPES only samples a shallow depth range of the sample $k_\perp$ conservation is 'smeared', and photoemission intensity from a small range of $k_\perp$ values is seen in the detected signal at each $k_\parallel$ value. In order to extract the true $k_\perp$ dispersion of the 3D bands it is therefore necessary to obtain data like the image of Fig. 6.9(c) at many different photon energies, providing a 3D data set of initial state binding energies corresponding to photoemission peaks as a function of both polar emission angle and photon energy. Assuming a free-electron final state (as in Equations (6.4) and (6.5)) it is then

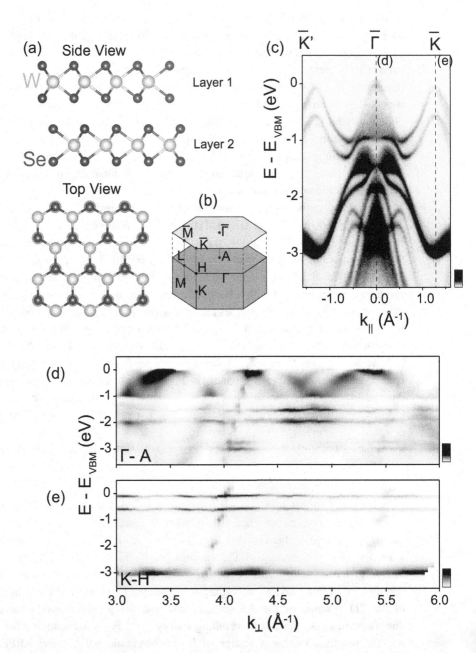

Figure 6.9 ARPES data recorded from a sample of WSe$_2$, courtesy Matt Watson and Phil King, University of St. Andrews. Adapted by permission from Nature (Riley et al., 2014), copyright 2014. See text for details.

possible to generate a grid of binding energy versus $k_\parallel$ and $k_\perp$. Fig. 6.9(d) and (e) show data extracted in this way, presented as maps of the intensity as a function of initial state binding energy and $k_\perp$, at two specific values of $k_\parallel$ corresponding to the $\bar{\Gamma}$ and $\bar{K}$ points of the surface Brillouin zone. At the $\bar{K}$ value of $k_\parallel$, mapping along the bulk K-H direction in reciprocal space shown in Fig. 6.9(e), the bands are flat. Here the electronic states show no $k_\perp$ dispersion, and must therefore be 2D localised. However at $\bar{\Gamma}$, with $k_\parallel = 0$, mapping along Γ-A (Fig. 6.9(d)) there is clearly significant dispersion in $k_\perp$, showing the states here to have 3D character. This surprising effect is a consequence of the different atomic orbital character of the uppermost electronic states at these two points in the surface Brillouin zone. At $\bar{\Gamma}$ the states are largely of W $d_z{}^2$ and Se p_z character, leading to significant interlayer hybridisation, whereas at $\bar{K}$ the states have mainly in-plane $d_{xy} + d_{x^2+y^2}$ character, leading to much less interlayer coupling and essentially 2D behaviour. Notice that the features that appear as 'dashed lines' with a steep gradient in Figs. 6.9(d) and (e) are an experimental artefact due to core level photoemission excited by small amounts of higher-order radiation from the monochromator.

In ARPES experiments the ability to control the polarisation state of synchrotron radiation can be used to determine the symmetry of electronic bands and of orbital states of molecules adsorbed on surfaces. For linearly polarised radiation, as produced close to the axis from bending magnets as well as from planar undulators, these experiments exploit the dipole selection rule that dominates most optical transitions. Specifically, for emission from atomic orbitals this selection rule is that the change in the orbital angular momentum quantum number, Δl, must equal ± 1. Thus, for example, photoemission from an initial s-state ($l = 0$) can only be into a p-state ($l = 1$) while photoemission from an initial p-state is into a mixture of s- and d-states ($l = 0$ and 2). This effect is also exploited in the EXAFS, NEXAFS and XANES techniques described in Chapter 5, and indeed a breakdown of this selection rule at higher energies leads to some complications in the interpretation of X-ray Standing Wave measurements (Chapter 4) monitored by core level photoemission. The same symmetry selection rule applies to photoemission from molecular orbitals and electronic band states of solids as discussed particularly in Section 5.6.2.

The general rule is a consequence of the matrix element for photoemission, namely $\langle \mathbf{f} | \mathbf{A}.\mathbf{p} | \mathbf{i} \rangle$ where $\langle \mathbf{f} |$ and $| \mathbf{i} \rangle$ are the final and initial state wavefunctions, $\mathbf{A}$ is the vector potential of the incident radiation and $\mathbf{p}$ is the electron momentum operator. The selection rules are most powerful if the system of interest, be it an oriented molecule or a crystalline solid, possesses a mirror reflection symmetry plane. In this case the initial and final state wavefunctions can only be either symmetric or antisymmetric with respect to this plane. If one then detects the photoemission in this mirror plane the detected intensity can only be non-zero if the final state is symmetric (an antisymmetric state has zero amplitude in the mirror plane). In order to excite into a symmetric final state the whole matrix element must be symmetric. This can be achieved if the initial state wavefunction is symmetric and the electromagnetic wave is symmetric (i.e. $\mathbf{A}$ lies in the mirror plane), or alternatively if the initial state wavefunction is antisymmetric and the electromagnetic wave is antisymmetric (i.e. $\mathbf{A}$ lies perpendicular to the mirror plane). Varying the direction of the polarisation vector of the incident radiation thus allows one to distinguish emission from symmetric and antisymmetric initial states. The discussion of Section 5.6.2 on NEXAFS from oriented molecules describes how symmetric (σ) and antisymmetric (π) states arise for planar molecules. For planes of atoms in 2D or 3D solid bands, for example, those largely derived from atomic s-p_{xy} states are symmetric relative to a mirror reflection lying in the plane, while those that are p_z-derived are antisymmetric. These selection rules help to identify specific initial state contributions to the observed photoemission spectra and, in particular, help to distinguish such features that may have almost the same energies. Circularly-polarised incident radiation can also provide significant information on electron spin states as described in Section 6.2.

6.1.3   HAXPES (Hard X-Ray Photoelectron Spectroscopy)

While the use of relatively low photon energies can have significant advantages for the study of shallowly bound valence states in terms of instrumental resolution and photoionisation cross-sections, much higher photon energies still lead to valence band photoemission from solids. Indeed, recording such data has been one traditional application of laboratory-based XPS instruments using photon energies ~1.2–1.5 keV. The relative intensities of features in XPS spectra at kinetic energies within typically no more than ~10 eV of the most energetic emission (corresponding to emission from states at the Fermi level in a conductor) correspond to the density of occupied valence states, albeit modified by different cross-sections for components of different orbital character. The fact that the kinetic energy of the photoelectrons in such a measurement are in excess of 1 keV clearly leads to some loss of energy resolution relative to that achievable at a much lower photon energy, assuming that the electron energy analyser has a fixed resolving power ($E/\Delta E$). Nevertheless, the idea of using much higher energy radiation (e.g. ~10 keV), available at high brightness from a synchrotron radiation source, has led to renewed interest in this method of probing valence electronic states, because the higher photoelectron energy leads to an increased sampling depth and thus to data being more representative of the bulk of a solid. This is one key rationale for the development of HAXPES – Hard X-Ray Photoelectron Spectroscopy. Notice that the electron inelastic scattering mean-free-path that determines the degree of surface specificity at higher energies scales as ~$E^{0.75}$ where E is the photoelectron kinetic energy, so increasing the energy by a factor of 10 only increases the probing depth by a factor of ~5, but this seems to be sufficient to very significantly suppress the component of the detected signal that is specific to the surface.

In practice, the technique of HAXPES is exploited in two very different ways. In core level HAXPES the main objective is generally to obtain compositional and chemical state information on buried interfaces below, or within, thin film structures. Experiments are performed at different sampling depths, using some combination of different photon energies and different photoelectron take-off angles, to try to separate information from above, below and within the buried interface. In valence level HAXPES the objective is generally to suppress the component of the signal arising from the surface, which is sensitive to contamination and surface reconstructions, and obtain true 'bulk' information on these electronic states. Very pronounced changes in the relative photoionisation cross-sections of different components of the valence band, at higher energies, can also help to identify the partial densities of states (PDOS) of these different components. Of course, extracting useful information from valence band photoemission at energies ~10 keV is extremely challenging. One problem concerns the photoionisation cross-section. Photoionisation cross-sections fall off steeply with increasing photon energy; this is typically described for 1s emission by a simple empirical polynomial expression, the Victoreen formula (Victoreen, 1943), in which the leading term that is commonly dominant is proportional to $(1/(h\upsilon)^3)$, so an increase in the photon energy by a factor of 10 can lead to a decrease in the cross-section by a factor of ~1,000. Moreover, detailed investigations of valence electronic structure ideally require a combined spectral resolution, arising from the photon monochromator and the electron spectrometer, of less than 100 meV. As the resolution of a Si(111) double-crystal monochromator used for the incident X-rays at a photon energy of ~10 keV typically has a resolving power $(h\upsilon/\Delta h\upsilon)$ of ~10^4, this part is achievable relatively routinely, but obtaining a similar resolving power for the photoelectron energy analysis (with kinetic energies also ~10 keV) is more challenging, and does lead to a further consequential loss in detected intensity. Despite these problems, HAXPES studies are being actively pursued at beamlines at a number of synchrotron radiation sources worldwide.

Notice that the valence electronic structure information obtained from ARPES (at low photon energies, typically no more than ~100 eV) and from XPS and HAXPES (with keV energies) are not equivalent. In ARPES it was shown that k-conservation was important, so even if the measurement at low photon energies is truly angle integrated what is measured is not simply the PDOS. At the higher energies of XPS and HAXPES most measurements are 'angle-integrated' in that they do average over one of more complete Brillouin zones. For example, for a material with a small real-space unit mesh (such as an elemental fcc metal) the Brillouin zone edge corresponds to a relatively large $k_\parallel$ value of ~1 Å^{-1}. The emission angle to detect states at the zone edge is therefore ~15° at a photoelectron energy of ~50 eV, but only ~1° at a photoelectron energy of ~10 keV. A detector with an aperture of ~2° thus averages over the complete first Brillouin zone of such a material. Most materials of interest have significantly larger real-space meshes, so the angular range corresponding to emission from a single zone is even smaller. Nevertheless, if k-conservation were still to be important, the resulting spectrum would still not be of the PDOS. However, the role of thermal vibrations does suppress the importance of k-conservation in determining the measured spectrum. This arises because coherent elastic electron scattering interference, like X-ray scattering, is attenuated by a Debye-Waller factor

$$\exp\left(-\Delta k^2 \langle u^2 \rangle\right) = \exp\left(-M\right), \tag{6.5}$$

where $\langle u^2 \rangle$ is the mean square vibrational amplitude, measured in the direction of the scattering vector Δk. As k (and thus also Δk) is proportional to the square root of the photoelectron energy, this attenuation of the k-conserving component is much stronger at high photon energies. The intensity that is lost from this coherent component is redistributed into incoherent non-k-conserving emission, so what is measured is even more nearly true PDOS in HAXPES than in conventional laboratory-source XPS. Indeed, in HAXPES the coherent k-conserving component is typically no more than ~1% of the total signal at room temperature. Of course, if experiments are performed at very low temperatures, of particular interest for some materials, this aspect has to be re-evaluated.

While HAXPES typically ensures that the valence spectrum probes the PDOS directly and avoids significant contributions from the surface electronic structure, the large variation in photon energy available in such experiments can also prove valuable in identifying the orbital character of different features within the PDOS. This is because the photon energy dependence of the photoemission cross-section from states of different orbital character can differ significantly. Fig. 6.10 shows an example of this effect in the valence band photoemission spectrum obtained from a CdO(100) sample recorded at a range of photon energies from 600 eV to 7,935 eV reported by Mudd et al. (2014). The main contributions to the valence PDOS arise from the O 2p and Cd 4d states and DFT calculations using a number of different functionals were used by the authors to attempt to identify these components.

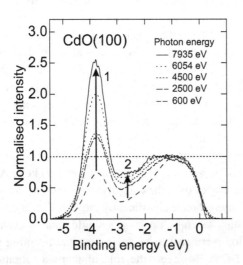

Figure 6.10 HAXPES valence band spectra recorded from CdO(100) at a range of different photon energies, normalised to the intensity of the feature at a binding energy of ~ -1 eV. Adapted from Mudd et al. (2014) under a CC BY 3.0 licence (https://creativecommons.org/licenses/by/3.0). The arrows labelled 1 and 2 highlight the main features showing significant relative intensity changes.

6.2   X-Ray Absorption and Magnetic Circular Dichroism

The use of X-ray absorption spectra to obtain structural information was described in Section 5.6. In particular, the modulations in the absorption cross-section from ~50 eV to several hundred eV above a core-level absorption edge form the basis of the EXAFS technique to determine the local structural environment of the absorbing atoms. Within ~50 eV of the absorption edge the modulations also provide structural information, particularly for molecular adsorbates through the NEXAFS technique, but the same measurements can also provide electronic structure information, identifying transitions from occupied to unoccupied states; this application of such measurements is usually described by the XANES acronym. XANES spectra are often used as a spectroscopic fingerprint of atomic species in a particular bonding environment without any detailed analysis of the features in the spectra. An example of this is an archaeological investigation by Cotte et al. (2006) to identify the materials involved in wall paintings at Pompei. Fig. 6.11 shows two S K-edge XANES spectra taken from this investigation illustrating the very substantial difference in both the energy of the edge (c.f. 'core level shifts' in XPS described in Section 6.1.1) and the spectral shape beyond it from two different S-containing bulk compounds, a sulphate and a sulphide mineral.

A rather different example is provided by the near-edge structure measured in the study of transition metal dopants in Bi_2Se_3 of Figueroa et al. (2015), part of the same study in which the EXAFS data of Fig. 5.17 were used to obtain structural information. The XANES results are shown in Fig. 6.12 for Cr, Mn and Fe dopants. The transition metal K-edge spectra are shown for these dilute alloys, compared with similar spectra

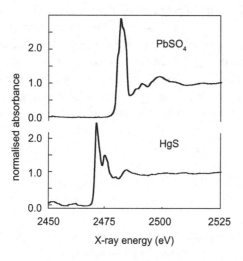

Figure 6.11 XANES recorded at the S K-edge in two bulk mineral materials, a sulphate (anglesite) and a sulphide (cinnabar), showing both the chemical shift of the absorption edge and the different near-edge spectra. Adapted with permission from Cotte et al. (2006). Copyright (2006) American Chemical Society

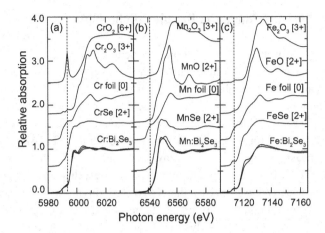

Figure 6.12 XANES spectra above the K-edge of Cr, Mn and Fe dopants in Bi_2Se_3 compared with a number of reference materials containing the same elements. The charge states of the metal ions in the reference compounds are shown in brackets. Reproduced with permission from Figueroa et al. (2015). Copyright American Chemical Society

from foils of each metal and from several different compounds of these atoms in known charge states. These spectra also show two main changes in the different compounds, namely shifts in the exact energy of the adsorption edge, and differences in the structure above the edge. Shifts in the edge energy (like core level shifts in XPS) are most commonly attributed to changes in the charge state of the ion, while the XANES structure gives some insight into the local valence bonding environment.

The K-edge XANES spectra of Fig. 6.12 correspond to excitations from the 1s core state, which are dominated by electric dipole transitions to final states of p symmetry, which in the case of condensed matter are εp states in the unoccupied continuum. In the case of the 3d transition metals being investigated in this study the excitations appearing in the XANES are to the 4p-derived unoccupied density of states, while the pre-edge structure is proposed to derive from 1s $\rightarrow$ 3d transitions. Such transitions are dipole forbidden, and although electric quadrupole cross-sections are generally very weak, the presence of these features for Cr is attributed to a non-centrosymmetric local environment of the ion, consistent with the slightly distorted octahedral environment indicated by the EXAFS data. A fuller discussion of these issues and their interpretation is given by Figueroa et al. (2015).

An important variant of the XANES technique, which exploits circularly polarised incident radiation to obtain local magnetic structure information, is X-ray Magnetic Circular Dichroism (XMCD).

Circular dichroism is the difference in the measured signal in an experiment (commonly absorption but also, for example, in photoemission) obtained with right- and left- circularly-polarised light. In the visible and UV spectral ranges it is characteristic of molecules and materials that are chiral; namely the mirror image of their structure cannot be superimposed on the original structure. It can be used, for example, to provide a spectral fingerprint of the folding of protein structures, with some experiments utilising synchrotron radiation to extend the measurements to shorter wavelengths. In a magnetic material, however, one can have circular dichroism due to the 'chirality' of the electronic states, but one important advantage of investigating magnetic CD in the XANES signal is that the information is local to the atomic species associated with the absorption edge. XMCD is a particularly effective way of investigating magnetism in an element specific way with sum rules, first developed in the early 1990s (Thole et al., 1992; Carra et al., 1993), that allow the spin and orbital magnetic moments of atoms in solids being determined independently.

XMCD can most simply be understood as a two-step process. Fig. 6.13 shows schematically the processes involved and the resulting spectra in XMCD from a ferromagnetic 3d metal (Fe, Co or Ni) recorded at the L_2 and L_3 edges (corresponding to the thresholds for photoionising the $2p_{1/2}$ and $2p_{3/2}$ states, respectively). The first step involves photoexcitation from these occupied 2p states, while in the second step these excited electrons must find empty states in the metal d-band. A characteristic of these materials is that the exchange interaction leads to an offset in energy in the valence d-band electronic states depending on their electron spin orientation relative to the direction of the magnetic field. The result is that there is a majority of occupied states of one spin orientation and a majority of unoccupied states of the opposite orientation. The minority spin direction, also the majority hole spin direction, is the direction of the magnetic field in Fig. 6.13. The 2p core states are spin-orbit split into two states depending on the relative orientation of the spin and angular momentum moments, with these being parallel in the $2p_{3/2}$ state $(j = l + s = 1 + 1/2)$ and antiparallel for the $2p_{1/2}$ state $(j = l - s = 1 - 1/2)$. Photon absorption is subject to the dipole selection rule that $\Delta l = \pm 1$ (and also $\Delta s = 0$), so electrons from a p-state can be excited into s or d final states, but the cross-section for transitions to the $l + 1$

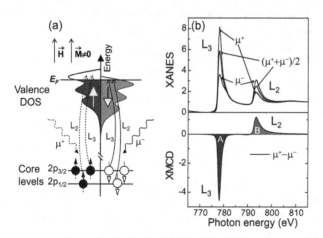

Figure 6.13 Schematic diagram showing (a) the excitations associated with XMCD measurements from a 3d magnetic transition metal and (b) the resulting XMCD spectra obtained from Co. Adapted from van der Laan & Figueroa (2014). Copyright 2014, with permission from Elsevier

d-state generally dominates strongly, so L-edge XANES is an effective probe of the unoccupied part of the 3d valence band. Indeed, measured across the 3d transition metals, the sum of the intensities of the XANES absorption peaks (the 'white lines' – a term that came from early absorption spectra recorded in a spectrograph on photographic emulsion) at the L_3 and L_2 adsorption edges, is proportional to the number of unoccupied 3d states (holes), N. Specifically,

$$I_{L_3} + I_{L_2} = CN. \tag{6.6}$$

Strictly this equation requires that the XANES intensities are averaged over all incidence angles for anisotropic materials, but for bulk fcc and bcc solids the high symmetry removes this requirement. C is the modulus squared of the matrix element for $p \rightarrow d$ transitions.

In addition, however, using circularly polarised radiation with its helicity vector parallel to the $2p$ orbital moment results in preferred ejection of spin-up electrons, while if the helicity vector is antiparallel to the $2p$ orbital moment the preferred ejection is of spin-down electrons. In this case the relative intensity of the L_3 and L_2 adsorption edge peaks is determined by the number of unoccupied 3d states (holes) of the appropriate spin. The opposite spin orientations of the $2p_{3/2}$ (L_3 edge) and the $2p_{1/2}$ (L_2 edge) thus allow circularly polarised radiation to probe empty states of the opposite spin; positive helicity of the incident radiation, μ^+, will enhance the L_3 edge peak, exciting 62.5% spin-up electrons, while radiation of negative helicity, μ^-, will enhance the L_2 edge peak, exciting 75% spin-down electrons. As the circular dichroism is defined as the difference between the absorption using the two opposite light helicities, the consequence is that the XMCD signal will be negative at one absorption edge and positive at the other. Important sum rules allow the orbital and spin magnetic moments of the absorbing atoms to be extracted separately from the XMCD measurements. These, too, are relatively simple for isotropic materials, measured in a magnetically saturated sample with a strong external field along the direction of the X-ray propagation, for which the spin moment is given by

$$m_s = -\frac{(A - 2B)}{C}\mu_B, \tag{6.7}$$

while the orbital moment along the field direction is

$$m_o = -\frac{2(A + B)}{3C}\mu_B \tag{6.8}$$

in units of Bohr magnetons, μ_B. A and B are the XMCD signals at the two L edges (see Fig. 6.13). For anisotropic materials, including ultrathin films and surfaces, more complex orientation-dependent sum rule equations are involved as given in a similar format, for example, in the review by Stöhr (1999).

Results from a very different materials application of XMCD are presented in Fig. 6.14, which shows XANES and XMCD measurements at the Os L_2 and L_3 edges from the filled skutterurite material, $SmOs_4Sb_{12}$, reported by Kawamura et al. (2009). 'Filled skutterudites' are compounds with the same crystal structure as the mineral

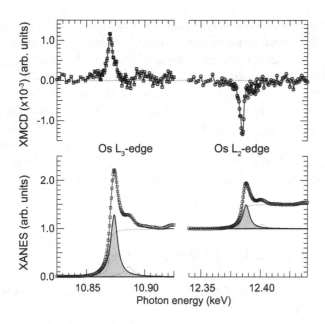

Figure 6.14 XANES and XMCD spectra from $SmOs_4Sb_{12}$ at the Os L_3- and L_2- edges. The extracted white line components of the XANES spectra are shown as shaded peaks. After Kawamura et al. (2009) under a CC BY 3.0 licence (https://creativecommons.org/licenses/by/3.0)

skutterudite ($CoAs_3$), but having two voids in this structure 'filled' with other atoms; their general formula is RT_4X_{12}. Those in which R is a rare earth, T is Fe, Ru or Os and X is P, As and Sb have a number of 'exotic' properties including magnetic ordering, heavy fermion behaviour and superconductivity, associated with strongly correlated electron behaviour due to the hybridisation between the (localised) rare earth f-electrons and the conduction electrons. $SmOs_4Sb_{12}$ shows weak ferromagnetic ordering at ambient pressure below a critical temperature of ~3 K and interesting changes at high pressures. The MXCD experiment was directed to investigate the size of the transferred magnetic moment from Sm $4f$ to Os $5d$, and did find evidence for antiferromagnetic coupling between the Sm $4f$ and Os $5d$ magnetic moments at a range of extreme pressures and temperatures. As shown in Fig. 6.14, a very weak dichroic signal of ~0.1% was rather clearly detected, while application of the sum rules revealed no detectable Os $5d$ orbital moment but a spin moment of 0.016 μ_B. This example highlights the ability of the XMCD to detect extremely small local element-specific moments in a complex material.

Notice that these magnetic effects on the X-ray absorption also lead to changes in the *elastic* X-ray scattering cross-section. In the discussion of structure determination using X-ray diffraction in Chapter 4 it was implicitly assumed that the diffracted amplitude was related to the Fourier Transform of the electron (charge) density distribution within the crystal, neglecting any magnetic scattering depending on the orbital and spin moments of the electrons. In general this is a reasonable assumption – the magnetic scattering cross-section at most energies is orders of magnitude weaker than the charge scattering. Close to appropriate absorption edges, however, one achieves *resonant magnetic scattering* with strongly enhanced cross-sections. This effect arises because a change in absorption corresponds to a change in the imaginary part of the form factor (and thus the refractive index), and this change in the imaginary part leads to a corresponding change in the real part through the Kramer-Kronig relation, as described in Section 4.3. An example of the application of the resulting change in scattering cross-section using circularly-polarised coherent X-ray diffraction from ferromagnetic domains is given in Section 8.6, while the general topic of both resonant and non-resonant magnetic X-ray scattering is covered in detail in the book by Lovesey & Collins (1996).

6.3   X-Ray Emission and Inelastic X-Ray Scattering

6.3.1   XES and RIXS – X-Ray Emission Spectroscopy and Resonant Inelastic X-Ray Scattering

When one creates a core hole in an atom, as in core level photoemission, XANES or EXAFS, this hole is subsequently refilled by an electron from a more shallowly bound occupied state in the atom, releasing energy. This energy release may be in the form of the emission of a photon or of an Auger electron, as described in Chapter 5 (also shown schematically in Fig. 5.16). For example, in K-edge experiments the resulting emitted K_α photons arise from decay from the occupied 2p states to the 1s core hole (note the dipole selection rule, $\Delta l = \pm 1$), so their energies are determined by these core level binding energies, which are characteristic of the atomic species. Measurement of these energies thus provides a method of determining the elemental composition of a solid. This X-ray fluorescence forms the basis of an extremely sensitive trace element analysis technique using incident synchrotron radiation to create the core holes. It is also exploited in electron microscopy (in which initial core ionisation is by energetic electron impact), leading to the standard compositional technique of EDAX (energy-dispersive analysis of X-rays). To achieve this elemental analysis by distinguishing the emitted photon energies from different atomic species the very modest energy resolution (~120 eV) of typical solid state detectors is sufficient.

While the X-ray fluorescence resulting from K-shell absorption is dominated by the 2p → 1s transitions, emission also occurs, at lower intensity, due to transitions from more shallowly-bound occupied states, both core levels and valence states, and the emission from the valence → core transitions carries information on the density of occupied valence states. Specifically, the spectral lineshape of X-ray emission resulting from transitions from these valence → core transition may be expected to reflect the density of occupied states in the valence band, albeit modified by transition matrix elements. This is the basis of the X-ray Emission Spectroscopy (XES) technique. Fig. 6.15 shows the two steps – initial core hole ionisation, followed by core

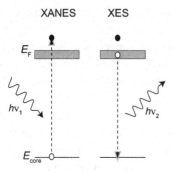

Figure 6.15 Energy level diagrams showing the electron transition in a solid associated with core hole creation in a XANES experiment, and the X-ray emission resulting from a valence → core hole transition. These are the two steps of the two-step interpretation of XES.

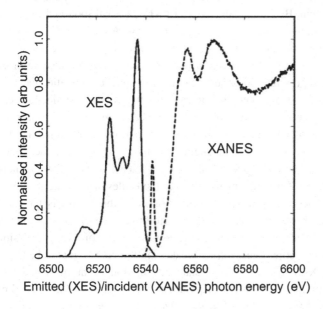

Figure 6.16 Mn K_β XES (full line) and K-edge XANES (dashed line) data obtained from a Mn(V) nitride complex plotted on a common energy scale to show the information content of occupied and unoccupied valence states. Adapted from Bergmann & Gratzel (2009) with permission from Springer, copyright (2009)

hole refilling and fluorescence emission – that underlie the XES process. This figure also shows rather clearly the potential complementarity of the XANES and XES technique; XANES probes the density of unoccupied states, XES probes the density of occupied states. This complementarity is illustrated rather clearly by the results shown in Fig. 6.16 of applying both techniques to a Mn(V) nitride complex, taken from the review of Bergmann & Glatzel (2009). By plotting the data from the two techniques in terms of the relevant photon energies (the emitted photons for XES, the absorbed photons for XANES) one can display information on the occupied and unoccupied states of this sample.

Evidently, the X-ray emission in such an experiment must be measured with very significantly higher resolution than that provided by the solid-state detectors routinely used in EDAX; detailed density of states information, from solid valence bands with widths of only few eV, clearly requires a detector with sub-eV resolution. This can be achieved using a wavelength dispersive spectrometer, particularly combined with a position sensitive detector to allow multiple wavelengths to be measured simultaneously; in effect, such an instrument is one of the monochromators, discussed in Chapter 3, operated in reverse. Of course, a key difference in the requirements of a spectrometer for XES compared with a monochromator for synchrotron radiation is the degree of intrinsic collimation of the source. The synchrotron radiation incident on a monochromator is a near-parallel beam, only very slightly divergent without pre-focusing, and the desired output radiation should also be near-parallel but generally slightly convergent. By contrast the source of XES, a small spot on the sample defined by the incident ionising beam, emits in all directions. A challenge in the spectrometer design is thus how to capture a reasonably large solid angle of this emission while also achieving a physically narrow wavelength-dispersed output. In the case of soft X-ray detection, curved gratings and/or curved mirrors in some combination can address this problem. At harder X-ray energies one can use curved (Bragg-reflecting) crystals or, essentially equivalently, an array of smaller crystals at slightly different orientations. The combination of low emission cross-sections and relatively small aperture detection does make XES quite a challenging technique, though the fact that this is a photon-in/photon-out technique ensures that, at least at hard X-ray energies, the resulting signal arises from the many atoms or molecules in the bulk of solid materials. At soft X-ray energies, however, the penetration of the X-rays is far less, and the detectable volume of emission is smaller. Moreover, the alternative energy release mechanism of Auger electron emission dominates over soft X-ray emission for the refilling of shallow core holes, further weakening the XES signal to be detected.

The use of XES to gain valence state information is far from new, with early investigations of chemical shifts in K_α fluorescence energies, akin to XPS core level chemical shifts, being conducted in the 1920s. However, measurements of valence band densities of states developed somewhat later. A review of soft XES By Fabian, Watson & Marshall in 1971 includes several historical citations. However, the huge increases in incident X-ray flux delivered to the sample by synchrotron radiation, plus advances in dispersive X-ray detection that can exploit the higher emitted intensities and provide higher spectral resolution, have enabled more recent significant improvements in the technique and its viability. Moreover, intense monochromatic incident synchrotron radiation provides an important new capability to XES, namely the selective excitation of particular states, and even allows studies of a very small amount of material involved in a single molecular layer adsorbed on a crystal surface.

An example of the benefit of the selective excitation and the sensitively to molecular adsorbates is provided by an XES investigation of the adsorption of molecular nitrogen, N_2, on the Ni(100) surface, by Sandell et al. (1993), subsequently described more fully in reviews by Nilsson & Pettersson (2004 and 2008). The results of this study are shown in Fig. 6.17, which also shows as inset schematic diagrams the adsorption geometry with the N-N molecular axis perpendicular to the surface, the molecule being bonded atop surface Ni atoms. This bonding geometry causes the two N atoms to be structurally and electronically inequivalent, leading to two different energetically resolved N 1s photoelectron binding energies in XPS, but also two energetically resolved $1s \rightarrow 2\pi^*$-resonances in the N K-edge NEXAFS. Using incident photon energies corresponding to these two NEXAFS peaks therefore allowed XES spectra to be obtained that were specific to the local electronic p-states (determined by the optical selection rule $\Delta l = \pm 1$ for transitions to the 1s hole) of each of the two distinct N atoms. The spectra in Fig. 6.17 corresponding to the outermost N atoms are shown as black continuous lines, whereas those corresponding to the innermost N atoms, bonded to a Ni atoms, are shown as dashed grey lines. In the N_2 molecule the N 2p electronic states of the two atoms hybridise to form two different types of molecular orbital states, those of σ-symmetry, which are totally symmetric about the N-N axis, and those of π-symmetry, which are antisymmetric relative to this axis. These are shown schematically in the inset diagrams in the upper and lower parts of Fig. 6.17, respectively. In order to separate the XES arising from transitions between the σ and π molecular orbital states and the 1s hole, one can exploit the

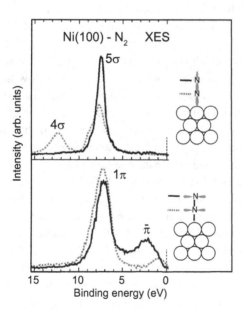

Figure 6.17 Chemical-state specific XES data recorded from N_2 adsorbed on Ni(100) recorded at photon energies close to the N K-edge. The X-ray emission energies have been converted to a binding energy scale of the initial states from which the emission transitions occurred. Different incidence and collection geometries were used to separate the molecular orbital states of σ and π symmetry as described in the text. After Nilsson & Pettersson (2004) copyright (2004) with permission from Elsevier

same polarisation selection rules that apply to the reverse transition in NEXAFS, described in Section 5.6.2. To do so it is not necessary to measure experimentally the polarisation of the emitted X-rays (which would be extremely challenging) but to exploit the fact that the polarisation vector **A** must always be perpendicular to the direction of X-ray emission. As a consequence XES measured for emission perpendicular to the surface can only arise from the $\pi \rightarrow 1s$ transitions with the molecular axis perpendicular to the surface, leading to the spectra shown in the lower part of Fig. 6.17. At off-normal emission XES from both σ and π states to the 1s hole can be detected, but by taking an appropriately weighted difference between the off-normal and normal emission spectra, the $\pi \rightarrow 1s$ contribution can be subtracted, leading to the pure $\sigma \rightarrow 1s$ spectra shown in the upper panel of Fig. 6.17. The resulting XE spectra are shown with the excitation energy subtracted to yield a binding energy scale for the initial valence hole state energies. The binding energies of the observed peaks have been previously assigned through valence band photoemission spectra to the 1π, 4σ and 5σ molecular orbitals of N_2, but the XE spectra show very clearly that the 4σ orbital is localised on the innermost (Ni bonding) N atoms, whereas the 5σ orbital is shared by the inner and outer N atoms, albeit more localised at the outermost N atom.

The description of the XES process associated with its representation in Fig. 6.15 assumes that the initial core hole ionisation and the subsequent core-hole refilling leading to the emission of the X-ray photon are separate processes (i.e. that the X-ray emission has no memory of the initial ionisation process). If the initial ionisation places the electron emitted from the core level into a delocalised state in the continuum, this two-stage description is appropriate. However, if the initial ionisation leads to this electron being excited to a localised state (to an unoccupied molecular bound state or an unoccupied state in a solid that is below the vacuum level), then this *resonant* XES process may be regarded as a one-step process. This resonant XES or RXES process is more commonly referred to as RIXS- resonant inelastic X-ray scattering or even X-ray Raman scattering (recall that the Raman process is inelastic photon scattering, albeit more commonly exploited in the visible or infrared part of the spectrum). This resonant one-step process has an important consequence, because (as discussed in the context of ARPES in Section 6.1.2), the fact that the photons carry very little momentum, relative to that of electrons of comparable energy, means that there must be electron momentum conservation between the initial and final states. Specifically, this k-conservation means that the momentum of the photoelectron and of the valence hole should be the same. Experimental evidence for this was first provided by a study of C K-edge X-ray emission from crystalline diamond using different photon excitation energies by Ma et al. (1992), the energy spectra of the emitted photons, reflecting the density of the valence hole states, showing a strong dependence on the initial excitation energy. RIXS therefore offers the possibility of an alternative approach to the use of ARPES for 'band mapping' and studies of complex spin, orbital and lattice excitations in solids are possible, as reviewed by Ament et al. in 2011.

An illustration of the different mechanisms that can be observed in a RIXS experiment is provided by Fig. 6.18, which shows the results of an investigation of RIXS

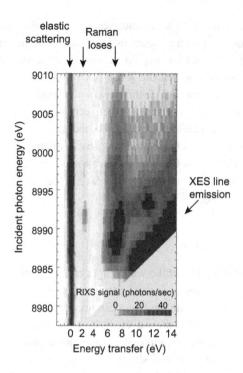

Figure 6.18 Grey-scale representation of the intensity of RIXS X-ray signal as a function of incident photon energy above the Cu K-edge recorded from CuB_2O_4. Reprinted with permission from Hancock et al. (2009). Copyright (2009) American Physical Society

spectra obtained from a CuB_2O_4 sample as a function of the incident photon energy above the Cu K-edge, reported by Hancock et al. (2009); intensities are represented on a grey scale (the original publication uses a false colour scale). The energy scale of the individual spectra corresponds to the difference between incident and detected X-ray energy and thus the size of the energy transfer. At zero energy transfer is the dark vertical line in Fig. 6.18 corresponding to elastic scattering – the scattered energy is the same as the incident energy. A second feature corresponds to emission at a fixed photon energy independent of the incident energy – this standard XES $K\beta_{2,5}$ 'line' emission appears in the figure as a dark diagonal line. Most significant in terms of RIXS are vertical dark streaks at specific values of energy loss, characteristic of a Raman type of process – the intensity of these features varies with incident photon energy, with resonant enhancement at certain incident energies. These spectra thus show both resonant and non-resonant processes coexisting, but they are readily distinguished by this type of data presentation.

6.3.2 Compton Scattering

A very different probe of electronic structure information, also based on X-ray inelastic scattering, is Compton scattering (e.g. Cooper et al., 2004), typically performed at significantly higher incident X-ray energies, and providing information on the momentum distribution of valence electrons in materials rather than their energies. The basic equation governing Compton scattering is generally derived assuming that the 'free' electron from which the X-ray photon scatters is stationary. One can then obtain the Compton formula from the conservation of energy and momentum, starting from an incident X-ray and a static electron and ending with an inelastically scattered X-ray and a recoiling electron. As the energies involved are large, it is necessary to assume the recoiling electron may have a relativistic energy, so the energy of the initial static electron is taken to be its rest mass energy, m_0c^2. This leads to an equation for the increase in wavelength, $\Delta\lambda$, of the emerging X-ray, inelastically scattered through an angle ϕ, of

$$\Delta\lambda = \frac{h}{m_0c}(1 - \cos\phi). \tag{6.9}$$

A striking feature of this equation is that the wavelength change is independent of the incident X-ray photon energy. Notice that its maximum possible value, when $\theta = 180°$, is $\Delta\lambda_{max} = 0.048$ Å. Thus, while the effect can occur at all incident photon energies, it can only be easily detected if the wavelength of the incident photons is no more than a few ångström units. For example, with an incident photon energy of ~1.2 keV $(\lambda \sim 10$ Å$)$, the wavelength change is ~0.5% of the incident wavelength; by contrast, with an incident photon of ~120 keV $(\lambda \sim 0.1$ Å$)$, the wavelength change is ~50%. Traditional laboratory-based Compton scattering experiments thus became most commonly based on incident gamma rays from radioactive sources rather than conventional X-ray sources used for standard diffraction measurements.

In reality, of course, studies of the initial electron momentum rely on the initially electron *not* being stationary. High densities of electrons are to be found in atoms and solids, but they are clearly not stationary, and indeed not 'free'. However, providing that the photon energies are large compared to the binding energies of the electrons to the atom or in a solid, the 'impulse approximation' allows them to be treated as free. The fact that they are not stationary means that there is a Doppler-like shift in the inelastically scattered photon wavelength and energy, so a measurement of this gives insight into this electron motion in these materials. In order to analyse the consequence of this effect it is helpful to re-express Equation (6.9) in terms of the energies (rather than the wavelengths) of the incident photon E_i and the scattered photon E_s to give

$$E_s = \frac{E_i}{1 + (E_i/m_0c^2)(1 - \cos\phi)}. \tag{6.10}$$

Scattering from a moving electron then leads to a shift in the scattered photon energy that depends on the component of the initial electron's momentum, p, in the direction of the scattering vector, $\Delta\mathbf{k}$. Notice that this Δk (i.e. $|\Delta\mathbf{k}|$) differs from that in *elastic* scattering, also referred to as q, in Chapters 4 and 5, because not only do the directions of the incident and scattered photon wavevectors differ, but in inelastic scattering their magnitudes are also different. Specifically, if the direction of $\Delta \mathbf{k}$ is defined as z, then

$$p_z = \frac{(E_i - E_s) + E_i E_s (1 - \cos \phi)}{c \left(E_i^2 + E_s^2 - E_i E_s \cos \phi \right)^{1/2}}. \tag{6.11}$$

The consequence of this is that at a fixed incidence direction and scattering angle the energy spectrum of the scattered photons comprises a broadened peak, the width and shape being determined by the momentum distribution of the scattering electrons. Specifically, this leads to what is known as the Compton profile, $J(p_z)$, symmetric about $p_z = 0$, that is determined by the electron momentum density of the scattering material, $n(\mathbf{p})$

$$J(p_z) = \int \int n(\mathbf{p}) dp_x dp_y. \tag{6.12}$$

Notice that, in the same way that electron density, $\rho(\mathbf{r})$ is related to the square modulus of the wavefunction $\psi(\mathbf{r})$, so $n(\mathbf{p})$ is related to the square modulus of the momentum wavefunction, $\chi(\mathbf{p})$; these two wavefunctions are simply Fourier transforms of each other. Ultimately, therefore, the information content here can be related to that obtained from conventional diffraction experiments, although in general the information is regarded as complementary. The technique is generally used to study the momentum distribution of the shallowly-bound valence electrons of a solid, but Compton scattering can also occur from electrons in the inner core levels of the constituent atoms. Fortunately, the wavefunctions of these states are rather well-known, so it is possible to calculate their contributions to an experimentally determined Compton profile and subtract them in order to extract the valence electron component of interest. Notice, though, that for these calculated corrections to be reliable the impulse approximation must also be valid for these core electron scattering contributions, a fact that leads to a need for much higher incident photon energies than one might otherwise expect. In practice, to ensure this approximation is valid, incident energies are typically ~100 keV or more. Extracting the required Compton profile from the original measurement of the scattered photon energy spectrum also requires corrections for the (calculated) influence of multiple scattering, for the effect of the energy dependence of the Compton scattering cross-section and for absorption in the sample, as well as a number of instrumental factors. One further reason for the need for the incident photon energy to greatly exceed the core level binding energies of the atoms that comprise the material being studied is to ensure that the Compton scattering peak in the scattered photon energy spectrum is well removed from absorption edges and their associated XANES and EXAFS. Although direct interpretation of a Compton profile is not really possible, comparison of the measured profile with theoretically predicted ones provides a valuable means of testing various aspects of current theoretical understanding of the electronic structure of solids.

While there are some benefits in using synchrotron radiation for general Compton scattering studies of valence electron momentum distributions in solids, notably the ability to deliver very high fluxes of monochromated high energy radiation to investigate a phenomenon that has a very low cross-section, a more significant advantage arises from the possibility of delivering partially circularly polarised radiation. Under these circumstances the scattering cross-section contains a term that is spin dependent, allowing the magnetic properties of a sample to be investigated. While this magnetic scattering cross-section is generally very weak relative to the charge scattering cross-section, its relative magnitude does increase with increasing photon energy. This spin-dependent term can be isolated by either reversing the helicity of the incident radiation or, more commonly, by reversing the direction of the sample magnetisation, with respect to the scattering vector. The difference between these two measurements allows one to extract the magnetic Compton profile, $J_{mag}(p_z)$, the 1D projection of the spin-polarised electron momentum density

$$J_{mag}(p_z) = \iint \left(n^\uparrow(\mathbf{p}) - n^\downarrow(\mathbf{p})\right) dp_x dp_y, \tag{6.13}$$

where $n^\uparrow(\mathbf{p})$ and $n^\downarrow(\mathbf{p})$ are the momentum densities of the majority and minority spin bands. At the high photon energies of Compton scattering studies the partially circularly polarised incident synchrotron radiation is taken from slightly above or below the orbital plane of the storage ring. This can be from a bending magnet, but in practice, to achieve high flux at high energy, a multipole wiggler is more appropriate. Notice that the difference measurement implied by Equation (6.13) removes the need for many of the corrections described above that must be applied in standard Compton scattering experiments to extract the Compton profile for the valence electrons of interest. In particular, all Compton scattering from electrons in closed shells is excluded from this difference as these contain equal numbers of spin up and spin down electrons. However, the need for high statistical accuracy to obtain reliable difference spectra does mean that the energy resolution in the detection of the scattered photons must be compromised through the use of solid-state detectors rather than energy dispersive spectrometers.

An example of a measured magnetic Compton profile, recorded by Butchers et al. (2015), is shown in Fig. 6.19, recorded from a sample of UCoGe, a material in which superconductivity and ferromagnetism are known to coexist. Notice that, as is typical in Compton scattering studies, the electron momentum is expressed in atomic units (1 a.u. $= \hbar/a_0 = 1.99 \times 10^{-24}$ kg m s^{-1}, where a_0 is the Bohr radius). In this investigation the determination of the magnetic Compton profile was complemented by measurements of bulk magnetisation and XMCD data to provide a more complete understanding of the magnetism of this material. In order to separate the atom site specific moments the experimental results are compared in Fig. 6.21 with the results of theoretical calculations performed in the local spin density approximation using a computational package based on the KKR (Korringa-Kohn-Rostoker) method. The scaled simulated contributions to the magnetic Compton profile, together with their sum, are superimposed on the experimental data. The results indicate that the spin moments at the U and Co atoms are ≈ -0.30 μ_B and ≈ 0.06 μ_B, respectively.

Because it is only the electrons with unpaired spins that contribute to the spin moment of the sample it is possible to determine the spin magnetic moment rather

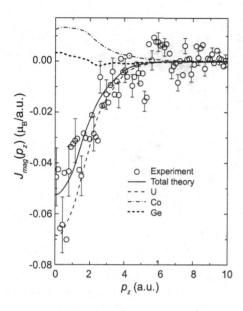

Figure 6.19 Experimental magnetic Compton profile from a sample of UCoGe along the c-axis at an imposed field of 5 T and a temperature of 1.5 K using an incident photon energy of 90 keV. Also shown are the KKR calculated profiles for each of the component atoms and the resulting sum of these. Adapted with permission from Butchers et al. (2015), copyright (2015) by the American Physical Society. Original figure kindly provided by Jon Duffy

directly from this measurement, comparing the results with those of similar measurements under the same condition from a sample of known spin moment. Specifically, the spin moment is proportional to the flipping ratio, R, defined as the ratio of the integrated magnetic and charge Compton profiles

$$R \propto \frac{\int J_{mag}(p_z)\,dp_z}{\int J(p_z)\,dp_z} \tag{6.14}$$

so by making a similar measurement under the same conditions of a sample of known spin moment the value of this moment for the unknown sample can be calculated. An example of the use of this approach is in an investigation of magnetic Compton scattering from a Fe-rich $NbFe_2$ sample (specifically $Nb_{(1-y)}Fe_{(2+y)}$ with $y = 0.015$) by Haynes et al. (2012). This is a complex material with very small changes in the exact composition at low temperatures leading to different structural phases. A particular question concerns the nature of the magnetism in the sample composition investigated in this study and whether the ordering is ferromagnetic or ferrimagnetic. Comparison of the measured magnetic Compton profile with theoretical calculations for these two states clearly favours ferrimagnetism, as shown in Fig. 6.20.

The experimental geometry used, with the scattering vector along the [0001] direction of the crystal and an imposed magnetic field of 2.5 T in this direction, was chosen as being the one found in linear muffin-tin orbital (LMTO) calculations, using the local

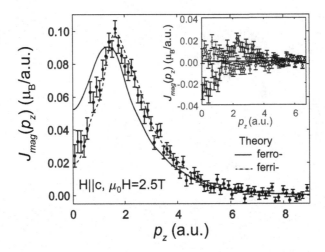

Figure 6.20 Magnetic Compton scattering profile measured from Fe-rich $NbFe_2$ along the c-axis ([0001] direction) compared with the results of LMTO calculations for the ferrimagnetic and ferromagnetic states. The inset shows theory/experiment difference plots (the data points for the ferrimagnetic state are shown as open circles with dashed error bars. Reprinted with permission from Haynes et al. (2012), copyright (2012) by the American Physical Society. Original figure kindly provided by Jon Duffy

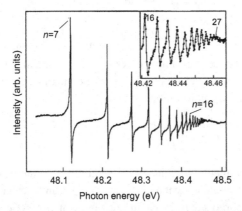

Figure 6.21 Photoabsorption spectra of Ne gas in the energy range of $2s \rightarrow np$ transitions. These spectra were recorded to determine the energy resolution of a monochromator, found to be 1.6 meV. After Blyth et al. (1999) copyright (1999) with permission from Elsevier

spin density approximation, to show the largest difference in predicted magnetic Compton profile between the ferri- and ferro-magnetic behaviour. Notice that in both Figs. 6.20 and 6.19 the amplitudes of the calculated profiles were normalised to those of the experiment to account for the overestimated saturation spin moments, due to a failure in the theory to account for spin fluctuations. Based on the measured flipping ratio the total spin moment of the sample was determined to be $0.245 \pm 0.004 \ \mu_B$.

6.4   Photoemission and Photoionisation Studies of Gas-Phase Atoms and Molecules

So far in this chapter, and indeed in all of the preceding chapters, the emphasis has been on the study of condensed phases of material, particularly crystalline phases but also less ordered solid and liquid phases. However, there is a long history of the use of synchrotron radiation to investigate gas-phase atoms and molecules. For example, photoemission and photoabsorption studies of gas-phase materials, particularly involving excitation of valence electronic states, were some of the earliest experiments performed using synchrotron radiation. In condensed matter these electronic states are commonly broadened into bands due to interaction between the constituent atoms, but in isolated atoms and molecules there is a rich landscape of spectroscopic phenomena with excited states coupling to vibrational and rotational modes. This commonly means that high spectral resolution is required to explore fully these phenomena. This is illustrated by the data shown in Fig. 6.21 which shows photoabsorption of Ne gas in the photon energy range corresponding to excitation of Ne 2s electrons to higher np Rydberg states, recorded to test the resolution of a monochromator on a gas phase beamline at the Elettra facility in Trieste. The observed spectral features arising from excitations in this free atom are clearly much narrower than comparable data from a solid sample. A further obvious challenge is presented by the low density of gases compared with those of solids. Vacuum requirements also define some constraints in experiment design. Electron (and ion) detectors (dispersive analysers, channeltrons and microchannel plates) need high vacuum for their operation, and synchrotron radiation experiments at low (vacuum ultraviolet) photon energies require windowless connection to the storage ring; this leads to a need for high vacuum in the measurement vessel, but also for intermediate differentially pumped chambers to avoid degrading the ultra-high vacuum of the storage ring. However, the use of local gas inlet at the point of interaction with the incident photon beam, or even a molecular beam source, can ameliorate this problem. A low density target necessarily leads to low detected photoemission signals, but, on the other hand, the high background signals arising from inelastic scattering in a solid are not present.

Photoemission and photoabsorption experiments on gas-phase atoms and molecules involve instrumentation very similar to those investigating condensed matter described in Section 6.3. One motivation is sometimes to provide a reference to gain a better understanding of the same species in and on condensed matter, for example in core level photoelectron binding energy chemical shifts and in associated shake-up structure (generally much richer than on the solid state). This also applies to core edge photoabsorption (NEXAFS or XANES) experiments. Studies of gas-phase clusters provide a valuable and potentially much simpler reference for understanding nano-particles that may be deposited on, or in, condensed matter. Fig. 6.22 shows the results of a core level photoemission study of the nucleobase molecule guanine in the gas phase. There are 8 possible tautomers of this molecule, so a key question is which are the most common and with what relative probabilities. In the study by Plekan et al. (2009) *ab initio* calculations were performed to determine the relative total energies of

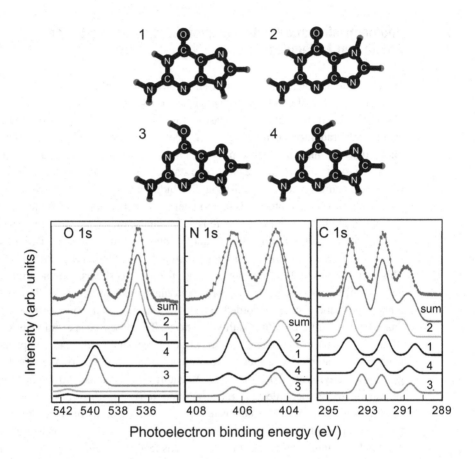

Figure 6.22 Schematic diagrams of the four most energetically favoured tautomers of the guanine molecule in the gas phase according to *ab initio* calculations. Below are shown comparisons of experimental O 1s, N 1s and C 1s photoemission spectra with predicted spectra based on these calculations. Adapted with permission from Plekan et al. (2009). Copyright (2009) American Chemical Society

these alternative configurations of the molecule. The four most favoured tautomers were found to be those shown in Fig. 6.24. Further calculations were performed to predict the relative photoelectron binding energies of the O1s, N 1s and C 1s states in the constituent atoms. Based on the expected relative occupation probabilities of these four tautomers it was then possible to predict the expected photoemission spectra from these three states, convoluting with Gaussian broadening to simulate the experimental resolution. In Fig. 6.22 the experimental photoemission spectra (shown by the individual experimental data points at the top of each panel) are compared with simulated spectra shown as the sum of the predicted contributions from each of the four individual tautomers. The generally good agreement supports the results of the calculations that identify the preferred tautomers and their relative occupations.

Early gas-phase photoemission experiments were typically focused on shallowly bound valence states and used laboratory-based sources, particularly He I radiation, a source that is intrinsically spectrally narrow, but has a fixed energy (21.2 eV), and the first such experiments exploiting synchrotron radiation used similar photon energies and instrumentation. However, an important benefit of the continuous tunability of monochromated synchrotron radiation is that it enables an alternative approach to measurement of these electronic states of atoms and molecules by determining ionisation thresholds with high precision and high efficiency. In this threshold photoelectron spectroscopy (TPES) experiment, the photon energy is incremented, and each time the photon energy passes an ionisation threshold, emission of almost zero kinetic energy electrons occurs. These emissions can be detected with high (~100%) efficiency by the application of modest extraction fields. An important feature of such a measurement is that the spectral resolution is determined almost entirely by the resolution of the photon source, unlike conventional photoemission experiments that measure the photoelectron kinetic energy using incident photon energies well above threshold. In these experiments the spectral resolution involves a convolution of the photon energy monochromacity and the resolution of the electron spectrometer. The large (essentially 4π steradians) detected collection angle of TPES also contrasts with a conventional photoemission spectroscopy (PES) experiment in which an electrostatically dispersing electron energy analyser collects only a small solid angle of the total emission. Both measurements generally use dispersive energy-selective detectors (though time-of-flight – TOF- energy detectors have also been used), but the application of an extraction field can only deliver nominally zero initial energy electrons to the analyser at a well-defined detection energy.

The relative absence of an inelastically scattered background in gas-phase photoemission also means that it becomes more realistic to explore electron-electron and electron-ion coincidence measurements; high backgrounds can lead to large numbers of spurious 'accidental' coincidences, rendering such experiments from condensed matter far more challenging. Fig. 6.23 shows an early example of TPES coincidence (TPESCO) experiment detecting the two emitted photoelectrons in double ionisation of O_2 molecules (by a single photon) by Hall et al. (1992). Fig. 6.23(a) shows a simplified schematic diagram of the experimental arrangement used for a two-electron coincidence TPES experiment. A flow of O_2 molecules is introduced into the vacuum chamber through a needle close to the crossed photon beam and the emitted near-zero energy photoelectrons are extracted by electrostatic fields and passed through a series of electron lenses and deflectors before passage through 127° cylindrical electrostatic electron energy analysers to channeltron detectors. Electrons detected by both detectors in coincidence are recorded as a function of photon energy, the resulting spectrum of coincidence events versus incident photon energy being shown in Fig. 6.23(b). This spectrum shows a sequence of clearly resolved peaks associated with the different vibrational states of the ground state O_2^{++} dication.

Doubly charged positive ions ('dications') of molecules are of interest as they can occur in plasmas, including those in the ionosphere and in interstellar clouds. They are generally unstable due to the Coulomb repulsion between the two atomic ions, but the exchange interaction between the electrons can shield the Coulomb

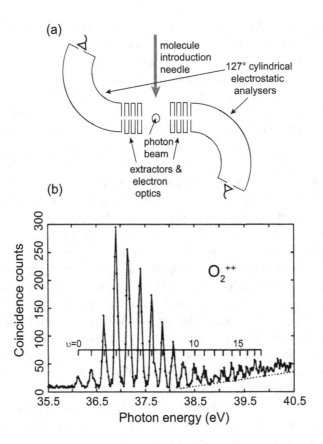

Figure 6.23 (a) simplified schematic diagram of a TPESCO spectrometer and (b) experimental data from a TPESCO investigation of O_2^{++}. Reprinted with permission from Hall et al. (1992), copyright (1992) by the American Physical Society

interaction and lead to a local energy minimum. The lifetime of dissociating molecular dications can be investigated by photoelectron-photoion coincidence (PEPICO) experiments. An example of such a study of doubly ionised CO_2 molecules by Alagia et al. (2012) is shown in Fig. 6.24. Fig. 6.24(a) shows a highly simplified schematic of the instrument used to conduct these experiments. The circle labelled s is the region of overlap of the incident photon beam and the molecular beam in which photoionisation occurs. The near-zero kinetic energy photoelectrons are extracted to the electron detector by a positive potential on grid g+ while the positively-charged ions are extracted in the opposite direction by grid g⁻ and further accelerated to a well-defined energy by the grid g⁻; they travel down a drift tube to the ion detector. The time of flight of these ions of known kinetic energy, relative to the start time at which the photoelectron is detected, defines their mass-to-charge ratio. The resulting ion arrival rate as a function of time-of-flight time is displayed on a grey scale in Fig. 6.24(b), overlaid in two orthogonal directions to allow correlations of the ions arising from the same photoionisation event to be identified.

(a)

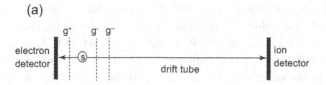

(b)

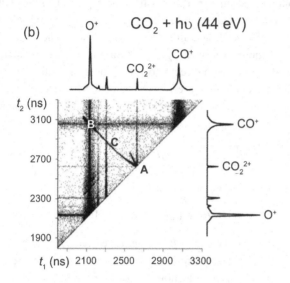

Figure 6.24 Example of PEPICO investigation of double ionisation of CO_2 molecules. (a) shows a highly simplified schematic diagram of the experimental arrangement while (b) shows the data obtained at a photon energy of 44 eV. Reprinted from Alagia et al. (2012), copyright (2012) with permission from Elsevier

Some of the CO_2^{2+} dications survive the journey to the ion detector intact and appear in the grey-scale 'image' at point A, some dissociate (into CO^+ and O^+) before being accelerated into the drift tube and are detected at point B, while others are metastable and dissociate during their acceleration, appearing along the line C. Evidently these data provide information on the lifetime of the metastable dications, although the quantitative analysis does depend on the detailed geometry and applied potentials of the instrumentation.

Notice that these PEPICO experiments, while intrinsically exploiting time-resolved aspects though the TOF ion detection, do not exploit (or need) the time structure of the synchrotron radiation. Even in single bunch operation, the interval between light pulses determined by the circulation time of the electron bunch in the storage ring is typically ~1,000 ns, while the ion flight times shown in the example of Fig. 6.24 extend above 3,000 ns. However, to perform these experiments it is only necessary to attenuate the intensity of the (pseudo-continuous multi-bunch) incident radiation such that the photoelectron count rate is only ~10 kHz, leading to an average separation of these detected events of ~100 μs. Even with the weak fields used for electron and ion extraction the electron velocity is very much higher than that of the ions, so electron detection is effectively instantaneous. When each photoelectron is detected the clock is started and the field for ion acceleration is pulsed on, the clock only restarting after all the ions have completed their travel down the flight tube and the full coincidence event has been recorded.

Angle-resolved photoemission experiments have also proved to be of interest in gas-phase photoemission, although the information content is quite different from that of ARPES experiments of crystalline solids to determine electronic band dispersion. Of course, the fact that molecules in the gas phase are randomly oriented in space means that it is only the polarisation vector of the incident radiation that defines a reference direction and the only angular effects to be observed depend on the angle θ between the emission direction and the polarisation vector. This dependence can be written as

$$\frac{d\sigma}{d\Omega} = \frac{\sigma}{4\pi}[1 + \beta P_2(\cos\theta)], \tag{6.15}$$

where $P_2(\cos\theta)$ is a Legendre polynomial of second order.

$$P_2(\cos\theta) = 0.5(3\cos^2\theta - 1) \tag{6.16}$$

For purely atomic emission the factor β is determined only by the orbital angular quantum number of the initial state from which the electron is photoemitted. For emission from an s state $(l = 0)$ the selection rule $\Delta l = \pm 1$ ensures that the outgoing electron is a pure p state and $\beta = 1$. For other initial states, a range of values between -1 and 2 can occur depending on the relative amplitudes of the $l - 1$ and $l + 1$ outgoing component waves. Many early experiments were directed to measurements of these atomic β factors as tests of theoretical predictions. For molecules, Equation (6.15) still defines the angular distribution, but the elastic scattering of the emitted photoelectrons by atoms provides an added complication influencing the value of β. This scattering is the same effect as photoelectron diffraction (and EXAFS and NEXAFS/XANES) in the solid state. As such the photoelectron diffraction could be used to extract information on the internal structure, but only if the orientation of the molecule in space were known. One way to achieve this is through an angle-resolved PEPICO experiment. Fig. 6.25 shows the results of an experiment demonstrating this effect for CO molecules, reported by Becker, Gessner & Rüdel (2000). As shown schematically in Fig. 6.25, the C 1s photoelectron emission was detected in coincidence with O^+ ion recoils in opposite directions, ensuring that the photoelectrons are detected from molecules having an orientation such that the emission directions corresponds to the O-C axis. In this geometry the detected emission comprises a coherent sum of the directly emitted component of the C 1s photoelectron wavefield and the component of the same wavefield backscattered from the O atom of the original intact molecule. The resulting photoelectron diffraction shows intensity modulations as the photoelectron energy, and thus wavelength, causing this interference to switch in and out of phase. The results of a simple scattering calculation of this

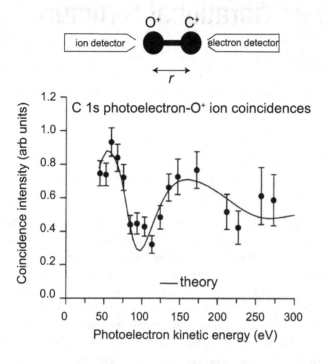

Fig. 6.25 Gas-phase photoelectron diffraction measurement of the interatomic bondlength, r, of the CO molecule obtained from coincidence measurements of C 1s photoelectrons and O^+ recoil ions measured in opposite directions. The superimposed line is derived from a scattering calculation using a value of $r = 1.13$ Å. After Becker, Gessner & Rüdel (2000), copyright (2000) with permission from Elsevier

effect are sensitive to the C-O intermolecular spacing, r, and the calculated curve for a value of $r = 1.13$ Å, superimposed on the experimental data, clearly gives a good fit to the experiment. The estimated precision of this measurement is ± 0.07Å, while the value agrees well with the known bond length.

More recently this basic idea is being explored using an X-ray FEL radiation source to determine structural changes in gas phase molecules in fs timescales (Boll et al., 2014; Rolles et al., 2014), albeit with active alignment of the molecules in space using laser alignment techniques (e.g. Stapelfeldt and Seideman, 2003).

7 Probes of Vibrational Structure

7.1 Introduction

The vibrational modes of molecules can provide a rich source of information, identifying structural components and the influence of chemical or physical environmental changes. The vibrational modes of crystalline solids comprise their phonon band structures. Typical energies of these modes are in the tens or hundreds of meV, although rotational modes in molecules can be even lower. While a primary method of investigating these modes is through direct optical excitation (providing there are dipole moments to couple to), these energies fall well outside the range of most applications of synchrotron radiation that typically go from tens to thousands of eV. Synchrotron radiation provides a unique broad band (continuum) source not only at these higher energies, but in the low energy (infrared) optical range there are readily available low-cost black body continuum sources. Recall, too, that the degree of intrinsic collimation of (bending magnet) synchrotron radiation is inversely proportional to the photon energy. Synchrotron radiation in the infrared energy range therefore does not emerge in cones with a divergence of only a few mrad or less, but nevertheless, it is very much brighter than a laboratory black body source, offering significant advantages for some types of experiment in this energy range. At least one infrared beamline is a feature of many synchrotron radiation facilities.

An alternative way of using synchrotron radiation to explore these low energy excitation modes is through inelastic scattering of X-rays. At first glance this seems to be rather perverse; why would one wish to try to measure energy losses in the meV energy range in photons of keV energies, clearly demanding extremely high spectral resolution? It transpires, however, that such experiments are possible and do have advantages for particular types of study.

7.2 Infrared Spectroscopy and Microspectroscopy

The spectral energy range corresponding to vibrational modes of molecules and solids corresponds to optical wavelengths in the infrared in the range up to ~1,000 μm. In infrared spectroscopy it is common to quote energies not in meV, but in cm^{-1}, the inverse of the wavelength. The approximate energy equivalence is 8 $cm^{-1} \approx$ 1 meV. Laboratory-based infrared spectroscopy commonly uses a black-body radiation source

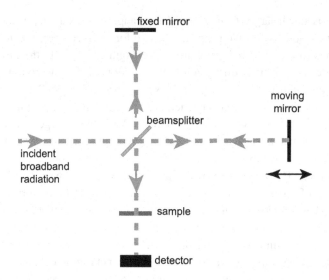

fixed mirror

moving
mirror

beamsplitter

incident
broadband
radiation

sample

detector

Figure 7.1 Schematic diagram showing the main components of a Michelson interferometer used in Fourier transform infrared (FTIR) spectroscopy.

(a 'Globar'), typically operating at a temperature of ~2,000 K. IR spectroscopy from this incident broadband radiation nowadays is most commonly achieved using a Michelson interferometer (Fig. 7.1), rather than a dispersive spectrometer in which different wavelengths are accessed sequentially. The incident radiation is split into two equally-intense beams that travel to plane mirrors, thereby being reflected back to recombine before passing through the sample. By moving one mirror along the direction of the impinging light the optical path length difference of the two component beams is varied, so the signal at the detector is the Fourier transform of the absorption spectrum. This improved approach has been made possible by the easy low-cost availability of computers powerful enough to perform the fast Fourier transforms required to extract the absorption spectra. The broadband character of black-body radiation is well suited to this measurement technique that not only provides a major multiplex advantage (known as Fellgett's advantage), measuring many wavelengths concurrently, but also offers enhanced throughput and higher achievable spectral resolution. However, one disadvantage with a black body source is that the emitted intensity falls off very steeply at long wavelengths (low energies).

The broadband continuum of synchrotron radiation is also ideally suited to this Fourier transform infrared (FTIR) spectroscopy approach, but has the major advantage in this far infrared region, corresponding to energies less than $50-100$ cm^{-1}, that the output of a typical synchrotron radiation (bending magnet) source, in terms of photons per second, shows a significant gain over a conventional (Globar) black body radiation source. Moreover, a more general advantage of synchrotron radiation relative to the black-body source is its spectral brightness, a factor of 2–3 orders of magnitude higher, which extends over the whole infrared range of interest in these experiments. As remarked in the introduction above, the divergence of synchrotron radiation in this spectral range is much larger than in the X-ray range. For example, at the critical wavelength of the source the divergence is ~$1/\gamma$, so for a source operating at ~3 GeV, this corresponds to less than 0.5 mrad at a wavelength of ~1 Å. In the infrared the photon emittance is essentially diffraction limited at all modern synchrotron radiation sources so the divergence is $\sigma'_{ph} \approx \sqrt{(\lambda/2L)}$, as described in Chapter 2. L is the length of the source, which for a bending magnet is a few mm, so at a photon energy of 200 cm^{-1} ($\lambda = 50$ μm) the divergence is ~70 mrad. While this is large compared with the behaviour of synchrotron radiation in the X-ray region, it still corresponds to a very bright source relative to the essentially 2π steradians emission of a conventional black body source. The magnitude of the practical advantage over the black body source depends, of course, on the nature of the experiment to be performed and, in particular, on what is referred to in optical systems as the étendue that the experiment can accept. Etendue is essentially the equivalence of emittance; namely, it is the product of the lateral dimensions and divergence of the source, a quantity that is conserved through any optical system. The bright synchrotron radiation source is therefore particularly advantageous for experiments requiring a small étendue.

Notice that these arguments are based on conventional (bending magnet) synchrotron radiation, as discussed in Chapter 2. However, it transpires that in the infrared spectral range a further mechanism producing radiation from a bending magnet arises in the region where the circulating electrons pass through the inhomogeneous magnetic field at the edges of the bending magnets. While the total intensity of this 'edge radiation' may be weaker than the synchrotron radiation emitted as the electrons travel through the bending magnet, it is actually significantly brighter; infrared beamlines can be designed to exploit this effect. One further special property of synchrotron radiation at very long wavelengths is coherence. In Chapter 2 it was pointed out that in general synchrotron radiation is not truly coherent in the same way that laser (and FEL) radiation involves emission of photons that are coherent with each other. The degree of 'coherence' of synchrotron radiation is a measure of the ability of the source to exploit phase interference phenomena due to its very low emittance. However, the statement that synchrotron radiation is not truly coherent is only correct if the wavelength is significantly shorter than the electron bunch length (typically a few mm) in the storage ring. A wavelength of 1 mm corresponds to 10 cm^{-1}, in the very far IR (usually referred to as THz radiation; 1 THz = 33 cm^{-1}), so true coherence in this spectral range can be achieved if the bunch length can be shortened, as has been shown in both early and more recent tests (e.g. Andersson, Johnson and Nelander, 1999; Cinque, Frogley and Bartolini, 2011). Characterisation of this THz radiation has been undertaken at many synchrotron radiation sources; the gain in intensity can be many orders of magnitude due to the coherence of the emission (as in a FEL) although it is commonly emitted in bursts due to a beam instability leading to microbunching; at sufficiently low currents this effect can be suppressed and stable coherent THz synchrotron radiation is emitted (Bartolini et al., 2011).

The IR spectroscopy experiments that gain particular benefit from the enhanced brightness of synchrotron radiation are those requiring particularly high spectral resolution, sampling of very small sample areas (and hence microspectroscopy) and reflection absorption IR spectroscopy (RAIRS) of ultra-thin molecular layers on surfaces, which can only exploit a narrow divergence of the incident radiation. The benefit of very high spectral resolution is of greatest benefit in studies of the vibrational and rotational modes of molecules in the gas phase, for which the intrinsic linewidths are narrow. The spectral resolution of a Fourier-transform interferometer is ultimately determined by the inverse of the maximum optical pathlength difference (OPD) between the interfering split components of the beam, but the low divergence, spatial coherence and associated high spectral brightness of synchrotron radiation allow OPD values of ~10 m or more to be used. The resulting resolution can be as high as 0.0008 cm^{-1} (or 0.1 μeV). This allows the determination of IR absorption bands associated with excitations of different combinations of vibrational and rotational (ro-vibrational) modes to be investigated. A short review by McKeller (2010) summarises activities in this field at different beamlines internationally at the time of publication.

Fig. 7.2 shows two examples of such measurements of two halogenated molecules of relevance to atmospheric pollution, the destruction of the ozone layer, and global warming, namely methyl chloride (CH_3Cl) and trifluoroamine (nitrogen trifluoride, NF_3). For example, the 100 year Global Warming Potential (GWP$_{100}$, referenced to CO_2) of NF_3 is 17,200, so understanding the IR absorption properties of these gases is potentially very important. Fig. 7.2(a) shows mid-IR transmittance spectra of ro-vibrational states of CH_3Cl recorded at three different gas pressures by Fathallah et al. (2020), while Fig. 7.2(b) shows absorbance spectra of pure rotational excitations in the far IR region for NF_3 recorded at two different pressures and gas temperatures by Bolotova et al. (2018). The high spectral resolution of these measurements (0.003 cm^{-1} and 0.0008 cm^{-1}, respectively) clearly reveals an exceptional degree of detail that could be compared to theoretical descriptions of the molecular modes in the two reported studies.

A rather different field of application of FTIR spectroscopy is in the study of vibrational modes of molecules adsorbed on surfaces at sub-monolayer coverages, typically under UHV conditions. A highly simplified diagram of the basic instrumental arrangement for such a reflection-absorption IR spectroscopy (RAIRS) experiment is shown in Fig. 7.3. Note that all components are in vacuum; UHV is essential for the beamline to preserve that of the storage ring, while UHV is also essential for the surface science study. Vacuum is also required for the optical path throughout to avoid gas-phase molecular absorption bands. The energies associated with these vibrational modes can be used to identify the chemical character of the adsorbed species (which may be the products of a heterogeneous catalytic reaction on the surface) but also may help to determine the local adsorption site (for example, whether the molecule is bonded to one, two or more atoms of the underlying surface). The main mechanism of coupling of the incident IR radiation to these vibrational modes is via their dynamic dipole moments. On conducting surfaces these moments perpendicular to the surface are enhanced by the image charges in the surface, the absorption of these modes being very strongly enhanced by using a very grazing incidence angle of the radiation (Fig. 7.3); this ensures such that the **A** vector of the radiation is also near-perpendicular to the surface and increases the area of the 'footprint' of the beam on

(a)

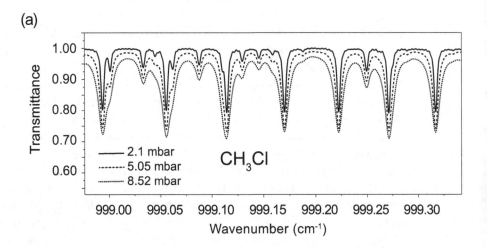

(b)

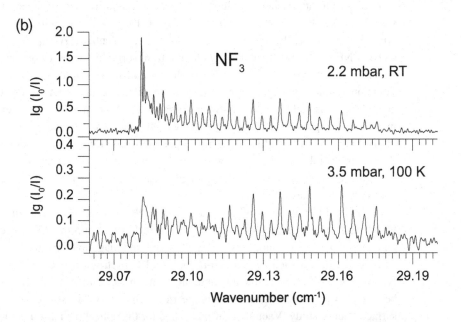

Figure 7.2 Two examples of high-resolution gas-phase FTIR spectroscopy. (a) shows mid-infrared ro-vibrational spectra from CH_3Cl at three different pressures, after Fathallah et al. (2020), copyright (2020) with permission from Elsevier. (b) shows far IR spectra of rotational modes in the ground vibrational state from NF_3, after Bolotova et al. (2018), copyright (2018) with permission from Elsevier.

the surface, interacting with more adsorbed molecules. In this geometry, the experiment can accept only a very narrow angular range of IR radiation, for which the high spectral brightness of synchrotron radiation offers a significant advantage. A further benefit is the enhanced flux in the longer wavelength, lower energy, IR range of the spectrum below ~600 cm^{-1}. Most laboratory-based (Globar source) RAIRS studies

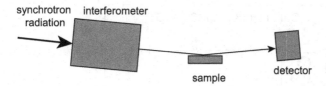

synchrotron   interferometer
  radiation

                                              sample   detector

Figure 7.3 Simplified schematic diagram of experimental set-up for a reflection-absorption IR spectroscopy (RAIRS) investigation of vibrational modes of molecular adsorbates on surfaces.

focus on the internal vibrational modes of the adsorbed molecule at higher energies, such as the C-O 'stretching' mode of adsorbed CO at ~2,000 cm^{-1}, and cannot access the lower energy modes associated with the movement of the molecule relative to the surface. An alternative approach that has been widely used to access all of these modes is high resolution electron energy loss spectroscopy (HREELS), but the 'high resolution' of the name is relative to EELS experiments to access electronic excitations; typically HREELS has a resolution of a few meV, and even the best instruments can barely achieve ~1 meV, an order of magnitude worse than standard RAIRS studies conducted with a resolution of 1–2 cm^{-1}. A further important advantage of the use of incident IR radiation relative to incident low energy electrons is the absence of radiation damage and the ability to perform the experiment not only under UHV conditions, but also under ambient gas pressures more relevant to real surface catalytic reactions; in this case the sample is contained in a small higher-pressure cell to minimise the gas-phase absorption, which is largely cancelled by comparing the FTIR spectrum with and without the absorbed layer and reactant gases.

Fig. 7.4 shows RAIRS data recorded using incident synchrotron radiation from a Cu(111) surface covered with an ordered adsorption phase of 0.33 ML (monolayers) of CO, under UHV conditions, based on the data of Hirschmugl, Chabal, Hoffmann & Williams (1994). Notice that in these UHV experiments, involving reflection from the sample surface, the reference signal used to determine the transmittance is the intensity of the reflected signal without the adsorbed layer, R_{clean}. At ~2,100 cm^{-1} there is a strong absorption band associated with excitation of the C-O stretching vibration. This is very well-known from conventional Globar source studies of CO adsorbed on many different metal surfaces. In addition, however, in the lower wavenumber range, inaccessible to such laboratory-based RAIRS studies, are two much weaker spectral features expanded in the inset of Fig. 7.4. The absorption band at 339 cm^{-1} is associated with the vibration of the whole molecule relative to, and perpendicular to, the surface. In addition, however, is an anti-absorption band at 282 cm^{-1}, which can be attributed to a frustrated rotational mode of the adsorbed molecule. The presence (and character) of this band was completely unexpected. Up until that time the established conventional wisdom was that in a RAIRS experiment of adsorption on a conducting surface only vibrational modes with a dynamic dipole moment perpendicular to the surface can be excited by the grazing incidence IR radiation. By contrast, the dynamic dipole moment of this frustrated rotational mode is parallel to the surface. The appearance of this feature led to significant new efforts to

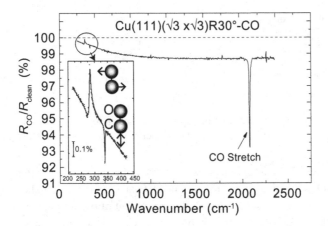

Figure 7.4 RAIRS spectra recorded from the Cu(111)($\sqrt{3}\times\sqrt{3}$)R30°-CO surface, reproduced after Williams (2001) with permission, copyright IOP Publishing. The inset shows the circled low wavenumber region on an expanded scale together with a schematic diagram of the two vibrational modes leading to the two spectral features in this region.

develop further the theory of RAIRS. The appearance of this band is explained (Persson and Volokitin, 1994) by the fact that scattering of the substrate electrons by the adsorbates leads to broad band absorption of the incident IR radiation, but when the IR frequency corresponds to that of the frustrated rotation mode there is a resonance with the collective drift motion of the electrons leading to a fall in the surface resistivity and thus a reduction of the broad band absorption. This leads to the observed anti-absorption peak.

The results of a similar study of CO adsorption, but in this case on the Pt(111) surface (Fig. 7.5), provide further support for this interpretation as well as demonstrating the utility for structural characterisation benefit of being able to perform RAIRS in this lower wavenumber range. Notice that Fig. 7.5 shows no anti-absorption band in this IR wavenumber range; however, a further important observation in these experiments by Surman et al. (2002) is that for this system there was no broad band absorption as a result of the CO adsorption, and thus no possibility of a feature due to the loss of this broad band absorption when the IR frequency coincides with the frequency of the frustrated rotational mode. Instead, the spectrum from this system shows two different conventional absorption bands at 464 cm^{-1} and 376 cm^{-1}, both due to vibration of the whole CO molecule against the surface, but associated with coadsorption in two different local sites, namely singly coordinated atop a Pt surface atom and bridging two nearest-neighbour Pt atoms, respectively.

While the desire to perform these low wavenumber RAIRS experiments motivated the original constructions of several of the earlier IR synchrotron radiation beamlines, much greater emphasis in recent years has been on exploiting the high spectral brightness in this spectral range to perform microspectroscopy/spectromicroscopy in the mid-IR ranges with a growing number of applications in both the physical and life sciences. The underlying scientific motivation of these studies is, nevertheless,

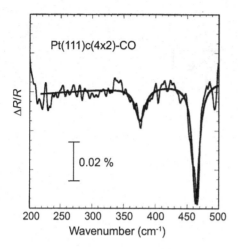

Figure 7.5 RAIRS data in the far IR from an ordered phase of 0.5 ML of CO adsorbed on Pt(111) showing absorption bands associated with the vibration of the CO molecule in two different adsorption sites against the metal surface. A smooth fit is superimposed on the raw data. After Surman et al. (2002), copyright (2002) with permission from Elsevier

essentially the same as that of RAIRS studies of surfaces. In particular, the FTIR measurements provide a spectral fingerprint of specific molecules and molecular components, while the high spectral brightness makes it possible to focus the incident beam to a small size, allowing spatial variations of these molecular characteristics to be determined. Notice, though, that the spatial resolution achievable is determined by the diffraction limit of Abbe theory to be $\lambda/2$, which for infrared radiation is ~$10-1,000$ μ, very much larger that the microfocus achievable for X-rays with wavelengths ~1 Å $\left(10^{-1} \text{ nm}\right)$.

These measurements are based on an IR microscope coupled to a FTIR spectrometer. In concept an IR microscope is similar to that of a standard optical microscope operating in the visible range of wavelengths. The key components of both are a condensing lens to focus the light onto part of the sample and an objective lens to image the illuminated part of the sample. In an optical (visible range) microscope these are generally refractive lenses, fabricated from some type of glass. However, to exploit the broad band capability of a FTIR spectrometer (interferometer) coupled to this microscope, one must use reflecting optics to avoid the chromatic aberration of refractive lenses. Fig. 7.6 shows a schematic diagram of these key optical components based on Schwarzchild reflecting optics.

By inserting an aperture in the appropriate part of the optical path a particular part of the image can be selected and an FTIR spectrum obtained from the radiation sampling this part. A complete FTIR microscopic image can be obtained by physically rastering the lateral position of the aperture. While the ultimate achievable spatial resolution is determined by the diffraction limit, in practice it is more limited by the amount of light that can be passed through the aperture and thereby by the achievable

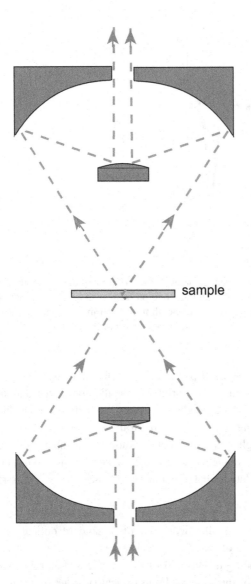

sample

Figure 7.6 Schematic diagram showing the key components of an IR transmission microscope based on Schwarzchild reflecting optics. An important feature of reflecting optics is the absence of chromatic aberration, which is important when using the broadband radiation source required for FTIR microspectroscopy and spectromicroscopy.

signal-to-noise ratio of the collected spectra. In this regard the small étendue of the emitted synchrotron radiation is a very significant advantage over a conventional black body source. A small aperture gives higher spatial resolution in an image scanned in this way, but the collection time required to accumulate it with an adequate signal-to-noise ratio may be very long. Fortunately, in recent years, 1D and 2D pixel detectors, known as FPAs (Focal Plane Arrays) have provided a far more efficient way to collect such images, with a complete FTIR spectrum being collected from each pixel in the image simultaneously, the image collection time being ultimately limited by the read-out time of the FPA.

FTIR spectromicroscopy and microspectroscopy have a wide range of applications but are particularly widely used to investigate biological samples including live cells. The very low photon energies appear to avoid the radiation damage that occurs not only with X-rays, but also with UV radiation used in fluorescence microscopy. For example, Holman, Martin & McKinney (2003) showed that they could collect FTIR spectra from individual live human cells during the cell-cycle with no evidence of damage or disturbance of the reproductive integrity. Spectra from this study (Fig. 7.7) show clear spectral differences for cells at different stages in the cell cycle including the G_1 (first gap) stage in which the cells grow physically larger, the S phase in which the cell synthesizes a copy of the DNA in its nucleus and the M (miotic) phase in which cell division occurs. The features in the spectra labelled amide (R-C=O acyl groups bonded to N atoms) are attributed mainly to C=O and C-N stretching vibrations (amide I) and to N-H bending and C-N and C-C stretching modes (amide II). Detailed analysis of these spectra revealed subtle changes in intensities and exact energies of different bands that could be interpreted in terms of biological function.

Notice that the schematic diagram of IR microscope optics in Fig. 7.6 is for a transmission geometry, which can be used for suitable samples, including those deposited on an IR-transparent substrate. However, the results of Fig. 7.7 were obtained in a different 'transflectance' mode in which the sample is deposited onto an IR-reflecting substrate and one analyses the reflected beam that has actually been transmitted through the sample twice, before and after the substrate reflection. Fig. 7.8 shows the various methods of sampling different types of sample. It includes the ATR

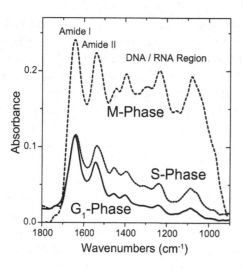

Figure 7.7 FTIR spectra from individual live human cells showing the spectral differences at various stages of the cell cycle. After Holman, Martin & McKinney (2003) under a CC BY-NC licence

transmission reflection transflectance   ATR

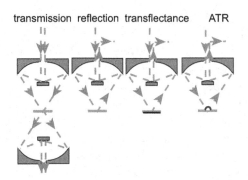

Figure 7.8 Simplified schematic diagram of four different modes of obtaining FTIR microspectroscopy spectra from different samples.

(attenuated total reflection) method, an alternative to the reflection mode for samples that cannot be prepared in a form thin enough for transmittance or transflectance mode measurements but also may be unsuitable for standard reflection measurements. In this case an ATR IR-transparent material (typically ZnSe, Ge or diamond), incorporated in the objective, is placed in contact with the sample, the refractive indices being such that one has total internal reflection at the ATR/sample interface. In this case an evanescent wave extends from the interface into the sample, leading to partial absorption in the sample. Evidently this method does require good physical contact between the ATR crystal and the sample.

A particularly important area of application of FTIR spectromicroscopy is in biological and medical applications (see, e.g. Marcelli et al., 2012; Pilling and Gardner, 2016), and indeed in medical screening for conditions such as cancer. Routine screening, however, generally requires very quick results, not easily compatible with the use of synchrotron radiation, but standard laboratory radiation sources can be used with lower spatial resolution, for which the high brightness of synchrotron radiation is less necessary. Rather different areas of application include forensic science, earth sciences, and studies of archaeological and cultural heritage materials (e.g. Margariti, 2019; Marcelli and Cinque, 2019). One such example is a study of 5,000-year-old bone found in the Jura in France reported by Reiche et al. (2010). The issue addressed by the FTIR spectromicroscopy study was to establish the state of preservation; a poorly preserved sample may influence the reliability of the results of investigating archaeological markers such as ^{14}C dating. The FTIR spectra identified absorption bands well known to be attributable to different species and mapping the spatial variation of these bands provides an indication of the state of preservation. Fig. 7.9 shows some of the results of this study. Fig. 7.9(a) shows an optical micrograph of part of the sample centred on one of the characteristic circular histological units (osteons) of such a bone. Fig. 7.9(b) and (c) show images constructed from the relative amplitudes of specific IR absorption bands. Specifically, Fig. 7.9(b) is based on the relative amplitudes of the collagen absorption band at 1,660 cm^{-1} and that of phosphate at 1,095 cm^{-1}, while Fig. 7.9(c) shows the relative amplitudes of the

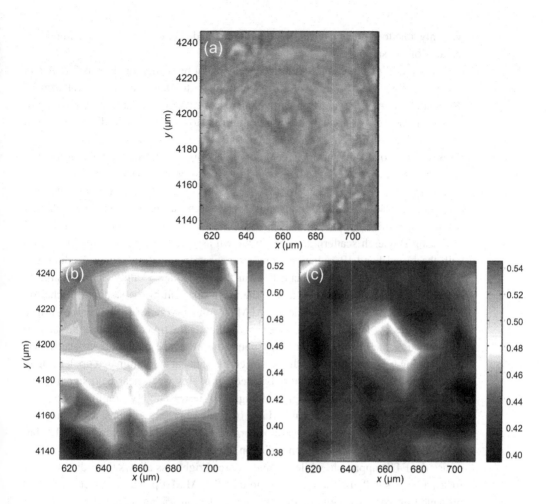

Figure 7.9 (a) shows an optical micrograph of a section of 5,000-year-old bone taken from a site in France, while (b) and (c) show 'heat maps' of the amplitude ratio of IR adsorption bands associated with different components of the bone material as described in the text. Adapted from Reiche et al. 2010. Reprinted by permission from Springer, copyright (2010).

carbonate adsorption band at $1,415 \text{ cm}^{-1}$ and that of the phosphate at $1,095 \text{ cm}^{-1}$. The images in the original paper are shown as 'false colour' heat maps, but conversion to the monochrome images of Fig. 9 does lead to a non-linear brightness scale as shown by the scale bar. Nevertheless, even in this form it is clear that these images show a higher value of the carbonate/phosphate ratio and a lower collagen/phosphate region in the same central region. The main conclusion of the study was that this very old bone sample was in an excellent state of preservation, with broadly similar results to much fresher bone sample. Notice that these images were obtained in transmission mode by physical scanning of the sample with a 15 μm beam focused in a FTIR microscope taking IR radiation from the BESSY II synchrotron radiation facility in Berlin. The authors report that an earlier similar investigation using a globar source was only able to achieve a resolution of ~50 μm, inadequate to resolve some important features of the sample.

One method of obtaining much improved spatial resolution, as low as 200 Å and thus well below the conventional diffraction limit, is to exploit Rayleigh scattering of IR radiation at the tip of an atomic force microscope (AFM). Atomic scale imaging of a solid surface can be obtained by bringing a tip to within atomic dimensions of a surface and measuring either the change in force between the tip and the surface as the tip is scanned over the surface (AFM) or measuring the electron tunnelling current between tip and surface (scanning tunnelling microscopy – STM), well-established techniques described in standard texts (e.g. Woodruff, 2016). A range of other technique with lower resolution also involve scanning a tip over a surface including scanning near-field optical microscopy (SNOM) and particularly, the form of SNOM exploiting Rayleigh scattering at the tip known as s-SNOM. By illuminating the tip with the broad band IR radiation from a synchrotron radiation source, leading to a strongly enhanced electromagnetic field in the immediate vicinity, it is possible to detect IR absorption bands by placing the tip in one arm of the standard Michelson interferometer of FTIR experiments. AFM instruments generally detect tip-surface force changes by 'tapping' the surface (i.e. oscillating the tip height above the surface) and detecting the change in frequency under the influence of the tip-surface forces. This tapping will also modulate the enhanced IR signal at an illuminated tip, leading to a modulated scattered signal that aids detection of this weak near-field signal in the far-field diffraction-limited background. Fig. 7.10 shows a highly simplified schematic of the experimental arrangement. Many of the earlier experiments exploiting this general IR s-SNOM technique were performed with tuneable lasers or IR FELs, but the use of broadband synchrotron radiation allows one to fully exploit the advantage of use the FTIR approach, while the high source brightness allows the radiation to be focused to a small, illuminated area around the AFM tip. Hermann et al. (2013) presented an early example of this technique, demonstrating a spatial resolution of 100 nm, while Khatib et al. (2018) have shown this technique can be extended down in 320 cm^{-1} in the far IR with a spatial resolution of ~30 nm.

7.3   IXS: (High-Resolution) Inelastic X-Ray Scattering

Inelastic X-ray scattering can be used to provide information on electronic states, as described in Chapter 6. In addition, however, this general technique can be used to investigate vibrational and related dynamic excitations, when the technique is typically known by the abbreviation IXS. As discussed particularly in Chapter 5, a range of techniques are concerned with measuring elastic scattering as a function of momentum transfer, q, by varying the scattering angle. If one could measure the small energy transfers due to vibrational excitations one could therefore, for example, map the phonon band dispersion in crystalline solids. The problem in doing so, of course, is that one needs to detect energy transfers in the meV range using X-rays in the keV range. For 1 meV resolution at a photon energy of 10 keV the resolving power required

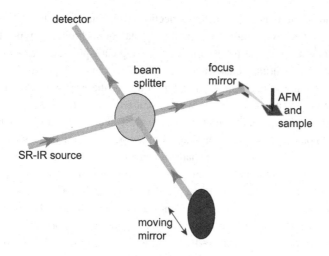

Figure 7.10 Simplified schematic of the instrumental arrangement for nano-scale FTIR *s*-SNOM experiments using synchrotron radiation.

$(hv/\Delta hv)$, 10^7, in both the incident and detected X-rays, is a huge instrumental challenge and for many years was regarded as unachievable or, at least, unrealistic in practice. In fact it has proved possible to build instruments that do operate effectively in this way, and beamlines to perform this type of IXS experiment have been in operation at several synchrotron radiation facilities for a number of years. Of course, before undertaking the required major instrumental developments one might ask if there are no other techniques available that can provide the same information in simpler ways. For example, to map the phonon bands of surfaces both high-resolution electron energy loss spectroscopy and helium scattering are established methods that provide this information (see, e.g., Woodruff, 2016). In bulk solids inelastic neutron scattering (INS) is generally regarded as the method of choice to provide this type of dynamical information on crystalline and amorphous solids and liquids. The low scattering cross-sections of neutron scattering ensure deep penetration, so that truly bulk information is provided, while the energy of neutrons with appropriate wavelengths of the order of interatomic distances (~100 meV) seems more naturally suited to such studies. However, IXS does offer certain advantages over INS for specific applications. Most obviously, the ability to focus X-rays to beams of ~μm (or less) lateral dimensions, coupled with the higher scattering cross-sections, means IXS can be performed on small samples (e.g. with volumes of ~10^{-6} mm^3). It is also possible, through the use of grazing incidence angles close to the critical angle for total reflection (c.f. the discussion of SXRD in Section 4.5), to obtain information on surface phonons and thus, by varying the incidence angle to obtain complementary surface and bulk information in the same experiment. IXS also offers access to data at low values of q in studies of disordered materials for which INS presents some problems.

A number of highly specific features characterise beamlines designed to achieve the exceptionally high resolution required to perform these experiments but in general a key ingredient is the use of higher order Bragg reflections from silicon crystals. The standard component of most X-ray beamlines is a double-crystal monochromator, typically using Si(111) reflections as described in Section 3.5; these typically provide a resolving power of $\sim 10^4$. Higher order Si(nnn) backscattering reflections (with n typically in the range 9–13) offers one route to much higher resolution due to the narrower Darwin rocking curve (see Section 3.5) as a result of the weaker scattering and deeper extinction length. In order to exploit this potential, however, one must address a number of technical problems, most notably those due to the impact of temperature variations, which change the Bragg energy (due to thermal expansion and contraction). Fig. 7.11 shows a simplified schematic representation of the main components of one beamline (ID28 at the ESRF) constructed to address to these problems. Two pre-monochromators are the first components that meet the intense incident (undulator) radiation. The first of these takes the main heat load of the high power incident radiation and consists of a channel-cut Si(111) monochromator cooled to ~ 100 K (at which temperature Si has a maximum in thermal conductivity and a minimum in the thermal expansion coefficient), providing a relatively low resolution output beam. The second pre-monochromator uses higher order Si(nnn) reflections leading to somewhat higher resolution ($\Delta h\nu / h\nu$) of $\sim 10^{-5}$, thereby further reducing the heat load on the main monochromator crystal which uses a higher order reflection in a backscattering geometry to achieve ultimate resolution in the 10^{-8} range. To do so, however, it is necessary to minimise temperature gradients within the crystal due to the remaining incident heat load, this crystal operating in the mK temperature range, and to maintain a constant crystal temperature to within ± 1 mK over significant time periods. Small, controlled changes in the temperature of this crystal are used to control the energy of the monochromatic beam delivered to the sample. The inelastically scattered X-rays emerging from the sample are also analysed by high order Si(nnn) backscattering, but in order to focus the wider range of emitted angles onto discrete detector components after dispersion, a large number (12,000) of small silicon diffracting crystals are glued onto a spherical sector silicon crystal, effectively producing a highly perfect spherically 'bent' focusing analyser crystal.

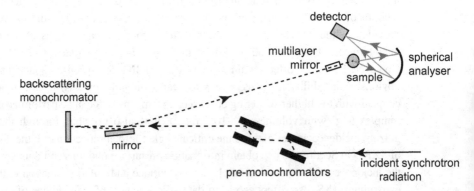

Figure 7.11 Simplified schematic diagram showing the main components of the ID28 beamline at ESRF used for high resolution IXS experiments. See www.esrf.eu/UsersAndScience/ Experiments/DynExtrCond/ID28/BeamlineLayout.

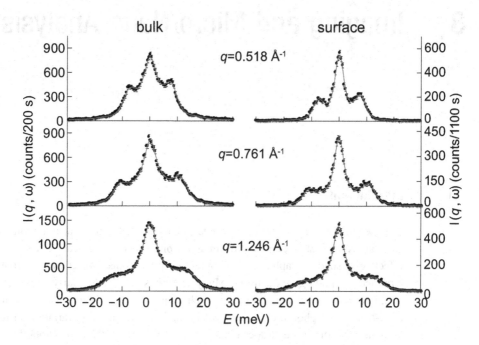

Figure 7.12 Representative IXS spectra obtained from liquid In obtained with bulk or surface sensitivity by using grazing incidence angles above and below the critical angle for total reflectivity. The experimental data (circles) are shown together with the best fit model (lines). After Krisch & Sette (2017). Reprinted with permission from Springer, copyright (2017)

Fig. 7.12 shows an example set of data obtained using the high-resolution IXS technique to investigate the near-surface and bulk dynamical modes of liquid indium reported by Krisch & Sette (2017) (see also Wehinger, Krisch and Reinchert (2011) for a detailed discussion of the underlying theory). Liquid indium has the big advantage for this study in that its vapour pressure is extremely low ($\ll 10^{-13}$ mbar at 170°C, just 14°C above its melting point) while its relatively high atomic number leads to a larger critical grazing incidence angle for total reflectivity (~0.16°), so it is possible to conduct measurements at half of this grazing angle when the effective scattering depth is only 40 Å. The near-surface region of liquid In is also known to possess a clear layering effect with fluctuations in the density as a function of depth that may be expected to modify the microscopic dynamics. The spectra shown in Fig. 6.20 were recorded at grazing incidence angle below ('surface') and above ('bulk') the critical angle and show clear differences, also as a function of the momentum transfer, q. Superimposed on the raw experimental data shown as circles are continuous lines that represent theoretical fits based on a generalised hydrodynamics model for the density fluctuations and, for the surface data, the influence of capillary waves. The review by Krisch & Sette (2017) summarises the results of a range of other relatively recent applications of this technique to crystalline solids, glasses including studies at high pressures.

8   Imaging and Micro/Nano-Analysis

8.1   Introduction

The idea that X-ray can be used for imaging is one that is familiar to anyone who has had 'an X-ray' at a dentist's or in a hospital. These medical X-ray images are effectively shadowgraphs, or radiographic projections, which were traditionally recorded on photographic films. Many materials, and particularly those composed mainly of low atomic number atoms (such as human bodies!) are relatively transparent to X-rays at the photon energies used in medical imaging but variation in the small amount of absorption that does occur in different parts of the body (notably bone, but also, more subtly, some organs and modifications of organs due to disease) allow this simple technique to be used for medical screening of a range of ailments and injuries. Of course, this medical imaging has now adopted digital electronic recording methods, and the availability of the resulting 2D digital data has enabled the development of what are commonly referred to as 'CT scans' ('computed tomography'), whereby the analysis of multiple 2D images recorded in different directions is used to construct 3D images.

These applications of X-ray imaging are clearly on a macroscopic scale, but the use of potentially highly focused synchrotron radiation X-ray beams enables microscopic imaging to be undertaken, while the ability to vary the photon energy above and below the absorption edges of different elements means that one can enhance the contrast in transmission images of specific materials. There is, however, also a much wider range of imaging modes that can be exploited. In recent years there has been an increasing trend in the application and development of synchrotron radiation sources and beam-times to investigate smaller and smaller samples, but also small regions of samples, and thus to produce imaging in a number of different modes with high spatial resolution. This has been a major driving force in the desire to develop 'diffraction-limited' sources, commonly interpreted as meaning diffraction-limited at an X-ray wavelength of ~1 Å, as discussed in Chapter 2.

There are several different approaches to imaging using synchrotron radiation. Conceptually, the simplest of these are various forms of scanning microscopy. The incident X-ray beam is focused to a small spot on the sample, and is rastered across the sample (most commonly by physically scanning the sample), constructing an image from some signal (that could be the transmitted intensity but may be one of a range of other detectable responses) as a function of the position of the beam on the sample.

This mode of operation is broadly similar to that using in scanning electron microscopy, although as electrons are charged particles rapid scanning can be easily achieved by time-dependent electrostatic or magnetic fields without the need for physical movement. A second approach is so-called full-field imaging. This is often based on the same principle as a standard optical microscope, namely illuminating a part of the sample with a beam focused by condensing optics, followed by a second objective lens converting the transmitted signal to an image of the illuminated area of the sample. A rather special variant of this general approach is in photoelectron imaging in which an electron objective lens is used to produce the image. 'Full field' images can also be constructed from measurements of the diffracted interference pattern produced by coherent X-ray scattering from an object, a form of 'lens-less' imaging. Further details and examples of individual techniques are described in the following sections.

8.2   Incident Beam Focusing

As described above, scanning microscopies require the incident synchrotron radiation to be focused to a small spot on the sample; the size of this spot ultimately determines the spatial resolution of the image obtained by scanning this spot over the sample. Several types of optics are used to achieve this goal.

The first of these is based on reflection, and in particular, the use of cylindrical or spherical surface mirrors, the most commonly used type being the Kirkpatrick-Baez or K-B pair of mirrors achieve focusing in both vertical and horizontal directions, described in Section 3.2 (see Fig. 3.5). Sub-micron spot sizes can be achieved routinely with such devices, with advanced designs achieving ~50 nm resolution (e.g. Matsuyama et al., 2017). Very different focusing devices, also based on reflection, are polycapillary optics. The underlying principle is essentially the same as that of fibre optics used in the visible and infrared for digital data transmission cables. In a single-bore glass capillary X-rays travelling very close to the axial direction of the capillary travel down the device by a succession of total external reflections on the internal walls of the capillary. In this form the capillary acts as a guide or collimator, rather than a focusing device, but if the capillary is tapered, or clusters of capillaries are arranged in groups with non-linear geometries, focusing will occur. Of course, the total reflection grazing angle of typical glass material for hard X-rays can be only ~0.1°, so to be effective the diameter of the capillary must be no more than a few μm. The development of this concept is the polycapillary optic, a bundle of several thousand individual capillaries, as shown schematically in Fig. 8.1. The concept was first developed by Kumakhov in the 1980s (see, e.g., Kumakhov, 1990; MacDonald, 2011), and the devices are commonly referred to as Kumakhov optics; they can be used to significant benefit with conventional laboratory-based X-ray sources as well as at synchrotron radiation beamlines and typically produce spot sizes of a few tens of μm. An important feature of all focusing devices based on reflection is that they suffer no chromatic aberration.

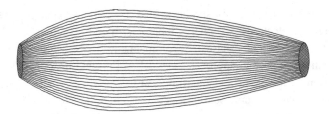

Figure 8.1 Sketch of the interior channels of a monolithic polycapillary optic with a short input and longer output focal length after MacDonald (2011) under a CC BY 4.0 license (https://creativecommons.org/licenses/by/4.0).

A second method of focusing the X-ray beam to small dimensions is to exploit refraction, although the fact that the refractive index of materials at X-ray energies is very close to unity does mean that combinations of such lenses, to create compound refractive lenses as described in Section 3.3 (Fig. 3.7), are required. Unlike devices based on reflection, they have strong chromatic aberration due to the energy-dependence of the refractive index. Specifically, the focal length scales as $1/\lambda^2$ so changing the incident wavelength may require changes in the number of component lenses. These devices are therefore most useful in experiments operating at a single X-ray wavelength, such a micro-diffraction.

The third method of focusing, using Fresnel zone plates (FZPs), is based on a quite different physical principle: namely, coherent interference of diffracted radiation. If one illuminates a planar surface with radiation from a point source, the contours of constant phase on this surface are circles centred on the point on the surface that is closest to the source (see Fig. 8.2(a)). Contours corresponding to phase differences of 2π (i.e. to pathlength differences from the source of one wavelength) become closer and closer as the radii of the circular contours increase. The areas between these adjacent contours are known as Fresnel zones. A Fresnel zone plate is a device that comprises alternate rings of transparent and absorbing material, with spacings corresponding to those of these Fresnel zones. It acts as a lens for coherent parallel incident radiation. The radiation that is diffracted from all of the transparent rings leads to coherent interference at a point on the axis of the device at a distance corresponding to that of the source-to-screen separation defining the radii of the Fresnel zones (Fig. 8.2(b)).

Of course, like all diffraction-based devices (diffraction gratings, diffracting crystals), FZPs also transmit and focus higher order radiation, but with different focal lengths. Specifically, the focal length for mth order radiation is

$$f_m \simeq D\Delta r/m\lambda, \tag{8.1}$$

where D is the diameter of the zone plate, Δr is the width of the outermost zone and λ is the incident wavelength. To suppress the higher orders (and the zero order) radiation one can use an axial beam stop and order sorting apertures. The ultimate (diffraction limited) theoretical spatial resolution that defines the limit of focusing is $1.22\Delta r$, so the

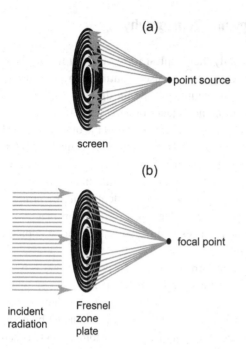

Figure 8.2 (a) shows the contours of constant phase on a screen illuminated by a point source. (b) shows how this can be exploited to create a Fresnel zone plate lens by creating a device that comprises alternate rings of transparent and adsorbing material with a spacing equivalent to those of the Fresnel zones shown in (a).

main challenge in fabricating FZPs is determined by the ability to produce sufficiently finely-spaced outermost rings, achieved using lithography. In practice, the overall diameter of FZPs is submillimetre, typically requiring the incident radiation to first pass through apertures, thereby limiting the incident (and thus transmitted) intensity. The fact that the focal length is inversely proportional to the wavelength effectively means that FZPs display strong chromatic aberration and are therefore only useful for experiments using incident monochromatic radiation; if the incident wavelength is changed, the focal length changes and the physical separation of the FZP and sample must be changed to maintain focus. These small devices also have a very short working distance (f) of, at most, ~mm. Despite these disadvantages, they have proved to be extremely effective for the highest resolution experiments (down to ~10 nm), particularly for soft X-rays, which have longer wavelengths; soft X-rays also have the advantage that it is relatively easy to ensure a high degree of absorption of the incident radiation in the nominally opaque rings of the FZP. The small dimensions (and low mass) of a FZP also means that it is sometimes possible to use coherent illumination of a larger area than this device (to 'overfill the optic') so that scanning of the beam over the sample can be achieved by physically scanning the FZP rather than the sample, allowing much faster scanning.

8.3   Transmission Microscopy and Tomography

Both conceptually and practically the simplest form of X-ray imaging is a transmission radiograph or 'in-line' geometry, as used routinely in conventional medical imaging for orthopaedics and screening for certain diseases. In the case of microscopy using synchrotron radiation the object of interest is placed in a (often parallel) beam of radiation and a detector is placed close behind the object. Image detection is by a direct X-ray pixel detector or using a scintillator, which converts the X-ray image to a visible light image, followed by some type of digital camera. Ultimate resolution is often determined by the pixel size of the final detector. This type of imaging can also be achieved using a divergent incident beam (by bringing the beam to a fine focus in front of the sample), leading to projection magnification of the image. By recording many such images at different sample orientations a computed tomograph can be constructed.

This very simple experimental arrangement also provides one way of obtaining phase contrast microscopy and tomography data. This mode of operation is of particular relevance in imaging materials containing only low atomic number elements, of which the most obvious is biological material, dominated by carbon, oxygen (notably in H_2O), hydrogen and nitrogen. These low atomic number elements have very weak absorption at hard X-ray energies, so absorption contrast in such material can be very poor. In traditional medical imaging heavier atom markers, such as organic iodine-containing molecules added to a patient's blood stream, or barium (sulfate) ingested to highlight the alimentary tract, provide a partial solution to this problem. One solution available using synchrotron radiation for microscopy is to use soft X-rays in the 'water window', having a photon energy above the carbon K-shell ionisation energy of 284 eV but below that of oxygen (543 eV). In this energy range there is very little absorption due to the water content of the material, but strong absorption (and thus contrast) in the carbon-containing material. Of course, the fact that there is strong absorption at carbon atoms does mean that there is significant energy deposited into the material, and for many of the materials of interest (biological matter, polymers) this can lead to radiation damage. To overcome this, one can work at low temperatures, the resulting cryo-soft X-ray tomography (cry-SXT) being the X-ray analogue of cryo-electron microscopy that has revolutionised transmission electron microscopy (TEM) as a probe of biological systems. An example of an image of cryo-preserved mouse primary neuronal cells obtained by cryo-SXT is shown in Fig. 8.3, taken from the work of Harkiolaki et al. (2018). These data were taken using full-field imaging using a capillary condenser and a FZP objective lens, with 500 eV photons. The spatial resolution of this instrumentation is 20–40 nm.

As discussed in Chapter 3, absorption is determined by the imaginary part, $i\beta$, of the refractive index of a material, $n = 1 - \delta + i\beta$ (Equation 3.1). The value of this component falls steeply with increasing photon energy above atomic absorption edges. However, the energy-dependent component of the real part of the refractive index, δ, has a weaker decline at high energies. The alternative approach to imaging soft matter exploits this fact by working at high photon energies, at which absorption

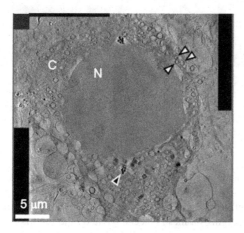

Figure 8.3 2D slice through a cumulative tomogram of cryo-preserved mouse primary neuronal cells slowing the nucleus (N) and surrounding vesicles in the cycloplasm (C), obtained by stitching together four tomograms from adjacent overlapping areas. Black arrows with white outlines point to lipid droplets, white arrows with black outlines point to mitochondria. Adapted from Harkiolaki et al. (2018) under a CC BY 4.0 license (https://creativecommons.org/licenses/by/4.0)

is very weak (leading to a further important advantage of less radiation damage), but exploiting the different values of δ in different material components of a sample in order to achieve phase contrast. Specifically, the real part of the refractive index determines the speed with which a wavefront passes though the material, so passage through different material within a sample leads to different rates of progression of the wavefront, while refraction can lead to changes in the direction of the propagation of different parts of the wavefront. The wavefront therefore becomes distorted if the sample is inhomogeneous (understanding the inhomogeneity is the objective of the microscopy!). In particular, variations in refraction in different parts of the sample lead to different propagation directions and thus the possibility of overlap and interference in different parts of the wavefront. These resulting phase variations can lead to intensity variations in the recorded image due to Fresnel diffraction. Exploiting this effect relies on the coherence of synchrotron radiation as defined in Section 2.7. A particularly notable example of this effect is 'edge enhancement' in the recorded image, effectively 'decorating' the boundaries between materials of different refractive index in the recorded image.

Fig. 8.4 illustrates one way that this can arise. Fig. 8.4(a) shows the classic situation when an opaque sharp-edged object is inserted to shadow part of the planar wavefront. Instead of a sharp shadow one gets interference fringes on the edge of the shadow. This example of Fresnel diffraction is commonly described in undergraduate textbooks on optics, and a quantitative analysis of the resulting diffraction pattern is generally derived by using the Huygens secondary wavelets description of diffraction. The amplitude on the distant screen is calculated by determining the optical pathlengths from different parts of the wavefront to each point on the detector, and summing coherently the

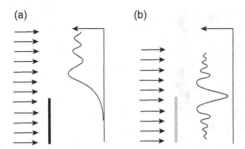

Figure 8.4 Highly simplified schematic diagram illustrating the effects of Fresnel diffraction at (a) a sharp edge of an opaque object and (b) at a sharp edge of an object that is transparent but has a different refractive index to the surroundings. On the left of each part is shown a planar wavefront impinging on these objects, while on the right is a sketch of the resulting interference intensity pattern on a more distant detector.

amplitudes of these wavelets from the unobscured part of the wavefront. Of course, in the inhomogeneous samples of interest here, there are no such obscuring (highly absorbing) objects, but there are sharp changes in refractive index associated with the sample inhomogeneity. Fig. 8.4(b) shows the consequence of replacing the opaque screen by a material that is transparent but has a different refractive index from that of the surroundings. The diffraction pattern shown in Fig. 8.4(b) is a simple sketched representation of one result from calculations by Tavassoly et al. (2012) of visible radiation incident on the edge of a thin glass plate surrounded by air; the detailed form of the diffraction pattern depends on the geometry and refractive indices of the two media, but this simple model case illustrates the general phenomenon.

Notice that this Fresnel diffraction is a 'near-field' effect, and as such must be detected at an intermediate distance behind the imaged object; too close and the image is dominated by a simple radiograph, too distant and this local diffraction is replaced by far-field Fraunhofer diffraction.

Of course what is detected in such an experiment is not a true real-space image, because it also contains significant features due to the Fresnel diffraction. Even in this form such an image may be valuable, as the edge enhancement brought about by the Fresnel diffraction can greatly improve the contrast between different parts of the sample, allowing them to be identified. However, it is also possible to reconstruct a 'true' quantitative image, particularly for tomographs, using a single distance phase retrieval method such as the Paganin filter (e.g. Mohammadi et al., 2015). A number of alternative experimental arrangements for synchrotron radiation phase contrast imaging, reviewed by Liu et al. (2013) are available, including the Zernicke type that includes the use of a FZP objective rather than a simple in-line projection.

Fig. 8.5 shows the result of applying the simple parallel beam arrangement to an *in situ* study of a model lithium ion electrochemical cell to investigate electrodeposition of metallic Li, which is known to be a major cause of failure in lithium batteries. As Li is the lowest atomic number element that is a solid under standard conditions, it is an extremely weak absorber of X-rays, even in the soft X-ray region (the K-shell

Figure 8.5 3D rendering of electrodeposited Li 'moss' microstructure formed in a model cell following galvanostatic cycling. After Eastwood et al. (2015) with permission from the Royal Society of Chemistry

ionisation energy is only ~55 eV) but using hard X-rays (in this case with an energy of 19 keV) allows easy penetration of the containing vessel but also phase contrast imaging of the cell. Fig. 8.5 shows a tomographically reconstructed image obtained from 1,800 separate projection images obtained during a 180° rotation of the sample. The 'mossy' non-faceted structures have the same X-ray attenuation as the metallic Li electrode, demonstrating that Li metal moss is formed after cycling the cell.

8.4 Spectroscopic Imaging

A number of synchrotron radiation based spectroscopic probes of materials have been described earlier in this book, most notably in Chapters 6 and 7. All of these techniques can be exploited using highly-focusing incident radiation beams, and scanning/rastering this beam over the sample provides one route to spectroscopic microscopy or micro-spectroscopy. In some cases, depending on the spectroscopic signal detected, alternative full-field imaging techniques can be exploited to record all the pixel components of a complete spectroscopic image concurrently rather than sequentially. One example already discussed in Chapter 7 is that of infrared microscopy.

8.4.1 X-Ray Fluorescence

A spectroscopic technique for elemental analysis that has not been discussed in earlier chapters is that of X-ray fluorescence (XRF), although the phenomenon has been described in Chapter 5 (see Fig. 5.16) and in discussing XES in Chapter 6. Irradiation of a sample with X-rays having an energy above a characteristic core level ionisation threshold leads to photoemission and the creation of a core hole, and one mechanism for its refilling (dominant for core level binding energies $> \sim 10$ keV) is X-ray fluorescence: an electron from a more shallowly-bound state falls into the core hole and the energy difference is released as an X-ray photon. The energy of this photon is characteristic of the atomic species and its energy-resolved detection provides a way of identifying the elemental composition of the sample. This effect is routinely exploited in electron microscopy (ionisation being achieved by energetic electron impact), a technique commonly referred to as Energy Dispersive X-Ray Analysis (EDX or EDAX). However, using incident X-rays rather than electrons does offer some important advantages. As a photon-in/photon-out experiment, a vacuum environment is not required, a particularly relevant factor in studies of biological material. Using incident photons rather than electrons can also mean that the fluorescence signal is even more bulk-sensitive, the depth of detection being determined only by the attenuation length of the fluorescent X-rays (the incident X-rays can be significantly more energetic than those of the fluorescence emission, and hence more penetrating). Radiation damage can also be much reduced by using high energy X-rays (that are essentially only absorbed by the photoionisation events that define the technique); by contrast an electron beam generates a cascade of low energy electrons, which typically produce more radiation damage in soft tissue. Synchrotron radiation XRF is therefore a particularly effective technique for trace element analysis. Until relatively recently EDAX in electron microscopy had the advantage over XRF of better spatial resolution due to the ease of focusing electron beams using electrostatic of magnetic lenses, but the use of synchrotron radiation and modern focusing facilities have changed this picture.

Fig. 8.6 shows an example of X-ray fluorescence microscopy from a study of plant samples, taken from the report of Kopittke et al. (2018). The Zn fluorescence image

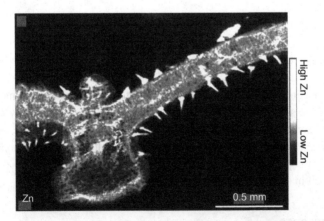

Figure 8.6 Distribution of Zn in a cross-section of a sunflower leaf following foliar fertilisation with Zn, by applying a droplet of $ZnSO_4$ solution on the upper leaf surface. Monochrome adaptation from Kopittke et al. (2018). Reproduced with permission from American Society of Plant Biologists

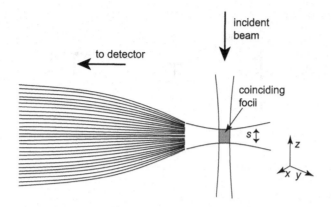

Figure 8.7 Basic experimental geometry of confocal XRF analysis using polycapillary focusing of the emitted fluorescent X-rays onto the detector. The parameter s determines the depth resolution of the sampling. Adapted with permission from Vincze et al. (2004), copyright (2004) American Chemical Society

shown is from an investigation of the distribution of foliar-applied Zn to sunflowers. The authors of this work comment that this image shows clearly enhanced Zn concentration in the treated sunflower leaf tissue. The technique has proved invaluable in a wide range of applications, such as in cultural heritage samples (e.g. Janssens et al., 2000). Although it is possible to extract XRF tomographs from multiple 2D raster scans of the focused X-ray beam from different sample orientations, an alternative approach for tomography is to insert focusing optics in front of the detector in a direction perpendicular to the incident beam so that the 3D information can be extracted directly from a 3D x, y, z scan of the sample. As shown in Fig. 8.7, the focusing optics (in this case a polycapillary array) ensure that the detected signal is emitted only from a particular volume within the sample where the incident beam focus and detector focus overlap.

While these scanning microscopy modes of XRF have been widely used for some years, the fact that the technique involves the emission of X-ray photons from the sample means that, as in transmission modes, full-field imaging using some type of objective lens is also possible. In this case a larger area of the sample is illuminated with the incident X-rays, and the location of the emitted fluorescence X-rays are identified by creating an image of the sample from these emitted X-rays. The simplest such arrangement is to use the imaging properties of a pinhole, as shown in Fig. 8.8(a); this is the same principle of operation as the *camera obscura* (though notice this Latin name means 'dark room' – 'camera' also means 'room' in modern Italian, though the device does have the properties of the meaning of 'camera' in English). Notice that pinhole 'focusing' produces an inverted image and has a magnification governed only by the relative distances of the object and the image from the pinhole. Of course, this device has an extremely small solid angle of acceptance, but it can nevertheless produce good resolution images in the X-ray fluorescence mode. Examples of this, and a presentation of the alternative modes of

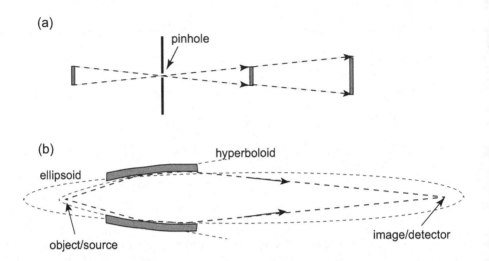

Figure 8.8 Schematic optical layouts two modes of full-field imaging of XRF microscopy: (a) pinhole camera and (b) Wolter optics.

full-field imaging in XRF, are discussed in a paper by De Samber et al. (2019). Other focusing optics that have been used include polycapillary arrays and Wolter optics such as that shown in Fig. 8.8(b). Wolter optics involve sequential reflections from hyperboloid and ellipsoidal mirror surfaces, that can be arranged as separate mirrors similar to the Kirkpatrick-Baez optic (see Chapter 3) or with rotational symmetry about the optical axis as shown in Fig. 8.8(b). Of course, the key requirement to exploit these full-field imaging modes is an efficient pixel detector with ability to discriminate the different emitted X-ray energies; significant developments of such devices have impacted on this field, as described by De Samber et al. (2019).

8.4.2 X-Ray Absorption Edge Spectroscopy and Photoemission

As described in Chapters 5 and 6, measurements of the core ionisation absorption edges and core level photoemission also provide information on both the elemental composition of a sample and the chemical state of these elemental components. In the case of photoemission this information is dominated by the near-surface region, whereas the bulk or surface sensitivity of X-ray absorption spectroscopy (XAS), also known as XANES or NEXAFS (see Chapter 5) depends on the mode of data collection. Evidently, if the absorption in XAS is measured via transmission, it is a bulk measurement. However, XAS thresholds and the associated fine structure can also be monitored by detecting the low energy electron emission from the absorbing sample, a method that does not require the samples to be sufficiently thin to allow a significant fraction of the incident X-rays to be detectable after transmission (see Section 5.6). XAS measurements using this low energy secondary electron emission lead to some degree of surface specificity, although on a depth scale much longer than in photoelectron spectroscopy (low energy secondary electrons can escape from deeper in the sample). Both of these techniques (and the different modes of detection) can yield 2D images by using a rastered focused X-ray beam. Indeed, scanning XRF and scanning XANES can be used to provide complementary elemental and chemical state information; one such example was a study of Cd distribution in a cadmium hyper-accumulating plant by Fukuda et al. (2008). Notice, though, that because the XANES measurements involve scanning the photon energy, FZP (or compound refractive lens) focusing of the incident beam is unsuitable, due to the need to refocus at each photon energy step, so reflective optics must be used. Of course, full-field XANES imaging can be achieved by simply measuring a standard transmission image as a function of incident photon energy around an absorption edge. An example of the application of this technique to investigate the progress of a heterogeneous catalytic reaction over a catalyst bed was reported by Grunwaldt et al. (2006).

Scanning photoelectron microscopy (SPEM) instruments typically involve the use of FZP focusing of soft X-radiation (in some cases down to the vacuum ultraviolet energy range) combined with an electrostatic dispersive electron energy analyser (generally a concentric hemispherical analyser – see Fig. 6.3). This combination is not only used for elemental and chemical-state imaging of surface composition with core-level photoemission, but also for ARPES studies of the electronic structure of small crystals. This need for highly-focused X-ray beams to allow analysis of very small crystals first emerged many years ago in macromolecular crystallography of materials from which large crystals could not be grown, but now appears in a very wide range of materials. Applications of 'nano-ARPES' (with beam sizes of a few 100 nm) include the need to measure the dispersion of electron bands of increasingly complex solids (such as novel superconductors, natural and metamaterial van der Waals solids) offering insight into fundamental aspects of condensed matter physics. Many of these materials can only be grown as small crystals.

Instruments employing a different approach of full-field imaging (but with the objective imaging of emitted electrons, not photons) are now widely used at most synchrotron radiation user facilities. These are mostly based on instruments that were originally developed for low energy electron microscopy (LEEM) rather that photo-electron emission microscopy (PEEM), and the full versatility of these PEEM instruments is best appreciated by first understanding the operation of the original LEEM component. To do so, it is helpful to first recall the basic Abbé theory of a conventional optical microscope. Specifically, such a microscope first produces a diffraction pattern, which can be described as a Fourier transform of the object, while the second objective stage produces a Fourier transform of the diffraction pattern, which corresponds to the real-space image. This same idea, of course, was discussed in Chapter 4 in examining crystalline structure determination through diffraction of X-rays. To illustrate this idea, Fig. 8.9 shows a schematic diagram of an optical microscope in which the object being imaged is a transmission diffraction grating. Incident (monochromatic) light is diffracted by the grating into different directions depending on the order of the diffraction, and the microscope objective leads to this diffraction pattern appearing in the back focal plane of the lens. These diffracted components then

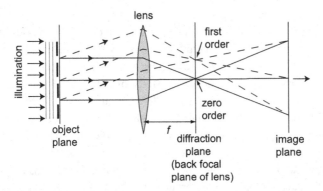

Figure 8.9 Schematic ray-tracing diagram of the operation of an optical transmission microscope with a diffraction grating as the object.

overlap to produce a real-space image of the grating in the image plane. The quality of the image depends on how many diffraction orders (Fourier components) are accepted by the optics.

The operation of a conventional transmission electron microscope follows the same principles, but in this case we may consider the case of a crystalline sample, which is effectively a 3D diffraction grating, being the object. However, a common feature of electron microscopy is to insert an aperture in the back focal plane so that only the zero-order diffracted beam is projected onto the image plane. As the zero order diffraction carries no information about the periodicity of the diffracting object, this would appear to be discarding the key information about the crystal, but in general the objective of these studies is not to determine this periodicity, but to identify the *defects* in this periodicity caused by dislocations, strain around vacancies, twin and grain boundaries, etc. These defects do show up in the image. Notice that these images based on the zero-order beam are referred to as 'bright-field' images; single higher-order diffracted beams can be deflected and expanded to produce 'dark-field' images.

A LEEM instrument is designed to obtain similar information from surfaces, but uses backscattered low energy electrons rather than forward scattered high energy electrons. However, fine focusing of low energy electrons (of energies no more than a few tens of eV) is far more difficult than doing this with high energy electrons (with energies of tens or even hundreds of keV). At low energies electron–electron Coulomb repulsion in the slow-moving electrons ('space charge effects') acts to spread the beam, while even the earth's magnetic field has a significant effect in deflecting the beam. The solution is to perform all the focusing and control of the electron beam at high energies, only slowing the electrons to low energies immediately in front of the sample to which a (high) retarding potential is applied. In this way low energy electrons diffracted back from the sample are then reaccelerated to high energy as they pass through the lens-sample electrostatic field (having a geometry that creates an electrostatic emersion lens) and on through the high-energy imaging electron optics. The only remaining complication is to separate the incident and scattered electron

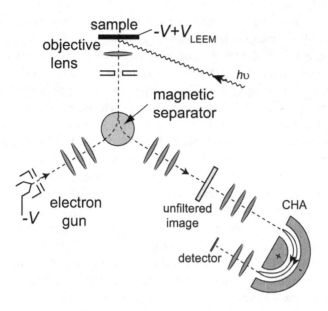

Figure 8.10 Simplified schematic illustration of the main features of a LEEM/PEEM instrument with photoelectron energy filtering. On the left is shown the incident electron gun and optics for LEEM, on the right the LEEM imaging optics but also the concentric hemispherical electron energy analyser used for energy filtering when operating in the PEEM mode with using incident photons instead of electrons.

trajectories (separating the condenser and objective lenses) which can be done by an applied magnetic field, as illustrated in Fig. 8.10. This is the basic arrangement in many of the LEEM/PEEM instruments that have been installed at synchrotron radiation beamlines, although some important newer refinements of this basic design have been used subsequently, including incorporation of aberration-correction systems, a development that is now widely used in transmission electron microscopes to improve the spatial resolution. Somewhat simpler 'bolt-on' instruments, with only the PEEM functionality, are also used, and continue to be refined, but all operate on the same basic principle of accelerating emitted electrons just in front of the sample to much higher energies, in order to achieve good resolution imaging with the electron optics. The electron gun for LEEM operation is shown on the left of Fig. 8.10, operating at an energy V eV – typically ~20–30 keV (the electron source having a potential of $-V$ volts relative to ground). In this design the electron path is diverted by 60° by a magnetic field and very strongly retarded electrostatically in the final lens to strike the sample at a kinetic energy of V_{LEEM} eV (often less than 10 eV). Scattered electrons are reaccelerated and diverted into the objective optics to provide an image of these elastically scattered electrons without energy filtering. In the PEEM mode the electron gun is not used, incident photons, $h\nu$, generating the emitted accelerated photoelectrons, which finally pass through a CHA dispersive electron analyser to produce one of three different modes of data display on the detector, depending on the operating mode of the analyser and lens system.

In the imaging mode, the analyser is operated as an energy band-pass microscope, providing a full-field energy-selected image, allowing images to be obtained at photoelectron energies characteristic of different elemental species or different occupied electronic states in the sample (Section 6.1.1). In the 'diffraction mode', the mode equivalent to that used to display the diffraction pattern of the sample when operating with incident electrons, the detected signal is a complete angular distribution of the photoemitted electrons at any specified kinetic energy. This is therefore an ARPES pattern that can be used for valence band mapping (Section 6.1.2) and for photoelectron diffraction (Section 5.7). Finally, in the spectroscopy mode the display is of a 1D photoemission intensity versus photoelectron energy spectrum, as described in Section 6.1. Notice that there is some variation in the use of the acronym for this technique, but in order to distinguish operation at higher photon energies used to investigate core level photoemission and photoabsorption, from applications directed more to valence electron emission, the former technique is often referred to as XPEEM. However, XPEEM is also sometimes used to distinguish synchrotron radiation experiments from near-threshold photoemission using a simple mercury discharge lamp as the photon source; this was the original much earlier version of PEEM, which has continued to prove valuable in its own right.

Fig. 8.11 shows the results of an investigation of the growth of graphene on an Ir (001) surface that exploits these different modes of operation of a PEEM instrument to explore both the electronic structure and the spatial distribution of features on the surface. The dispersion of occupied electronic bands are mapped in Fig. 8.11(b) (cf the ARPES data of Figs. 6.8 and 6.9) using incident photons of energy 40 eV. This shows the characteristic linear dispersion of the Dirac cones, the bands crossing the Fermi level (E_F) at a k-space location displaced from the Brillouin zone centre (the Γ point: $k_\parallel = 0$); the band crossing at E_D lies above the Fermi level. Fig. 8.11(a) shows the ARPES angular distribution of electrons emitted from the Fermi level from a small area of the film, showing the k-space distribution of these band crossings. Figs. 8.11(c) and (d) show real space images of the surface recorded using photoelectrons emitted from the Fermi level at two different emission directions corresponding to emission from the K and Γ symmetry points of the Brillouin zone. Normal emission from the zone centre is equivalent in electron emission direction to imaging in the zero-order (specular) diffraction in a LEEM experiment, while that recorded from the K point is equivalent to imaging using a non-specular diffracted beam in LEEM, generally referred to as a 'dark field' ('df') image.

The resulting graphene film is partly flat-lying graphene ('FG') on the Ir surface, but partly buckled ('BG'); the microstructure of these two regions shows up in images recorded at a photoelectron energy corresponding to emission from the states just below the Fermi level, for emission (Fig. 8.11(e)) along the surface normal (i.e. electrons from the Γ point of the Brillouin zone) and (Fig. 8.11(d)) at an angle corresponding to emission from the K point (a 'dark-field' geometry). Fig. 8.11(a) and (b) show the results of operating in the µARPES mode, from a small area of the sample, showing the angular distribution of the electron emission from the Fermi level (a 2D Fermi surface plot) and the 1D dispersion of the

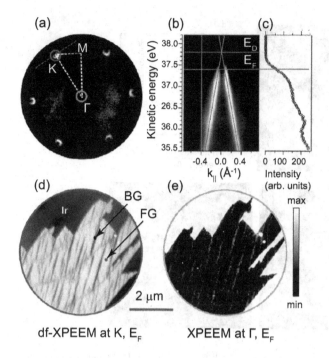

Figure 8.11 Results of a PEEM study, using 40 eV photons, of graphene growth on an Ir(001) surface showing the complementary information obtained by the different operating modes, namely (a) and (b) μARPES, (c) spectroscopy, (d) and (e) PEEM imaging. Adapted with permission from Locatelli et al. (2013). Copyright (2013) American Chemical Society.

graphene occupied band in the Γ-K direction. Fig. 8.11(c) shows a representative intensity-energy photoelectron spectrum showing the cut-off at the Fermi level.

Results from a rather different example, in this case exploiting the chemical imaging feature of XPEEM, are shown in Fig. 8.12. This experiment, reported by Zakharov, Mikkelsen & Andersen (2012), was concerned with the fabrication of quantum wires. To produce these 1D objects on a 2D surface some kind of local inhomogeneity is required for their nucleation, commonly achieved by depositing foreign (particularly Au) particles, but the incorporation of such atoms into the resulting quantum wire can sometimes be undesirable. One way to overcome this is to exploit 'selective area epitaxy' to suppress growth in two directions, thereby creating a 1D structure. The XPEEM experiment was aimed at exploring the idea that a thin SiO_2 film on a III–V semiconductor surface may serve as a mask layer, at the temperatures needed for nanowire growth, by characterising the initial nucleation stage. For this purpose the effect of heating a SiO_2-coated InAs(111) sample was investigated. Figs. 8.12(a), (c) and (d) show XPEEM images of an area of the surface after heating to 600°C recorded, using the Si 2p, In 4d and As 3d photoemission signals. The clear correlation of the dark features in the Si image with the bright features in the In and As images, shows that the heating has produced holes, of dimensions from less than 100 nm to a few μm, in the silica film, through which the

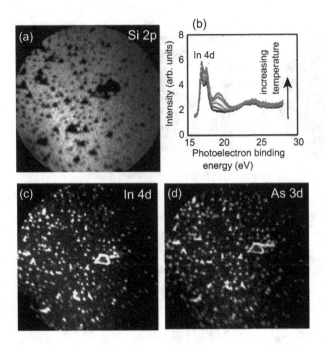

Figure 8.12 XPEEM images recorded with the (a) Si 2p, (c) In 4d and (d) As 3d photoemission intensities from a SiO$_2$-covered InAs(111) surface after heating to 600°C for 10 minutes (field of view 25 μm × 25 μm). (b) shows the In 4d spectra as the sample is heated from 135°C to 510°C. Reprinted from Zakharov, Mikkelsen & Andersen (2012) with permission of Elsevier

underlying InAs layer is exposed. A notable feature of this study was that if an InAs surface without the silica overlayer was heated in a similar way, mobile metallic In droplets formed on the surface. With the silica overlayer, however, small non-metallic In droplets that were well-dispersed, and not mobile, were formed. This was confirmed by the selected area XP spectra shown in Fig. 8.12(b). The well-resolved In 4d$^{5/2}$/4d$^{3/2}$ doublet characteristic of InAs is clearly present throughout the heating cycle with no chemically shifted component for metallic In being seen, although a broader component shifted to higher binding energy (~19 eV), attributed to In bonding to the oxide, does appear (reversibly) at the highest temperatures.

The fact that X-ray absorption spectra can be obtained by measuring the intensity of low energy secondary electron emission as one scans the photon energy through an absorption edge means that a PEEM instrument can also be used to obtain XANES images of a sample (recording the difference in images measured at photon energies below and above an absorption edge), and indeed this application of these instruments is widespread. For example, elemental and chemical state imaging of single particles in Li-ion battery composite electrodes was investigated in this way by Mirolo et al. (2020).

However, a rather different but particularly important application of PEEM operating in this mode, imaging the low energy secondary electron emission, is to vary the polarisation of the incident radiation to obtain images based on the XMCD signal from different parts of the surface (see Section 6.2). This can provide key information on the magnetic heterogeneity (domains) of a surface. These XMCD-PEEM experiments now account for a significant fraction of the applications of PEEM instruments at synchrotron radiation facilities. An important variant of this application is imaging of X-ray *linear* dichroism (XLD), which distinguishes the orientation of ferroelectric polarisation (and also the magnetic properties of antiferromagnets). Fig. 8.13 shows an example of a study that exploits both of these functionalities of a PEEM instrument. In this experiment by Ghidini et al. (2018) an array of 10 nm-thick, 1 μm-diameter (ferromagnetic) Ni discs were deposited onto a 0.5 mm thick (ferroelectric) BaTiO$_3$ (BTO) substrate in the sandwich structure shown in Fig. 8.13(a), in order to investigate the changes in the ferromagnetic magnetisation rotations induced by voltage-induced ferroelectric domain switching. Fig. 8.13 presents a composite XLD/XMCD PEEM image of the sample in which the lighter and darker stripes correspond to the two different directions of the ferroelectric polarisation directions of the substrate. Ni discs that fall entirely within a stripe of one specific ferroelectric polarisation direction were found to have relatively high XMCD asymmetries, implying that the local magnetisation lies approximately parallel (white) or antiparallel (black) to the in-plane directions of the incident X-ray beam direction (shown as a large arrow in Fig. 8.13). These discs are therefore seen to be single-domain structures. However, discs that straggle two different ferroelectric polarisation directions show two different magnetic domain orientations, consistent with the fact that the size of the Ni discs

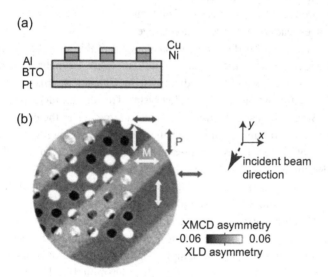

(a)

(b)

Figure 8.13 Results of XMCD/XLD-PEEM measurements taken from an as-prepared array of 1 μm-diameter Ni discs deposited on a BTO (BaTiO$_3$) substrate. (a) shows a schematic side view of the 'sandwich' structure of the sample.(b) shows an XMCD image (20 μm field of view) of the Ni discs combined with an XLD image of the BTO substrate. Double-headed arrows show the directions of the ferroelectric polarisation **P** and local magnetisation, **M**. Reproduced with permission from Ghidini et al. (2018), copyright IOP Publishing

exceeds the critical diameter (400 nm) of single domains for a Ni film of this diameter. Full details of the results of polarisation switching and their significance are reported in the original publication.

8.5   X-Ray Diffraction Topography

A rather different imaging technique, which predates most of the techniques described so far apart from basic radiography (and indeed also predates the recognition and exploitation of synchrotron radiation), was for many years known as X-Ray Topography or X-Ray Diffraction Topography (XRT), is now sometimes referred to as X-Ray Diffraction Imaging, but to avoid confusion with the topic of the next section, the original XRT name will be used here. The technique is directed particularly to obtaining images of various types of defects (e.g. dislocations, stacking faults, local strain) in (nominally) single crystal materials and exploits the same physical processes that allow transmission electron microscopy (TEM) to achieve the same objective, namely that these defects produce local modification of the intensity in crystal diffraction beams. Unlike TEM, however, no lenses are used in XRT. An important advantage of XRT, relative to TEM, is that the strongly penetrating character of X-rays as opposed to electrons means that there is no need to reduce the thickness of bulk samples to ensure adequate transmission. The need to reduce the thickness of samples in TEM can lead to this technique failing to truly characterise the bulk behaviour of some materials. On the other hand, the stronger elastic scattering of high energy electrons of TEM, relative to that of X-rays, does lead to higher resolution images.

While there is a range of different methods that exploit this same basic effect, the simplest (and widely used) variant uses 'white' (or 'pink') synchrotron radiation light in a transmission geometry. As discussed in Chapter 4, this is commonly referred to as Laue diffraction from a single crystal. The chosen incidence direction determines the Bragg angles to the different scattering planes in the crystal, and the broad band incident radiation ensures that the required wavelengths to match the Bragg condition at these angles are present to generate the resulting diffracted beams. An XRT image corresponds to the spatial intensity variation of a single diffracted beam. Fig. 8.14(a) shows a simple schematic representation of the configuration for this experiment. In this example a broad incident beam is incident on the sample and the detector is arranged to collect the beam diffracted from a set of Bragg planes perpendicular to the surface of the sample. Of course, this transmission geometry can only be exploited if the sample is sufficiently thin or if the absorption in the sample is sufficiently weak to allow the transmitted and diffracted beam intensities to be measurable. If these conditions are not met a reflection geometry can be used, although the diffracted intensity can then only detect the near-surface region of the sample within a depth defined by the extinction length of the X-rays. Traditionally (even until relatively recently), the detector took the form of a photographic emulsion, capable of recording images with a resolution down to ~1 μm (and with large lateral dimensions), but direct X-ray-detecting CCD devices are now most

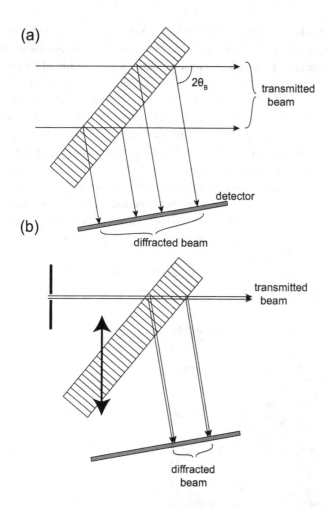

Figure 8.14 (a) Simplified schematic of the experimental geometry of a basic 'white' beam XRT experiment using the diffracted beam corresponding the Bragg reflection from scattering planes perpendicular to the surface of the sample. (b) shows a variant of this for 'sectional topography' in which the incident beam is reduced to a thin sheet of radiation, successive topographs of different sections of the sample being obtained by a relative displacement of the sample and beam.

commonly used, some enhanced resolution being possible with a scintillator and magnification to a visible-range light-detecting CCD.

One limitation of the arrangement of Fig. 8.14(a) is that, in common with radiographic imaging of 3D objects, the resulting image is only a 2D projection. Fig. 8.14(b) shows how this limitation can be overcome by inserting slits to provide a thin sheet of incident X-rays which thereby produce an image of only this very thin slice of the sample. By moving the relative position of the sample and incident beam a succession of these 'section topographs' allows one to build up a complete 3D image of the sample.

While XRT can be performed with laboratory X-ray sources (as it was in all of its early development) there are clear and important advantages to the use of synchrotron radiation. The need for an intense near-parallel incident beam, of course, ideally matches a key feature of synchrotron radiation, and by using a long source-to-sample distance (many tens of metres) one can obtain a wide (vertically and horizontally) beam to illuminate larger samples. The sample-to-detector distance is generally much shorter (tens of cm); this distance needs to be large enough to achieve a good separation of the diffracted and transmitted beams (in part to avoid overloading the detector) and while an increased distance can reduce background noise it has the disadvantage of degrading the image resolution. Moreover, recent advances in synchrotron radiation sources mean the higher energy radiation (tens of keV and more) are available to penetrate large samples, allowing, for example, assessment of the degree of perfection of as-grown semiconductor crystal ingots with multi-cm-scale diameters.

Fig. 8.15 shows the results of an investigation by Tanner et al. (2019) that exploits this ability, using a photon energy of 31 keV to penetrate samples of significant

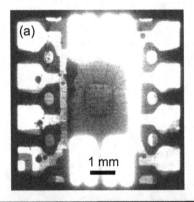

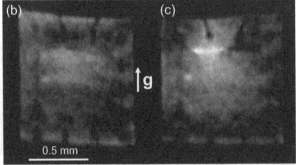

Figure 8.15 Results of an *in operando* study of a fully packaged silicon device. (a) is a radiograph of the complete device showing as dark lines the connecting wires to some of the 14 pins. (b) and (c) are XRT images of the chip under low load (b) and under a load of 2.4 W (c). The direction of the [220] scattering vector **g** is shown by the arrow. Reprinted from Tanner et al. (2019) with permission from Elsevier copyright (2019)

thickness to study technological problems. This was an *in operando* investigation of the strain induced by internal heating in a working fully packaged silicon device, specifically a widely commercially available LM3046 Small Outline Integrated Circuit transistor array. Operating such devices at power levels that are too high can cause failure due to internal heating, leading to local strain and distortion of the device, but there is no simple way of measuring this internal temperature. However, XRT can reveal the associated local strain while operating the device at different power levels. Fig. 8.15(a) shows a transmission radiograph of the complete device including the external contacts and packaging; the Si wafer and contact wires to it are clearly seen at the centre. Figs. 8.15(b) and (c) show the XRT images of this wafer under low load, and under a high load, respectively. The bright region towards the top of the image in Fig. 8.15(c) is due to pronounced strain due to local heating of the transistor component in this part of the device. Increasing the load to 2.4 W led to failure of the device, after which the strained region extended significantly, indicating that the wafer had ultimately bowed to a radius of curvature of 0.9 m.

These XRT images appear rather diffuse, which at first sight would seem to imply poor resolution. However, using the same imaging conditions, a much sharper image of a SiC wafer shows dislocations only a few microns across (Fig. 8.16), clearly indicating that this diffuseness of Figs. 8.15(b) and (c) is a result of the strain distribution in the sample, and is not due to the instrumentation. The bright features in Fig. 8.16 arise from the strain field around individual dislocation lines. The varying widths of these features in the image arise because the width depends on the scalar product of the scattering vector, **g**, and the varying values of the dislocation Burgers vectors, **b**

8.6   Coherent X-Ray Diffraction Imaging

The mode of operation of a conventional optical transmission microscope in the visible range of the spectrum is illustrated in Fig. 8.9 and is commonly described by

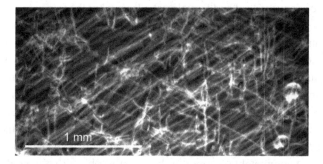

Figure 8.16 XRT image of dislocations in a SiC wafer, diffraction vector $[11\bar{2}0]$ vertical, 20 s exposure, sample-detector distance 173 mm. Courtesy B. K. Tanner, P. N. McNally, A. Danilewsky, R. Vijayaraghayan and B. Roarty (unpublished work)

the Abbé theory of the microscope. In this theory light scattering of the sample leads to a diffraction pattern that can be described as a Fourier transform of the sample, and the microscope image is a Fourier transform of the diffraction pattern, which reproduces the sample. Notice, though, that the image can only perfectly reproduce the sample if all the Fourier components can be captured, which in this case means capturing all components of the diffraction pattern. In the special case in which the sample is a diffraction grating, as in Fig. 8.9, the periodicity of this grating will only be reproduced in the image if the first-order diffracted beams are captured; capturing higher order beams will reproduce more detailed features of the shape of the grating. More generally, of course, a sample does not generate well-separated diffraction orders, and the quality of the image depends on the numerical aperture of the imaging optics. Abbé concluded that that the limit of resolution is $0.61\lambda/\sin\theta$, where the numerical aperture is $\sin\theta$, θ being the half-angle of acceptance of the imaging optics. Evidently, for any specific wavelength of radiation, good resolution requires high quality optics with a large numerical aperture, something that is not really possible to achieve in the hard X-ray range. In practice full-field X-ray microscopes that do exist typically achieve resolutions of a few tens of nm, but this is very much worse than the wavelength limit defined by the Abbé theory.

An alternative approach to this problem is to record the diffraction pattern directly and then invert this to an image numerically, sometimes referred to as 'lensless imaging'. There are two challenges to this approach. One is the need for the illumination of the sample to be coherent over its dimensions. The other is that one can only record the *intensities* of the diffraction pattern, not the *amplitudes*, as required for the Fourier inversion. This is the same phase problem that is familiar in solving complex crystal structures by crystal diffraction, as discussed in Chapter 4 and can be overcome by first guessing the phases and then refining these through many (often thousands) of iterations between real space and diffraction space, applying some constraints such as that the scattering electron density must be positive. An important requirement for the success of this approach is that the initial experimental data set is sufficiently complete. The problem of coherence is described in Section. 2.7. Synchrotron radiation does not have the complete coherence properties of conventional and free electron lasers, but the transverse coherence length is inversely proportional to the degree of collimation. This coherence can therefore be enhanced by passing the synchrotron radiation beam arriving at an experimental station though a 'pinhole' aperture defined by appropriately separated slits. The longitudinal coherence length is determined by the degree of monochromacity of the incident radiation, a property of the monochromator optics rather than the source design. Reducing the size of the aperture increases the transverse coherence but has the obvious disadvantage of reducing the transmitted flux. A key requirement for this Coherent X-Ray Diffraction Imaging (CXDI) technique is that the sample must be completely enclosed in the beam's coherent volume, defined by the transverse and longitudinal coherence lengths, which are typically in the low μm length scale. As such the technique is well-suited to investigate the structure of samples on the scale of tens to hundreds of nm.

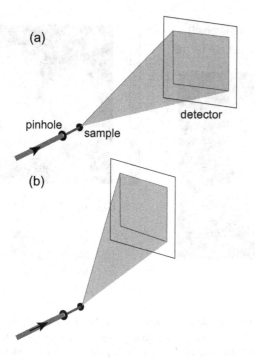

Figure 8.17 Schematic diagram showing the experimental arrangement for (a) PCXDI and (b) BCXDI investigations.

The simplest version of the technique, known as plane-wave CXDI (PCXDI) involves simply a pinhole, a sample, and a detector, all in line (see Fig. 8.17(a)), also including a beam stop to avoid saturating the detector with the direct beam. Fig. 8.18 shows the results of an investigation of the structure of 400 nm colloidal gold particles dispersed on a silicon nitride membrane by Takayama et al. (2018), using essentially this arrangement.

Notice that the grey-scale diffraction pattern shown in Fig. 8.18, recorded using a photon energy of 8 keV, has been converted from a false colour representation in the original article; this leads to a non-linear intensity grey scale. The Pilatus pixel detector used to record this pattern could only cover diffraction patterns with spatial frequencies up $65.2\,\mu m^{-1}$, so the pattern shown is a composite of three separate detector images covering different ranges. The spatial resolution in CXDI is determined by the largest recorded spatial frequencies, consistent with the earlier discussion regarding the effective numerical aperture of a measurement. In this case the image recovered from this diffraction pattern, with an effective resolution of 29.1 nm, is shown in Fig. 8.18(b) and is compared with a conventional scanning electron micrograph (SEM) of the same region in Fig. 8.18(c). The PCXD image clearly shows the presence of 100–160 nm internal voids in the two cubic and one spherical particles, not evident in the SEM images (and also not easily detectable by TEM (Nakasako et al., 2013)). An important feature of the CXDI technique, which makes it

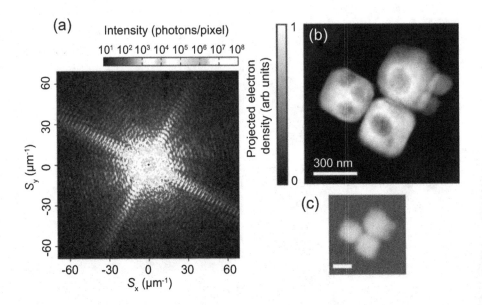

Figure 8.18 Visualisation of voids in colloidal Au particles by PCXDI. (a) shows the recorded diffraction pattern, (b) the PCXD image and (s) a scanning electron micrograph of the same sample. Reproduced from Takayama et al. (2018) with permission of the International Union of Crystallography (https://journals.iucr.org/)

possible to achieve the phase retrieval necessary to construct a reliable image from the recorded diffraction pattern, is that (unlike a crystal diffraction pattern from a large single crystal), the pattern is continuous (rather than discrete), so it is possible to oversample the pattern relative to the spatial Nyquist frequency. Of course, the use of a small pinhole to extract the coherent part of the synchrotron radiation, itself leads to diffraction of the radiation that is then incident on the sample, which needs to be taken into account in interpreting the recorded diffraction pattern. In this case Takayama et al. (2018) report the results of calculations of the effect of this aperture diffraction on the variation in the phase over the area of the sample, selecting an optimum sample-to-detector distance to minimise its impact.

Groups of gold nanoparticles have been used as a test sample in the development of this technique at a number of facilities, including 50 nm particles imaged with a resolution of 10 nm by He et al. (2003). By nanofocusing the coherent part of an incident beam to a diameter of ~100 nm, Schroer et al. (2008) showed it was possible to image a single Au particle with a size of less than 100 nm at a resolution of 5 nm. Of course, PCXDI does not require the sample to be crystalline, and indeed, in the paper by Takayama et al. (2018) described above, the authors also present results from a single macroporous silica particle, investigating the effect of changing the ambient humidity on its structure; the related study by Nakasako et al. (2013) also presents results for a frozen-hydrated biological sample (spinach chloroplast) at 66 K. Of course, the requirement for a high intensity coherent beam means that a FEL source has some significant advantages over even the latest synchrotron radiation beamlines in exploiting CXDI, collecting single-shot diffraction data from a single FEL pulse of radiation; Nakasako et al. (2013) presented preliminary data from such an experiment. A review by Miao et al. (2015) describes a range of variants of the general CXDI method including Fresnel CXDI in which the sample is placed in front of, or beyond, the focal spot of a focused coherent beam, leading to a curved wavefront. Miao et al. also discuss the significance of continuing developments of coherent X-ray sources, and particularly FELs, for CXDI experiments.

However, if the sample *is* crystalline, the alternative Bragg CXDI technique (Fig. 8.17(b)) can provide valuable information on strain and defects in sub-micron samples, the diffraction pattern now being recorded in a direction corresponding to a Bragg reflection. Notice that this technique also offers a way to separate the diffraction pattern from a single particle that is within a group of particles, because if they are randomly oriented their Bragg-diffracted beams emerge in different directions (as in a conventional powder diffraction pattern). To illustrate the utility of this technique, Figs. 8.19 and 8.20 show results, taken from the work of Xiong et al. (2014), of an investigations of Cu diffusion into a single ~300 nm gold nanocrystal (one of many deposited on a silicon substrate). The spatial resolution of the recovered images was ~20 nm, but because of the underlying use of the Bragg crystalline diffraction, BCXDI is sensitive to the longer-range strain fields that accompany atomic level defects such as dislocation loops, thereby detecting very small lattice distortions with 'picometre sensitivity'. Cu and Au are known to form well-defined bulk ordered alloy phases with stoichiometries of Au_3Cu, $CuAu$ and Cu_3Au, and the diffusion coefficient of Cu in bulk Au is well-known, but significant differences in behaviour are expected in nanocrystalline material, a general issue of increasing practical importance as components of devices become smaller and smaller. This experiment involved continuous measurement of the BCXDI diffraction pattern around a (111) reflection, recording the pattern while rocking the sample over a 0.5° range in order to allow 3D reconstructed images to be obtained. The sample was held at 300°C (to achieve a measureable amount of diffusion in a few hours) while slowly evaporating Cu onto the initially pure Au nanocrystals.

Fig. 8.19 shows the diffraction patterns, recorded at the midpoint of the Au(111) rocking curve, as cut-off XZ plane views, after a succession of different Cu diffusion times from 0–10 hours. The initial pattern (Fig. 8.19(a)) shows the modulated fringes characteristic of coherent illumination of a nanocrystal; the pattern is essentially symmetric and the dominant fringes correspond to the faceted directions of the nanocrystal. With increasing diffusion time the pattern becomes asymmetric and the fringes are distorted, attributable to inhomogeneous lattice distortions as the Cu atoms diffuse into the Au lattice. As shown by the upper values of the scale bars, the maximum intensity also decreases with increasing diffusion time, a consequence of the reducing volume of the Bragg diffracting Au nanocrystal as some Au crystalline parts are converted to an alloy with a different structure. In fact the authors report that after 10 hours of diffusion the many Au nanoparticles gave rise to a powder pattern characteristic of Cu_3Au; as this crystalline phase has a different lattice

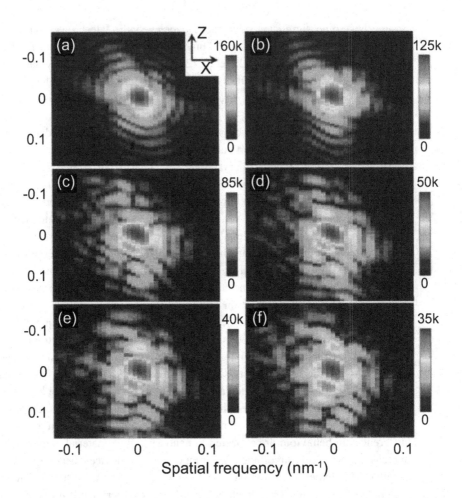

Figure 8.19 BCXDI diffraction patterns, taken from a gold nanocrystal under continuous Cu deposition, at the centre position of the Au(111) rocking curve. The XY plane view patterns were at different times after Cu deposition was initiated, namely: (a) 0 hours, (b) 2 hours, (c) 4 hours, (d) 6 hours, (e) 8 hours and (f) 10 hours. After Xiong et al. (2014) (https://doi.org/10.1038/srep06765) under a Creative Commons Attribution 4.0 International License. Note that conversion of the published false-colour patterns to monochrome leads to a non-linear grey-scale

parameter to pure Au it does not contribute to the Au(111) Bragg reflection of the imaged nanoparticle.

Reconstructions of these complete BCXDI data yield 3D maps of both the amplitude and phase of the complex electron charge density, and these properties, mapped on slices through the centre of the nanoparticle, are shown in Fig. 8.20. Specifically, Figs. 8.20(a)–(f) show the reconstructed phase images corresponding to the same Cu diffusion times from 0 to 10 hours, as in Figs. 8.19(a)–(f), while Figs. 8.20(g)–(l) show the corresponding amplitude images. The phase is a projection of the lattice distortion onto the Au (111) **g** vector (see arrow in Fig. 8.20(b)). Notice that the phase

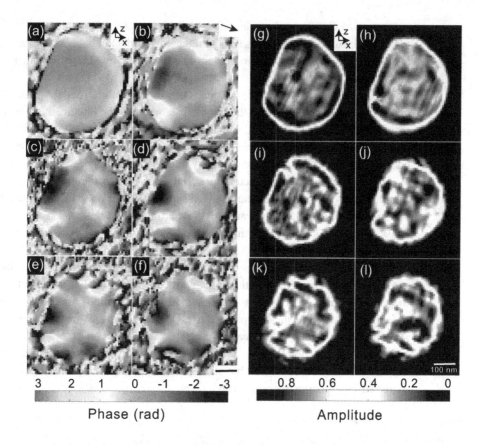

Figure 8.20 Phase (a)–(f) and amplitude (g)–(l) images extracted from the BCXDI diffraction patterns of Fig. 8.19, as cut-off views at the centre of the gold nanocrystal, after increasing Cu diffusion times. The arrow inset in (b) shows the direction of the (111) **g** vector. After Xiong et al. (2014) under a Creative Commons Attribution 4.0 International License. Note that conversion of the published false-colour patterns to monochrome leads to a non-linear grey-scale.

images are very noisy outside the particle, because here the amplitude falls to zero. Before diffusion (Fig. 8.20(a)) the nanocrystal is almost strain free, but with increasing diffusion significant strain develops. After 6 hours small regions near the top and lower-left of the crystal show a clear phase wrap where the phase changes from $-\pi$ to $+\pi$ over a small volume, indicating atom displacements equal to the lattice spacing; these can be attributed to dislocation loops consisting of edge dislocations due to the insertion of a single plane of atoms in a small region, with a consequential surrounding strain field. Whether this extra plane of atoms is Cu, Au or a mixture, is unknown. The amplitude images also show significant changes, the initial well-defined smooth edges of the crystal breaking up while 'channels' open up in the crystal, which probably comprise misoriented material rather than voids. More quantitative analysis of the development of these images, based on a 'shell' model with an inner core of Au and an outer shell of Cu, allowed a value of the diffusion coefficient D, of 8.7×10^{-9} μm/s to be extracted, some two orders of magnitude higher than in bulk crystalline Au. This is consistent with very much higher values found by TEM in much smaller nanocrystals.

One limitation of the CXDI technique is that the algorithms used for phase retrieval generally lead to ambiguities if the sample is larger than the illuminating beam due to their failure to handle the 'soft' boundaries of the diffracting volume. This problem can be overcome by the technique of X-ray ptychography, which is effectively a combination of CXDI and scanning X-ray microscopy. In this case the incident beam is rastered over the sample, recording CXDI diffraction patterns at each point. The sampled areas in this process must have a significant overlap, which ensures that the combination of patterns significantly over-determines the problem and allows reliable phase retrieval. The use of the different modes of CXDI leads to different associated modes of ptychography.

A particularly interesting modification of standard ptychography allowing it to provide information on magnetic domain structures is to exploit the effect of X-ray magnetic circular dichroism, as discussed in the absorption mode in Section 6.2. Recall that a change in absorption causes a change in the imaginary part of the scattering facto, which in turn leads to a change in the real part of the scattering factor as discussed in Section 4.3. The total diffracted intensity is therefore determined not only by the charge distribution, but also by a weaker magnetic component that can be extracted in a dichroism experiment.

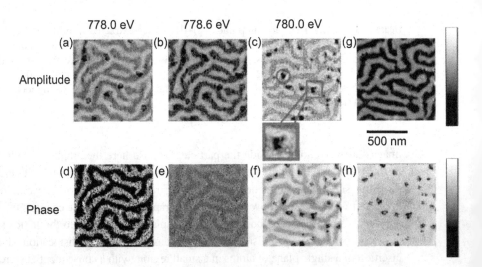

Figure 8.21 Reconstructed amplitude (a–c) and phase (d–f) from ptychography measurements at three different photon energies of a thin film of $SmCo_5$ using left-circularly polarised radiation. (g) and (h) show the difference and sum, respectively, of data using left- and right-circularly polarised radiation at the sample position of (c) and (f). After Shi et al. (2016), with the permission of AIP publishing

Fig. 8.21 shows the results of a soft X-ray transmission ptychography study by Shi et al. (2016) to exploit this effect, the sample being a thin film of $SmCo_5$ that displays perpendicular magnetic anisotropy; circularly polarised radiation at different energies around the Co L_3 edge was used to image the magnetic and structural texture. Diffraction patterns were recorded from a 20×20 grid of overlapping positions of a FZP-focused incident beam, from which the amplitude and phase images of Fig. 8.21 were extracted. Measurements at a photon energy of 778.0 eV (Figs. 8.21(a) and (d)), just below the absorption edge, provide the dominant magnetic phase contrast, while at an energy of 778.6 eV (Figs. 8.21(b) and (e)), corresponding to the absorption peak, one obtains the maximum magnetic absorption contrast. Using a photon energy of 780.0 eV (Figs. 8.21(c) and (f), on a different part of the sample), above this absorption 'white line' peak, the phase component of the refractive index passes through zero and switches sign, leading to an inverted magnetic phase contrast.

Overall, the images show the labyrinthine magnetic domain structure expected for a film with perpendicular magnetic anisotropy, with an average domain width of 76.5 nm. In addition, 'inclusions' are seen as dark spots in the amplitude images (e.g. those highlighted by the circle and square in Fig. 8.21(c)) and in the phase image of Fig. 2.21(f). In Fig. 8.21(d) the inclusions are seen as bright, resulting from the phase contrast inversion. Fig. 8.21(g) shows an amplitude image of the same region as Fig. 8.21(c) constructed from the difference between images obtained with left- and right-circularly polarised radiation at a photon energy of 780.0 eV, producing an XMCD image that isolates features associated with a component of magnetisation perpendicular the surface (the photon incidence direction). Notice that the inclusions are absent from this image clearly indicating that these inclusions image through charge scattering alone; other, lower resolution, imaging techniques indicate a probable composition of Sm_2Co_{17}. Fig. 8.21(h) shows an image obtained from the sum of the two different polarisation images; this effectively corresponds to the image one would obtain using linear polarisation, for which the magnetic contrast should be weak, and indeed the labyrinthine pattern is barely discernible but the inclusions are clearly imaged.

Of course, imaging of magnetic domain structures can be achieved by other techniques (e.g. XMCD-PEEM as in Fig. 8.13), but the spatial resolution achievable by CXDI and ptychography (in this case ~10 nm) is typically superior. While these 'lens-less' coherent imaging techniques using synchrotron radiation have advanced very considerably in recent years, a trend likely to continue, it also seems certain that they will advance further in the future through the use of FEL radiation and its greatly superior coherence properties.

References

Alagia, M., Candoni, P., Falcinelli, S. et al. (2012). *Chem. Phys.*, **398**, 134.

Allegretti, F., Polcik, M. & Woodruff, D. P. (2007). *Surf. Sci.*, **601**, 3611.

Amann, J., Berg, W., Blank, V. et al. (2012). *Nat. Phot.*, **6**, 693.

Ament, L. J. P., van Veenendaal, M., Devereaux, T. P., Hill, J. P. & van den Brink, J. (2011). *Rev. Mod. Phys.*, **83**, 705.

Andersson, A., Johnson, M. S. & Nelander, B. (1999). *Proc SPIE*, **3775**, 77.

Andruszkow, J., Aune, B., Ayvazyan, V. et al. (2000). *Phys. Rev. Lett.*, **85**, 3825.

Aquila, A., Hunter, M. S., Doak, R. B. et al. (2012). *Optics Express*, **20**, 2706.

Attwood, D. (1999). *Soft X-Rays and Extreme Ultraviolet Radiation: Principle and Applications*, Cambridge, UK: Cambridge University Press.

Bahena, D., Bhattarai, N., Santiago, U. et al. (2013). *J. Phys. Chem. Lett.*, **4**, 975.

Bartolini, R., Cinque, G., Martin, I. P. S. et al. (2011). *Proc IPAC2011*, THPC068, 3050.

Barton, J. J. (1988). *Phys. Rev. Lett.*, **61**, 1356.

Barton, J. J. (1991). *Phys. Rev. Lett.*, **67**, 3106.

Batterman, B. W. (1964). *Phys. Rev.*, **133**, A759.

Becker, U., Gessner, O & Rüdel, A. (2000). *J. Electron Spectrosc. Rel. Phenom.*, **108**, 189.

Beguiristain, H. R., Cremer, J. T., Piestrup, M. A., Gary, C. K. & Pantell, R. H. (2002). *Optics Lett.*, **27**, 778.

Bergamaschi, A., Cervellino, A., Dinapoli, R. et al. (2010). *J. Synch. Rad.*, **17**, 653.

Bergmann, U. & Glatzel, P. (2009). *Photosynth. Res.*, **102**, 255.

Bernardini, C., Corazza, G. F., Di Giugno, G. et al. (1963). *Phys. Rev. Lett.*, **10**, 407.

Biesinger, M. C., Lau, L. W. M., Gerson, A. R & Samart, R. St. C. (2010). *Appl. Surf. Sci.*, **257**, 887.

Bilderback, D. H., Brock, J. D., Dale, D. S. et al. (2010). *New J. Phys.*, **12**, 035011.

Billardon, M., Ellaume, P., Ortéga, J. M., Bazin, C., Bergher, M., Velghe, M., Petroff, Y., Deacon, D. A. G., Robinson, K. E. & Madey, J. M. J. (1983). *Phys. Rev. Lett.*, **51**, 1652.

Blanc, F. (2018). *ACS Cent. Sci.*, **4**, 1081.

Blewett, J. P. (1946). *Phys. Rev.*, **69**, 87.

Blewett, J. P. (1998). *J. Synchrotron. Rad.*, **5**, 135.

Blyth, R. R., Delaunay, R., Zitnik, M. et al. (1999) *J. Electron Spectrosc. Relat. Phenom.*, **101–103**, 959.

Boll, R., Rouzée, A., Adolph, M. et al. (2014) *Faraday Disc.*, **171**, 57.

Bolotova, I. B., Ulenikov, O. N., Bekhtereva, E. S. et al. (2018). *J. Mol. Spec.*, **348**, 87.

Bonifacio, R., Pellegrini, C. & Narducci, L. M. (1984). *Opt. Commun.*, **50**, 373.

Bourgeois, D., Schotte, F., Brunori, M. & Vallone, B. (2007). *Photochem. Photobiol. Sci.*, **6**, 1047.

Bragg, W. H. & Bragg, W. L. (1915). *X-Rays and Crystal Structure*, London: G. Bell & Sons. (www.archive.org/stream/xrayscrystalstru00braguoft#page/202/mode/2up)

Brown, F. C., Bachrach, R. Z. & Lien, N. (1978). *Nucl. Instrum. Methods*, **152**, 73.

Brown, F. C., Stott, J. P. & Hulbert, S. L. (1986). *Nucl. Instrum. Methods A*, **246**, 278.

Burmeister, W. P. (2000). *Acta Cryst. D.*, **56**, 328.

Butchers, M. W., Duffy, J. A., Taylor, J. W. et al. (2015). *Phys. Rev. B.*, **92**, 121107.

Carra, P., Thole, B. T., Alterelli, M. & Wang, X. (1993). *Phys. Rev. Lett.*, **70**, 694.

Cheng, L., Fenter, P., Bedzyk, M. J. & Sturchio, N. C. (2003). *Phys. Rev. Lett.*, **90**, 255503.

Cinque, G., Frogley, M. D. & Bartolini, R. (2011). *Rendiconti Lincei.*, **22**, 33.

Cocco, D., Bianco, A., Kaulich, B. et al. (2007). *AIP Conf. Proc.*, **879**, 497.

Cooper, M. J., Mijnarends, P. E., Shiotani, N., Sakai, N. & Bansil, A. (2004). *X-Ray Compton Scattering*, Oxford University Press.

Cotte, M., Susini, J., Metrich, N., Moscato, A., Gratzui, C., Bertagnini, A. & Pagano, M. (2006). *Anal. Chem.*, **78**, 7484.

Curbis, F., Allaria, E., Danailov, M. et al. (2005). In R. Reitemeyer, ed., *Proc. 27th Internat. FEL Conf*, Document SLAC –R-791. www-public.slac.stanford.edu/sciDoc/docMeta.aspx?slacPubNumber=slac-r-791

Deacon, D. A. G., Elias, L. R., Madey, J. M. J. et al. (1977). *Phys. Rev. Lett.*, **38**, 892.

De Samber, B., Scharf, O., Buzanich, G. et al. (2019). *J. Anal. At. Spectrom.*, **34**, 2083.

Diamon, H. (2018). *J. Phys. Soc. Jpn.*, **87**, 061001.

Dietz, E., Braun, W., Bradshaw, A. M. & Johnson, R. L. (1985). *Nucl. Instrum. Methods A*, **239**, 359.

Eastwood, D. S., Bayley, P. M., Chang, H. J. et al. (2015). *Chem. Commun.*, **51**, 266.

Egami, T. & Billinge, S. J. L. (2012). *Underneath the Bragg Peaks: Structural Analysis of Complex Materials*, 2nd ed. Oxford: Pergamon.

Eggl, E., Dierolf, M., Achterhold, K. et al. (2016). *J. Synchroton Radiat.*, **23**, 1137.

Eggl, E., Schleede, S., Bech, M. et al. (2015). *Proc. Nat. Acad. Sci.*, **112**, 5567.

Einstein, A. (1905). *Ann. Phys.*, **17**, 132.

Elder, F. R., Gurewitsch, A. M., Langmuir, R. V., & Pollack, H. C. (1947). *Phys. Rev.*, **71**, 829.

Elder, F. R., Langmuir, R. V., & Pollack, H. C. (1948). *Phys. Rev.*, **74**, 52.

Elias, L. R., Fairbank, W. M., Madey, J. M. J., Schwettman, H. A. & Smith, T. I. (1976). *Phys. Rev. Lett.*, **36**, 718.

Elleaume, P., Ortéga, J. M., Billardon, M. et al. (1984). *J. Physique*, **45**, 989.

Emma, P., Akre, R., Arthur, J. et al. (2010). *Nature Phot.*, **4**, 641.

Eriksson, M. (1997). *J. Synchrotron Rad.*, **4**, 111.

Fabian, D. J., Watson, L. M. & Marshall, C. A. W. (1971). *Rep. Prog. Phys.*, **34**, 601.

Fara, P. (2015). *Phil. Trans. Roy. Soc. A*, **373**, 20140213.

Fathallah, O., Manceron, L., Dridi, N., Rotger, M. & Aroui, H. (2020). *J. Quant. Spec. Rad. Transfer*, **242**, 106777.

Feynman, R. P., Leighton, R. B. & Sands, M. (1964). *The Feynman Lectures in Physics*, vol. 3, Ch. 2. Boston, MA: Addison-Wesley. (www.feynmanlectures.caltech.edu/)

Figueroa, A. I., van der Laan, G., Collins-McIntyre, L. J. et al. (2015). *J. Phys. Chem. C*, **119**, 17344.

Fisher, C. J., Ithin, R., Jones, R. G. et al. (1998). *J. Phys.: Condens. Matter*, **10**, L623.

Flavell, W. R., Quinn, F. M., Clarke, J. A. et al. (2005). *Proc. SPIE 5917, Fourth Generation X-Ray Sources and Optics III*, 59170C. Bellingham, WA.

Follath, R. (2001). *Nucl. Instrum. Methods A*, **467–468**, 418.

Fukuda, N., Hokura, A., Kitajuma, N. et al. (2008). *J. Anal. At. Spec.*, **23**, 1068.

Gaarenstroom, S. W. & Winograd, N. (1977). *J. Chem. Phys.*, **67**, 3500.

Garman, E. F. (2010). *Acta Cryst. D*, **66**, 339.

Ghidini, M., Zhu, B., Mansell, R. et al. (2018). *J. Phys. D: Appl. Phys.*, **51**, 224007.

Gianoncelli, A., Kourousias, G., Merolle, L., Altissimo, M. & Bianco, A. (2016). *J. Synchrotron Rad.*, **23**, 1526.

Gog, T., Len, P. M., Materlik, G. et al. (2014). *Phys. Rev. Lett.*, **76**, 3132.

Gold, S. H., Hardesty, D. L., Kinkead, A. K., Barnett, L. R. & Granatstein, V. L. (1984). *Phys. Rev. Lett.*, **52**, 1218.

Goward, F. K. & Barnes, D. E. (1946). *Nature*, **158**, 413.

Green, G. K. (1976). *Spectra and Optics of Synchrotron Radiation*, Brookhaven National Lab. Report BNL 50522.

Grunwaldt, J.-D., Hannemann, S., Schroer, C. G. & Baiker, A. (2006). *J. Phys. Chem. B.*, **110**, 8674.

Guinier, A. (1939) *Ann. Phys.*, **12**, 161.

Guinier, A. & Fournét, G. (1955). *Small-Angle Scattering of X-rays*, New York: Wiley.

Gurman, S. J., Binstead, N. & Ross, I. (1984). *J. Phys. C: Solid State Phys.*, **17**, 143.

Gurman, S. J., Binstead, N. & Ross, I. (1986). *J. Phys. C: Solid State Phys.*, **19**, 1845.

Hall, G. (1995) *Quarterly Rev. Biophys.*, **28**, 1.

Hall, R. I., Dawber, G., McConkey, A., MacDonald, M. A. & King, G. C. (1992). *Phys. Rev. Lett.*, **68**, 2751.

Hancock, J. N., Chabot-Couture, G., Li, Y. et al. (2009). *Phys. Rev. B.*, **80**, 092509.

Harkiolaki, M., Darrow, M. C., Spink, M. C., Kosier, E., Dent, K. & Duke, E. (2018). *Emerg. Top. Life Sci.*, **2**, 81.

Haynes, T. D., Maskery, I., Butchers, M. W. et al. (2012). *Phys. Rev. B*, **85**, 115137.

He, H., Marchesini, S., Howells, M. et al. (2003). *Phys. Rev. Lett.*, **67**, 174114.

Heinmann, P. A., Koike, M. & Padmore, H. A. (2005). *Rev. Sci. Instrum.*, **76**, 063102.

Henderson, R. (1990). *Proc. Roy. Soc. Lond. B*, **241**, 6.

Henderson, R. (1995). *Q. Rev. Biophys.*, **28**, 171.

Henke, B. L., Gullikson, E. M. & Davis, J. C. (1993). *At. Data Nucl. Data Tables* **54**, 181.

Hermann, P., Hoel, A., Patoka, P. et al. (2013). *Optics Express*, **21**, 2914.

Hirschmugl, C. J., Chabal, Y. J., Hoffmann, F. M. & Williams, G. P. (1994). *J. Vac. Sci. Technol. A*, **12**, 2229.

Holman, H. -Y. N., Martin, M. C. & McKinney, W. R. (2003). *Spectroscopy*, **17**, 139.

Hu, W., Hayashi, K., Ohwada, K. et al. (2014). *Phys. Rev. B.*, **89**, 140103.

Huang, J., Günther, B., Achterhold, K. et al. (2020). *Sci. Rep.*, **10**, 8772.

Itou, M., Harada, T. & Kita, T. (1989). *Appl. Optics*, **28**, 146.

Ivanenko, D. & Pomeranchuk, I. (1944). *Phys. Rev.*, **65**, 343.

Janssens, K., Vittiglio, G., Deraedt, I., Aerts, A. et al. (2000). *X-ray Spectrom.*, **29**, 73.

Jensen, K. M. Ø., Juhas, P., Tofanelli, M. A. et al. (2016). *Nature Commun.*, **7**, 11859.

Jones, N., Norris, C., Nicklin, C. L. et al. (1998). *Surf. Sci.*, **409**, 27.

Jonsson, G. K., Ulama, J., Johansson, M. Z. & Bergenholtz, J. (2017). *Colloid Polym. Sci.*, **295**, 1983.

Kawamura, N., Tsutsui, S., Mizumaki, M. et al. (2009). *J. Phys. Conf. Series*, **190**, 012020.

Kerst, D. W. (1940). *Phys. Rev.*, **58**, 841.

Khatib, O., Bechtel, H. A., Martin, M. C., Raschke, M. B & Carr, G. L. (2018). *ACS Photonics*, **5**, 2773.

Kingslake, R. (1994). *Opt. Photonics News*, **5**, 20.

Kirkpatrick, P. & Baez, A. V. (1948). *J. Opt. Soc. Am.*, **38**, 766.

Kitamura, H (1980). *Jap. J. Appl. Phys.*, **19**, L185.

Knight, M. J., Allegretti, F., Kröger, E. A. et al. (2008). *Surf. Sci.*, **602**, 2524.

Kopittke, P. M., Punshon, T., Paterson, D. J. et al. (2018). *Plant Physiology*, **178**, 507.

Kossel, W. (1920). *Z. Phys.*, **1**, 119.

Krafft, G. A. & Priebe, G. (2010). *Rev. Accel. Sci. Tech.*, **3**, 147.

Kraft, P., Bergamaschi, A., Broennimann, Ch. et al. (2009). *J. Synchrotron Rad.*, **16**, 368.

Krisch, M. & Sette, F. (2017). *Crystal. Rep.*, **62**, 1.

Kronig, R. L. (1931). *Z. Phys.*, **70**, 317.

Kumakhov, M. A. (1990). *Nucl. Instrum. Methods*, **B48**, 283.

Larmor, J. (1897). *Phil. Mag.*, **44**, 503.

LaShell, S., McDougall, B. A. & Jensen, E. (1996). *Phys. Rev. Lett.*, **77**, 3419.

Lee, J. (2002). PhD thesis, University of Warwick.

Lee, J., Fisher, C., Woodruff, D. P. et al. (2001). *Surf. Sci.*, **494**, 166.

Lee, P. A. & Pendry, J. B. (1975). *Phys. Rev. B*, **11**, 2795.

Lee, P. A., Citrin, P. H., Eisenberger, P. & Kincaid, B. M. (1981). *Rev. Mod. Phys.*, **53**, 769.

Lee, T.-L., Bihler, C., Schoch, W. et al. (2010). *Phys. Rev. B*, **81**, 235207.

Liénard, A. (1898). *L'Eclairage Elect.*, **16**, 5.

Liu, Y., Nelson, J., Holzner, C., Andrews, J. C. & Pianetta, P. (2013). *J. Phys. D: Appl. Phys.*, **46**, 494001.

Locatelli, A., Wang, C., Africh, C. et al. (2013). *ACS Nano*, **7**, 6955.

Lovesey, S. W. & Collins, S. P. (1996). *X-ray Scattering Absorption by Magnetic Materials*, Oxford: Oxford University Press.

Lytle, F. W. (1999). *J. Synchrotron Rad.*, **6**, 123.

Ma, Y., Wassdahl, N., Skytt, P. et al. (1992). *Phys. Rev. Lett.*, **69**, 2598.

MacDonald, C. A. (2011). *X-Ray Optics. Instrum.*, **2010**, 867049.

Madey, J. M. J. (1971). *J. Appl. Phys.*, **42**, 1906.

Marcelli, A., Cricenti, A, Kwiatek, W. M. & Petisbois, C. (2012). *Biotech. Adv.*, **30**, 1390.

Marcelli, A. & Cinque, G. (2019). *EMU Notes in Mineralogy*, **20**, 411

Margariti, C. (2019). *Herit. Sci.*, **7**, 63.

Mariedahl, D., Perakis, F., Späh, A. et al. (2018). *J. Phys. Chem. B*, **122**, 7616.

Marks, L. D., Erdman, N. & Subramanian, A. (2001). *J. Phys.: Condens. Matter*, **13**, 10677.

Masadeh, A. S. (2016). *J. Exp. Nanosci.*, **11**, 951.

Masadeh, A. S., Božin, E. S., Farrow, C. L. et al. (2007). *Phys. Rev. B*, **76**, 115413.

Matsui, F., Eguchi, R., Nishiyama, S. et al. (2016). *Sci. Rep.*, **6**, 36258.

Matsushita, T., Muro, T., Matsui, F. et al. (2018). *J. Phys. Soc. Jpn.*, **87**, 061002.

Matsuyama, S., Yasuda, S., Yamada, J. (2017). *Sci. Rep.*, **7**, 46358.

McKellar, A. R. W. (2010). *J. Mol. Spec.*, **262**, 1.

Merminga, L. (2020). In E. J. Jaeschke, S. Khan, J. R. Schneider & J. B. Hastings, eds, *Synchrotron Light Sources and Free-Electron Lasers, 2nd ed.* Switzerland: Springer Nature, 439–477.

Miao, J., Ishikawa, T., Robinson, I. K. & Murmane, M. M. (2015). *Science*, **348**, 530.

Michette, A. G. (1993). In A. G. Michette & C. J. Buckley, eds, *X-Ray Science and Technology*. London: IOP Publishing Ltd.

Mills, D. M., Helliwell, J. R., Kvick, Å. et al. (2005). *J. Synch. Rad.*, **12**, 385.

Mirolo, M., Leanza, D., Höltschi, L. et al. (2020). *Anal Chem.*, **92**, 3023.

Miyahara, T., Kitamura, H., Sato, S. et al. (1976). *Particle Accelerators*, **7**, 163.

Mohammadi, S., Larsson, E., Alves, F. et al. (2014). *J. Synch. Rad.*, **21**, 784.

Moretti, G. (1998). *J. Electron Spectrosc. Relat. Phenom.*, **95**, 95.

Morowe, Ch., Carau, D. & Peffen, J. -Ch. (2017). *Proc. SPIE*, **10386**, 1038603.

Morris, D., Schmidt, A., Acosta, R. E. et al. (1995). *Proc. SPIE*, **2437**, 134.

Mudd, J. J., Lee, T. -L., Muñoz-Sanjosé, V. et al. (2014). *Phys. Rev. B*, **89**, 165305.

Murray, C. A., Potter, J., Day, S. J. et al. (2017). *J. Appl. Cryst.*, **50**, 172.

Nakasako, M., Takayama, Y., Oroguchi, T. (2013). *Rev. Sci. Instrum.*, **84**, 093705.

Namioka, T. (1959). *J. Opt. Soc. Amer.*, **49**, 951.

Neil, G. R., Bohn, C. L., Benson, S. V. et al. (2000). *Phys. Rev. Lett.*, **84**, 662.

Newton, I. (1671). *Phil Trans. Roy Soc.*, **6**, 3075.

Nilsson, A. & Pettersson, L. G. M. (2004). *Surf. Sci. Rep.*, **55**, 49.

Nilsson, A. & Pettersson, L. G. M. (2008). In A. Nilsson, L. G. M Pettersson, J. K. Nørskov, eds. *Chemical Bonding at Surfaces and Interfaces*. Amsterdam: B. V. Elsevier. 58.

Patterson, A. L. (1934). *Phys. Rev.*, **46**, 372.

Patterson, A. L. (1935). *Z. Krist.*, **90**, 517.

Persson, B. N. J. & Volokitin, A. I (1994). *Surf. Sci.*, **310**, 314.

Pilling, M. & Gardner, P. (2016). *Chem. Soc. Rev.*, **45**, 1935.

Porod, G (1951). *Kolloid Z.*, **124**, 83

Prins, J. (1934). *Nature*, **133**, 795.

Puschmann, A., Haase, J., Crapper, M. D., Riley, C. E. & Woodruff, D. P. (1985). *Phys. Rev. Lett.*, **54**, 2250.

Okasinski, J. S., Kim, C. -Y., Walko, D. A. & Bedzyk, M. J. (2004). *Phys. Rev. B*, **69**, 041401.

Orzechowski, T. J., Anderson, B., Fawley, W. M. et al. (1985). *Phys. Rev. Lett.*, **54**, 889.

Phillips, R. M. (1988). *Nucl. Instrum. Methods A*, **272**, 1.

Plekan, O., Feyer, V., Richter, R. et al. (2009). *J. Phys. Chem. A* **113**, 9376.

Pollack, H. C. (1983). *Am. J. Phys.* **51**, 278.

Poole, M. W. (2017). Private communication.

Rasado-Colambo, I., Avila, J., Vignaud, D. et al. (2018) *Scientific Rep.*, **8**, 10190.

Ravel, R. & Newville, M. (2005). *J. Synch. Rad.*, **12**, 537.

Rehr, J. J. & Albers, R. C. (1990). *Phys. Rev. B.*, **41**, 8139.

Rehr, J. J. & Albers, R. C. (2000). *Rev. Mod. Phys.*, **72**, 621.

Rehr, J. J., Albers, R. C., Natoli, C. R. & Stern, E. A. (1986). *Phys. Rev. B*, **34**, 4350.

Reiche, I., Lebon, M., Chadefaux, C. et al. (2010). *Anal Bioanal. Chem.*, **397**, 2491.

Richardson, J. S. (2000). *Nature Struct. Bio.*, **7**, 624.

Rietveld, H. M. (1969). *J. Appl. Cryst.*, **2**, 65.

Riley, J. M., Mazzola, F., Dendzik, M. et al. (2014). *Nature Phys.*, **10**, 835.

Rolles, D., Boll, R., Tamrakar, S. R., Anielski, D. & Bomme, C. (2014). *Ultrafast nonlinear imaging and spectroscopy II Book Series: Proc. SPIE* **9198**, 919800.

Sandell, A., Björneholm, O., Nilsson, A. (1993). *Phys. Rev. Lett.*, **70**, 2000.

Sasaki, S. (1994). *Nucl. Instrum. Methods A*, **347**, 83.

Sayers, D. E., Stern, E. A. & Lytle, F. W. (1971). *Phys. Rev. Lett.*, **27**, 1204.

Schott, G. A. (1912). *Electromagnetic Radiation*, Cambridge, UK: Cambridge University Press.

Schroer, C. G., Boye, P., Feldkam. J. M. et al. (2008). *Phys. Rev. Lett.*, **101**, 090801.

Schwinger, J. (1949). *Phys. Rev.*, **75**, 1912.

Seah, M. P. & Dench, W. A. (1979). *Surf. Interface Analysis*, **1**, 2.

Seddon, E. A., Clarke, J. A., Dunning, D. J. et al. (2017). *Rep. Prog. Phys.*, **80**, 115901.

Senf, F., Eggenstein, F., Flechsig, U. et al. (2001). *Nucl. Instrum. Methods A*, **467–468**, 474.

Senf, F., Eggenstein, F. & Peatman, W. (1992). *Rev. Sci. Instrum.*, **63**, 1326.

Shi, X., Fischer, P., Neu, V. et al. (2016) *Appl. Phys. Lett.*, **108**, 094103.

Shirley, D. A. (1973). *Advan. Chem. Phys.*, **23**, 85.

Smedh, M., Beutler, A., Ramsvik, T. et al. (2001). *Sur. Sci.*, **491**, 99.

Stapelfeldt, H. & Seidemen, T (2003). *Rev. Mod. Phys.*, **75**, 543.

Stöhr, J. (1999). *J. Magn. Magn. Mater.*, **200**, 470.

Stöhr, J., Sette, F. & Johnson, A. L. (1984). *Phys. Rev. Lett.*, **53**, 1684.

Surman, M., Hagans, P. L., Wilson, N. E., Baily, C. J. & Russell, A. E. (2002). *Surf. Sci.*, **511**, L303.

Suzuki, Y., Uchida, F. & Hirai, Y. (1989). *Jpn. J. Appl. Phys.*, **28**, L1660.

Szöke, A. (1986). *AIP Conference Proceedings*, **147**, 361.

Takayama, Y., Takami, Y., Fukuda, K., Miyagawa, T. & Kagashima, Y. (2018). *J. Synchrotron Rad.*, **25**, 1229.

Tanaka, T. & Kitamura, H. (2001). *J. Synchrotron Rad.*, **8**, 1221.

Tanner, B. K., Vijayaraghavan, R. K., Roarty, B., Danilewsky, A. N. & McNally, P. J. (2019). *Mocroelect. Reliability*, **99**, 232.

Tavassoly, M. T., Hosseini, S. R., Fard, A. M. & Naraghi, R. R. (2012). *Appl. Optics*, **51**, 7170.

Tegze, M. & Faigel, G. (1996). *Nature*, **380**, 49.

Thole, B. T., Carra, P., Sette, F. & van der Laan, G. (1992). *Phys. Rev. Lett.*, **68**, 1943.

Tomboulian, D. H. & Hartman, P. L. (1956). *Phy. Rev.*, **102**, 1423.

Tsutui, K., Matsushita, T., Natori, K. et al. (2017). *Nano Lett.*, **17**, 7533.

Valegård, K., Hasse, D., Andersson, I. & Gunn, L. H. (2018). *Acta Cryst.*, **D74**, 1.

van der Laan, G. & Figueroa, A. I. (2014). *Coord, Chem. Rev.*, **277–278**, 95.

Vartanyants, I. A. & Zegenhagen J. (1999). *Solid State Commun.*, **113**, 299.

Verbeni, R., Sette, F., Krisch, M. H. et al. (1996). *J. Synchrotron Rad.*, **3**, 62.

Victoreen, J. A. (1943). *J. Appl. Phys.*, **14**, 95.

Vinze, L., Vekemans, B., Brenker, F. E. et al. (2004). *Anal. Chem.*, **76**, 6786.

Walker, R. P. (1993). *Nucl. Instrum. Methods A*, **335**, 328.

Walker, R. P. (1998). *CERN Accelerator School: Synchrotron Radiation and Free Electron Lasers, Grenoble, 1996*, CERN 98-04, 129.

Walker, R. P., Clarke, J. A., Couprie, M. E. et al. (2001). *Nucl. Instrum. Methods. A*, **475**, 20.

Walker, R. P. & Diviacco, B. (1992). *Rev. Sci. Instrum.*, **63**, 392.

Wehinger, B., Krisch, M. & Reinchert, H. (2011). *New J. Phys.*, **13**, 023021.

Westendorp, W. F. & Charlton, E. E. (1945). *J. Appl. Phys.*, **16**, 581.

Williams, G. P. (2001). *J. Phys. Condens. Matter*, **13**, 11367.

Williamson, G. K. & Hall, W. H. (1953). *Acta Met.*, **1**, 22.

Wilson, M. N., Smith, A. I. C., Kempson, V. C. et al. (1993) *IBM J. Res. Develop.*, **37**, 351.

Woodruff, D. P. (2005). *Rep. Prog. Phys.*, **68**, 743.

Woodruff, D. P. (2007). *Surf. Sci. Rep.*, **62**, 1.

Woodruff, D. P. (2016). *Modern Techniques of Surface Science, 3rd ed.* Cambridge, UK: Cambridge University Press.

Woodruff, D. P., Cowie, B. C. C. & Ettema, A. R. H. F. (1994). *J. Phys. Condens. Matter*, **6** 10633.

Woodruff, D. P., Seymour, D. L., McConville, C. F. et al. (1988). *Surf. Sci.*, **195**, 237.

Wu, H., Lustbader, J. W., Liu, Y., Canfield, R. E. & Hendrickson, W. A. (1994). *Structure*, **2**, 545.

Xiong, G., Clarke, J. N., Nicklin, C., Rawle, J. & Robinson, I. K. (2014). *Sci. Rep.*, **4**, 6765.

Yabashi, M., Tamasaku, K., Kikuta, S. & Ishikawa, T. (2001). *Rev. Sci. Instrum.*, **72**.

Zakharov, A. A., Mikkelsen, A. & Andersen, J. N. (2012). *J. Elect. Spect. Rel. Phenom.*, **185**, 417.

Zegenhagen, J. (1993). *Surf. Sci. Rep.*, **18**, 199.

Zhang, Z., Fenter, P., Cheng, L. et al. (2004). *Surf. Sci.*, **554**, L95.

Index

Printed in the United States
by Baker & Taylor Publisher Services